Sustainable Corrosion Inhibition:

Nanobiotechnological Solutions and Challenges

Shrikaant Kulkarni
Sanjivani University, Kopargaon, India

Susanta Das
https://orcid.org/0000-0002-9314-3988
Ajeenkya D.Y. Patil University, India

IGI Global
Scientific Publishing
Publishing Tomorrow's Research Today

Vice President of Editorial Melissa Wagner
Director of Acquisitions Mikaela Felty
Director of Book Development Jocelynn Hessler
Production Manager Mike Brehm
Cover Design Jose Rosado

Published in the United States of America by
IGI Global Scientific Publishing
701 East Chocolate Avenue
Hershey, PA, 17033, USA
Tel: 717-533-8845 | Fax: 717-533-7115
Website: https://www.igi-global.com E-mail: cust@igi-global.com

Library of Congress Cataloging-in-Publication Data

LCCN: 2025019855 (CIP Data Pending)
ISBN13: 979-8-3373-2617-7
Isbn13Softcover: 979-8-3373-2618-4
EISBN13: 979-8-3373-2619-1

British Cataloguing in Publication Data
A Cataloguing in Publication record for this book is available from the British Library.

Table of Contents

Detailed Table of Contents

production of these nanomaterials by green synthesis techniques, which reduce dependency on hazardous chemicals and promote more general sustainability objectives, has been the focus of recent study. This review highlights the various materials that are used in automobile industry that face the problem of corrosion, the nano composites that are used to inhibit corrosion in them and the challenges that are yet to be addressed by the industry in the implementation of these composites.

Soil microorganisms assumes in the corrosion of agricultural implements and irrigation systems, a significant economic issue for the agricultural sector. Soil corrosion is primarily electrochemical, driven by variations in soil properties that create potential differences on metal surfaces. Microbial Induced Corrosion (MIC), facilitated by bacteria and fungi such as Sulfate-Reducing Bacteria, accelerates this process by producing corrosive metabolites like acids, leading to localized pitting and structural degradation. The problem is exacerbated in agricultural environments by factors including soil salinity, sodicity, and exposure to agro-chemicals like fertilizers, pesticides, and acidic manure slurries. These conditions not only directly corrode metals and concrete but also alter soil microbiology, further influencing corrosion dynamics. This chapter also discusses mitigation strategies, including the use of corrosion-resistant materials like stainless steel, protective coatings, and system maintenance practices such as acid treatment for drip irrigation to prevent clogging.

Biotechnological solutions represent an environmental friendly and viable alternative for managing Microbiologically Influenced Corrosion (MIC), which is a ubiquitous issue in the industrial infrastructure, such as pipelines, marine ships, and water systems. The conventional procedures of corrosion control that use toxic compounds and synthetic finishes are harmful to the environment and have limited long-term efficacy. In contrast, biotechnology offers self-regulating, bio-inspired solutions. Key biotechnological strategies include the use of beneficial microorganisms as probiotic corrosion inhibitors, enzyme systems to degrade corrosive metabolites and biofilm disruption mechanisms to limit biofouling. The biosurfactant and biopolymers usage can give natural corrosion shield, while biogenic nanoparticles produced via green synthesis exhibit promising antimicrobial and anticorrosive properties.

Microbiologically Influenced Corrosion (MIC) presents a substantial threat to the integrity and efficiency of steel in marine and non-marine settings. This occurs due to microorganisms that, under specific conditions, alter their electrochemical environment on the metal's surface. Sulfate-Reducing Bacteria, Iron-Oxidizing Bacteria, and biofilm-forming species in aquatic settings catalyze redox processes that result in pitting, scaling, and localized corrosion of submerged steel structures. In contrast, microbial communities in soil systems react to corrosive metabolite such hydrogen sulphide and organic acids, that are intensified by moisture and organic matter under different redox conditions, thereby markedly affecting corrosion processes. The generation of biofilm was considered an important aspect in these processes because it creates zones of varying oxygenation that concentrate the corrosive chemicals in the interface of the metal.

Trace amounts of heavy metals pose serious threats to aquatic environments and corrosion inhibition in the water due to traditional purification technologies. It impacts human health and requires modern, sustainable purification technologies.

Water purification is essential for ensuring safe drinking water, as natural sources often contain physical, chemical, and microbial contaminants that pose health risks. Organizations like the EPA and WHO set quality standards, promoting technologies such as electrodeionization, reverse osmosis, ultrafiltration, sedimentation, biosorption, ion exchange, and UV treatment. These methods improve water quality by removing pollutants and pathogens. However, rising pollution and water scarcity demand innovative approaches. Nanobiotechnology offers promise by integrating nanomaterials and biological systems for efficient treatment. Techniques like nanofiltration, nano-adsorption, and nano-catalytic degradation remove bacteria, viruses, heavy metals, and pesticides effectively. Despite their benefits, challenges like cost, scalability, and regulation limit widespread use. Research, investment, and sustainable strategies are key to expanding access and ensuring global water security.

The governance of advanced nano biomaterials is failing due to obsolete regulations (like REACH and TSCA), which transforms technological risk into legal hazard and impedes the adoption of materials that could save $2.5 trillion annually in corrosion costs and cut waste by 60-70%. These frameworks suffer from definitional ambiguity and lifecycle myopia. The proposed solution is the 4-R Governance Framework (Responsibility, Risk, Reflexivity, Regeneration). Crucially, this involves the Doctrine of Perpetual Regulatory Design (PRD), featuring the Obligatory Reflexivity Clause and Trans-Generational Liability Ledger. This design mandates continuous legal adaptation and perpetual manufacturer accountability, providing a Model International

Preface

"Sustainable Corrosion Inhibition: Nanobiotechnological Solutions and Challenges" explores the innovative convergence of nanotechnology and biotechnology to address the global challenge of corrosion. The book emphasizes eco-friendly and sustainable approaches, presenting cutting-edge advancements in nanomaterials, microbial solutions, and bio-inspired strategies for corrosion control. It delves into self-healing coatings, smart materials, and AI-driven monitoring techniques, offering a comprehensive understanding of sustainable corrosion inhibition in various industrial applications.

The book contributes significantly to materials science and sustainable engineering by promoting green and innovative solutions for corrosion control. Highlighting interdisciplinary approaches fosters the development of environmentally conscious and cost-effective practices, reducing corrosion's economic and ecological impact. The book catalyzes further research and innovation, addressing critical industrial needs while aligning with global sustainability goals.

The book is intended for:

1. **Researchers and Academics**: In fields such as nanotechnology, biotechnology, materials science, and environmental engineering.
2. **Industry Professionals**: Engineers and scientists in aerospace, automotive, marine, and infrastructure sectors.
3. **Students and Educators**: Graduate and postgraduate students, as well as educators in interdisciplinary STEM fields.
4. **Policy Makers and Sustainability Advocates**: Interested in promoting green practices in industrial applications.

This edited volume has nine chapters, each one of them is unique. Together, they provide a comprehensive overview of how corrosion, sustainability, and nanotechnology intersect across various fields — from metallurgy and automotive engineering to textiles, agriculture, and environmental protection.

Chapter 1 sets the stage by unpacking corrosion in metal nanoparticles, highlighting the delicate balance between the unique reactivity that makes them valuable and the intrinsic vulnerabilities that demand innovative stabilization strategies. Building on the theme of applied nanosystems, Chapter 2 examines corrosion inhibition in the automobile industry through nano-bio composites, emphasizing green synthesis, sustainability, and the technological hurdles that still limit industrial implementation.

Chapter 3 discusses the role of soil microorganisms in the degradation of agricultural tools and irrigation systems, revealing the complexity of microbial–metal interactions within soil ecosystems.

Chapter 4, which presents biotechnological approaches for managing microbiologically influenced corrosion (MIC), positioning biological agents, enzymes, and biogenic nanomaterials as viable, environmentally friendly alternatives to toxic chemical inhibitors.

Chapter 5 offers this by detailing MIC in steel across marine and soil systems, clarifying biofilm dynamics and microbially driven redox processes that compromise structural integrity. Extending this environmental perspective.

Chapter 6 examines nanobiotechnological solutions in water purification—where corrosion inhibition, heavy metal removal, and sustainable materials converge to support cleaner, decentralized water systems.

Chapter 7 addresses the promise of biodegradable nanomaterials in the textile sector, illustrating how natural polymers and smart nanostructures can mitigate environmental pollution while enhancing fabric performance.

Chapter 8 introduces contemporary water purification technologies and highlights the growing relevance of nanobiotechnology in addressing water scarcity, contamination, and treatment inefficiencies.

Finally, Chapter 9 talks all preceding scientific innovations within a forward-looking policy and regulatory framework. The proposed 4-R Governance Model and Doctrine of Perpetual Regulatory Design challenge current paradigms and offer a blueprint for aligning innovation with responsibility, risk management, and sustainable global trade.

Together, these chapters collectively demonstrate that sustainable corrosion inhibition is no longer merely a material science problem—it is an ecological, technological, economic, and regulatory endeavor. The fusion of nanotechnology and biotechnology holds transformative potential, but its success depends on thoughtful design, interdisciplinary collaboration, and adaptive governance.

Shrikaant Kulkarni
Sanjivani University, Kopargaon, India

Susanta Das
Ajeenkya D.Y. Patil University, India

Acknowledgement

We are grateful to all the contributors and reviewers for their valuable input. We also extend our sincere thanks to our family members, friends, and colleagues for their constant support throughout the development and submission of this project.

Chapter 1
Corrosion in Metal Nanoparticles

Aditya Shrivastav
https://orcid.org/0009-0000-8461-8171
Ajeenkya D.Y. Patil University, India

Susanta Das
https://orcid.org/0000-0002-9314-3988
Ajeenkya D.Y. Patil University, India

ABSTRACT

MNPs are particles having unique nanoscale characteristics, including high surface-to-volume ratios & increased reactivity, that distinguish them over bulk materials & make them very effective in catalysis, biomedicine, and environmental remediation. Concurrently, these characteristics likewise make MNPs susceptible to corrosion, which is caused by electrochemical reactions, structural defects, & environment (pH & ionic strength). Gold, silver & platinum are noble metals with higher resistance than transition metals such as iron & copper and alloying and bimetallic structures are more stable. This chapter explains the underlying corrosion mechanisms & environmental factors in MNPs, compares between noble & transition metals, provides bibliometric information regarding corrosion in MNPs, and mitigation approaches such as coatings, core-shell architectures, and green inhibitors. It also points out the new uses in the field of catalysis, biomedicine & environmental remediation with a particular focus on the balance between stability & functionality in order to deploy it sustainably.

DOI: 10.4018/979-8-3373-2617-7.ch001

1. INTRODUCTION

Nanomaterials which are technologically important such as metal nanoparticles (MNPs) with the characteristics and properties fundamentally different than the bulk counterparts due to a large surface-volume ratio, quantum interactions, and great chemical reactivity (Shahalaei et al., 2024). The physicochemical peculiarities of the MNPs with their high surface energies and tuneable versatile optical and catalytic activity are what make it possible to apply these nanoparticles towards transforming the electronics, energy, medicine, and environmental technologies (Shahalaei et al., 2024). Gold and silver nanoparticles have a unique surface plasmon resonance, stability, and bio-compatibility, which allows adapting them to the diagnostics, sensing platform, and targeted drug delivery platform (Boruah et al., 2021).

An engineering challenge and a scientific opportunity is offered by metal nanoparticles for corrosion. It is not only the high surface energy of MNPs which renders them such spectacularly reactive but also causes a strong sensitivity of degradation mechanisms, especially in the conditions of operation or physiology. Compared to the bulk metals, the corrosion processes are more rapid on the nanoscale materials, which may have adverse effects on the stability of the device, its functionality as well as biocompatibility (Hiromoto, 2010; Ndukwe et al., 2025). A careful study of the mechanisms of corrosion is also essential in enhancing the design of nanomaterial-based technologies and its long-term stability. In the biomedical context, such as in the nanoparticles case, the implantable nanoparticles biocompatibility, possible toxicity, and safety profile depend directly on the corrosion resistance of nanoparticles (Hiromoto, 2010). Development of additional protective measures such as nanocoating's, alloying, and surface modification is in progress to reduce the effects of corrosion and increase the serviceability of MNPs (Ndukwe et al., 2025).

Corrosion applications in MNPs are cost-effective due to the application in the coating and catalysis, sensors, electronic parts. Platinum and palladium nanoparticles are vital in the operation of the fuel industry and in purifying exhaustive gases, however the existence in unfavourable gaseous or liquid settings causes corrosion of nanoparticles, which defines the duration and performance of the catalyst. Anti-corrosive finishes using silver and zinc oxide nanoparticles have been applied to the metals of the infrastructure and transport industry to provide them with self-healing, antimicrobial, and protective properties (Zeng et al., 2024). The corrosion behaviour can be utilized to determine the release kinetics, cellular uptake, and biosafety of nanoparticles, which can be used in biomedical systems in imaging, controlled drug delivery, and hyperthermia therapies, such as gold and iron oxide (Yaraki et al., 2022). Indicatively, the uses of gold nanoparticles in the wound dressings and tissue engineering have been widely applied because of their chemical stability, yet their corrosion against the biological conditions predetermines the performance in the long-term and patient consequences (Hiromoto, 2010). It is also reflected by the use of MNPs in dental amalgam, dental implants and stents which are made of orthopaedics that a complete corrosion analysis should be used to render the functionality and mitigate the adverse biological response (Hiromoto, 2010). The chapter discusses corrosion in MNPs starting with the foundation of the nanoscale reactivity and influencing factors of corrosion including size, shape and environment.

2. FUNDAMENTALS OF CORROSION IN METAL NANOPARTICLES

The corrosion behaviour of MNPs is significantly different than their bulk counterparts and this difference is mainly attributed to their sizes which are in the nanoscopic range, the high surface reactivity of MNPs, and the effects of both structural and environmental effects. The high surface to volume ratio of MNPs leads to high surface energies that makes MNPs chemically more reactive as compared to bulk metals. Such high reactivity is due to the large fraction of atoms on nanoparticles being on the surface, where atoms are not strongly coordinated and hence very susceptible to reaction with the corrosive substances of oxygen, water, and acidic species. This reactivity is also increased by the presence of inherent defects, edge sites and grain boundaries, which provide numerous active sites of nucleation of oxidation layers and localized corrosion (Dwivedi et al., 2017; Jeong et al., 2017; Peng et al., 2017). Besides, the agglomerative or aggregative tendency of the nanoparticles can dynamically alter the effective surface exposure of the nanoparticles and hence alter the rate and mechanism of corrosion. The high surface energy also boosts rapid exchange of electrons and ionic flow accelerating the electrochemical reaction which leads to the oxidation and dissolution of metallic cores (Peng et al., 2017).

At the nanoscale, the fundamental nature of the corrosion processes is dictated by the electrochemical reaction, with electrochemical reaction species cathodic reduction and base metal anodic oxidation reactions being catalysed by an increased reaction surface. Electrochemical analyses (cyclic voltammetry and electrochemical impedance spectroscopy) are also demonstrating that MNPs dissolve much faster than bulk metals, and that they tend to disperse or restructure the nanoparticle structure (Jeong et al., 2017; Shi et al., 2024). Another dimension that causes potential-dependent restructuring effects is the nanoscale dimension where the morphological integrity is lost and the ions are rapidly released by the exposure to the strong electrochemical potentials. The processes can be modulated in terms of deposition of protective passive layers or in core-shell. An example can be the resistance to corrosive attack of palladium nanoparticles, which can be enhanced by deposition of a platinum shell on the nanoparticles proving that surface chemistry and heterostructure design can be viable in corrosion mitigation strategies (Jeong et al., 2017).

The corrosion of MNPs is further determined by an interaction between intrinsic and extrinsic factors which include particle size, shape, surface area, environmental conditions and surface chemistry. The size effects, in particular, are rather high, as small nanoparticles contain more surface atoms, which are increasingly active and can dissolve (Peng et al., 2017). The morphological variations are also significant with diverse crystallographic planes surface in the geometry of a sphere, rod, octahedron, or hexagon presenting dissimilar susceptibilities to corrosion (Dwivedi et al., 2017). Accordingly, it is not only that the dominant fraction of the interaction of nanoparticles with aggressive ionic species is enhanced by the large effective surface area of the nanoparticles, but also the larger the area of the nanoparticles the easier it is to form more uniform protective coatings under conducive conditions. A significant effect of environmental factors is the pH, temperature, humidity, ionic strength, and the presence of chlorides or nitrates, and typically, acidic media increase the rate of dissolution of metal ion. Simultaneously, the surface modifications such as the formation of oxide shells or organic passivating layers may also be very effective in suppressing corrosion by inhibiting surface oxidation and the movement of ions. It is necessary to mention that these aspects do not only barely work in isolation but also relate to each other quite intricately. One such example is that not only their size and shape define the aggregation behaviour of nanoparticles, but also influence the level to which they are exposed to corrosive conditions (Peng et al., 2017). Combined, these results imply that corrosion in MNPs cannot be readily attributed to the derivation of bulk metal behaviour but instead, a consequence of nanoscale-related behaviours as well as the complexity of their interactions with the surrounding environment (Dwivedi et al., 2017; Peng et al., 2017). Table 1 presents the overview of the influencing factors of corrosion of MNPs.

Table 1. Factors Influencing Corrosion of MNPs.

Factor	Effect on Corrosion	References
Size	Smaller size increases corrosion rate due to higher surface reactivity	(Peng et al., 2017)
Shape	Distinct morphologies possess unique corrosion susceptibilities	(Jeong et al., 2017)
Surface Area	Higher area accelerates corrosion, enhances protection by coatings	(Dwivedi et al., 2017)
Environment	Acidic media, temperature, chemicals increase corrosion rate	(Peng et al., 2017)
Surface Chemistry	Passivating layers/coatings reduce corrosion and ion release	(Jeong et al., 2017)

3. CORROSION MECHANISMS IN METAL NANOPARTICLES

In MNPs, the corrosion processes are basically controlled by the distinct structural and chemical properties which are quite different to bulk metallic systems. They have a very high surface-to-volume ratio, high-confined nanoscale, and high density of surface defects, as well as very often metastable crystal structures, which combine to produce accelerated, spatially localized, and structurally complicated corrosion behaviours. The most common and frequent route is oxidation, which is very fast because of the increased reactivity of the surfaces of nanoparticles (Kadyrzhanov et al., 2022). Direct exposure to either oxygen or aqueous conditions usually results in the growth of multiple oxides and hydroxides, but the passivation layers formed are usually thinner, less uniform, and structurally less robust than in bulk metals and hence not sufficient to protect over time. A literature on iron oxide-based nanostructures, such as has shown a three-step corrosion cycle including an initial period of relative stability, then the development of oxide and hydroxide outgrowths on the surface, and, finally, completion of the corrosion process by pitting and amorphization of the material (Kadyrzhanov et al., 2022). The shell thickness and atomic ordering are very critical during the passivation process in bimetallic nanoparticles. As an example, in Fe@Au core-shell systems, with only three layers of gold, complete coverage of the core has demonstrated efficient oxygen-induced corrosion protection of the iron core, and incomplete coverage leads to the growth of onion-like Fe@Au@Fe-oxide structures or Janus-like configurations of the system suggesting partial and selective oxidation (Lasserus et al., 2019). Crystallinity, as well as oxide phase structure, is also a significant factor in the efficiency of passivation, and more resistant oxides, including $NdFeO_3$, have better long-term resistance than less resistant phases like Fe_2O_3 (Kadyrzhanov et al., 2022).

In addition to oxidation, another important degradation mechanism in a bimetallic or core shell nanoparticles is the galvanic corrosion, which is caused by the electrochemical potential difference between the constituent metals. Under aqueous conditions, galvanic coupling normally causes selective dissolution of the less noble metal resulting in hollow or partially eroded nanostructures. This is the case with Ag-Cu hetero-nanostructures, in which copper is dissolved rapidly in fractal mode, and silver is comparatively stable until its passive barrier ultimately fails (Du et al., 2021). Particle architecture is a very sensitive factor in determining the kinetics and morphology of galvanic corrosion, since thin oxide barriers may offer temporary defence but fail to cover completely or contain point defects on the surface can tend to localize anodic reactions and cause a selective leaching process (Bayat et al., 2021). Also, when localized corrosion processes like pitting, crevice attack and stress corrosion cracking are visible at the nanoscale. Atomic-scale inhomogeneities, grain boundaries and impurity sites provide preferential nucleation sites of pits, which are manifested as micron or submicron-sized voids in which the structural damage becomes severe. Specifically, chloride environments have been known to destabilize protective surface films to promote pit growth and lead to the formation of amorphous inclusions and ulcerative surface features which can be observed with complex electron microscopy (Kadyrzhanov et al., 2022). Stress corrosion is another cause of degradation that is caused by high density of surface defects and stress concentration at nanoscale junctions where dislocation slip and crack propagation cause mechanical breakdown of nanoparticles. Equally, crevice corrosion is formed at interfaces, junctions of particles, or agglomerated assemblies, which are motivated by a limited mass transport and the formation of local electrochemical gradients (Du et al., 2021).

The last experimental works and imaging studies allow explaining the decisive role of surface defects and grain boundaries in the control of nanoparticles corrosion. These structural features emerge as high-energy irregularly packed arrangement of atoms; which are favourable pathways of chemical influences and atom motions. Single-particle plasmonic nanoimaging has also demonstrated that grain boundaries are not only initial locations of concentration of vacancies but also promote the emergence of Kirkendall voids in the oxidation process and thereby augment the transformation of crystalline nanoparticles into porous and defective grains (Bayat et al., 2021). In addition to this, fast or non-equilibrium prepared nanoparticles are usually disordered or amorphous and have readily available sites of corrosion beginning. Enrichment of the grain boundaries leads to an enhancement of the capability to remove vacuity sinks, the revisability of the surface reactivity, and acceleration of the pitting and cavity formation mechanism with the end result of a high-volume loss and structural disintegration (Liu and Zhao, 2025). It is worth noting that such weaknesses can be prevented with the help of microstructural engineering techniques. It is shown that the ability to control grain size, the surface distribution of the material, and the application of gradient nanograined surfaces can be employed to enhance the corrosion resistance of the material through a reduced number of defects and stabilized passivating films (Beura et al., 2022). Collectively, these results indicate that corrosion of MNPs is not a reduced version of bulk corrosion, but a complicated interaction between the reactivity of nanoscale and structural instability as well as the aggressiveness of the environment and more complicated strategies should be used when creating a material to guarantee lengthy-term stability and functional viability.

4. CORROSION BEHAVIOUR OF DIFFERENT METAL NANOPARTICLES

Noble metals such as gold (Au), silver (Ag), platinum (Pt), palladium (Pd), etc., are known for their phenomenal resistance to corrosion that is an occurrence brought about by low chemical reactivity and high standard electrode potentials. Direct oxidation of the nanoparticle surface and the formation of cations is usually the main corrosion mechanism in such materials, but this mechanism is considerably subdued compared to transition metals, and it gives rise to much higher stability in the long term (Yanardag, 2019; Luciano et al., 2022; Liu et al., 2015). An example of such nanoparticles, namely, gold nanoparticles (AuNPs) exhibit high levels of inertness and can withstand even very adverse conditions like strong acids or saline solutions. However, in very oxidizing conditions or when the gold is reduced to very small sizes of a few nanometres, it can oxidize, but these reactions are inhibited by kinetic factors and seldom pose a threat to bulk stability (Yanardag, 2019). Although similarly stable over time, silver nanoparticles (AgNPs) are more susceptible to surface oxidation and sulfidation, especially in biological or marine conditions, in which silver oxide (Ag_2O) and silver sulphide (Ag_2S) coating and phase formation of the nanoparticles occur and change the physicochemical and optical properties over time (Luciano et al., 2022; Yanardag, 2019). The nanoparticles of platinum and palladium (PtNPs and PdNPs) are also resistant, and they can restructure their surfaces under the conditions of repeated electrochemical cycling, particularly in bimetallic systems (e.g. Pd@Pt) where the movement of atoms through adsorbates can potentially take place. Interestingly, this reorganization may result in a stability improvement in some cases because the creation of thin coatings of oxide on the edges or corners of nanoparticles creates extra passivation and inhibits catastrophic dissolution (Liu et al., 2015). It is due to these excellent corrosion-resistant behaviours that the applications of noble metal nanoparticles have been extensively used in demanding applications such as biosensing, corrosive environment catalysis, and medical devices, in which the degradation of the materials would negatively affect the performance and safety (Luciano et al., 2022).

In transition metal nanoparticles such as iron, nickel, copper and cobalt, however, they are significantly lower corrosion resistant due to the higher reactivity and will produce an oxide, hydroxide or other products of corrosion much more easily (Chen et al., 2022; Ahmad et al., 2010). There is a quick oxidation of FeNPs in wet or oxygenated environments, to form either Fe_2O_3 or Fe_3O_4 to lead to alteration of their magnetic and catalytic properties, and to achieve a rapid degradation. To address this, encapsulation techniques which are protective are commonly employed to delay the oxidation process and to extend the life of operation, including carbon coating and polymer shells (Chen et al., 2022). Nickel and cobalt nanoparticles (NiNPs and

CoNPs) may offer temporary resistance as passive oxide films (NiO or CoO) that are typically dissolved by acidic or highly oxidizing environments and as such, these films lose their resistance rapidly, resulting in poor stability (Shaheen et al., 2020). The copper nanoparticles (CuNPs) are more prone because it readily oxidizes to form Cu_2O and CuO, which are only prone to lower the electrical conductivity but at the same time, diminish its application. Some methods such as alloying metals that are more stable, functionalization of surfaces, and incorporation of organic stabilizers are only a few of the widely applied techniques that have been put in place to mitigate such vulnerabilities (Shaheen et al., 2020). This means that transition metal nanoparticles are required in studies in the business sector regarding catalysis, environmental remediation, and biomedicine to be deliberately stabilized to counteract the potential undesirable effect of their own corrosive degradation (Chen et al., 2022).

Alloyed and bimetallic nanoparticle's introduction has proven a strong solution to merge the functionality of the reactive transition metals with the stability of the noble metals to form hybrid systems that would be both highly active yet exhibiting improved corrosion resistance (Ahmad et al., 2010). Synergistic effects can be enabled by composition and architecture (e.g. better electronic interactions, as well as the creation of protective shells, protecting fewer stable cores) through meticulous tune-up. As in Au-Pt or Pt-Ni, Cu-Fe, and so on, not only does the stabilization of the structural frame, but also the increase in the catalytic activity of the alloy, the more resistant element to corrosion will change the surface energies and prevent early dissolution of the active form (Goswami et al., 2025). Nanoparticle microstructure can greatly determine the corrosion processes of these materials, with core-shell structures allowing the shell metal to serve as a protective layer, and disordered or complete alloy systems tending to have the same level of resistance since the elements are not mixed uniformly. These bimetallic and alloy nanoparticles are also becoming desired in the next generation applications where they should be capable of withstanding extreme electrochemical or environmental factors, such as fuel cell electrocatalysts, eco-friendly catalysts and biomedical imaging agents (Huynh et al., 2020). The current investigation efforts are focused on the optimization of these nanostructures to facilitate the stability of the associated operations during the long-term, the optimization of the oxide layers formation, and the implementation of self-protective mechanisms to make sure that the corrosion is reduced in even the cases of the intensive electrochemical cycling (Goswami et al., 2025). All of these pioneer developments indicate that rational design of alloys and bimetallic nanoparticles can serve as an effective remedy to the nature of the weaknesses of the transition metals and successfully utilize the nature of the stability of the noble metals to develop corrosion resistance nanomaterials in future technologies.

5. BIBLIOMETRIC ANALYSIS

The conceptual framework of the research into the corrosion in metal nanoparticles (MNPs) has experienced an impressive metamorphosis during the last thirty years as the study of nanoscience is becoming more mature, and its use is becoming more and more varied. A search using Lens.org as a provider of bibliometric data that was further analysed with VOSviewer offers a broad overview of the way in which this domain has evolved in terms of size and depth. Combined with journal and country-level analyses, temporal, network, overlay, and density visualizations can be used to understand the interaction between fundamental chemistry, applied materials science, biomedical engineering, and energy research to develop this research area.

Figure 1 illustrates that there is a definite trend of publications over time covered by the past five periods. The next stage (1996-2005) and the emergent phase had very little activity in it because the research on nanoparticles as a matter was still in its early stages. The nanoscale corrosion processes were sometimes recognized but not yet existed as a specific research area. The period of slow growth (2006-2012) saw a gradual rise of publications, which was associated with the global exploration of the issues of nanotechnology and the identification of nanoparticle stability and degradation as essential issues. By the boom period (2013-2019) corrosion in MNPs was now recognized by most as a central factor in defining the performance of nanoparticles, and the number of publications on the topic shot up as more characterization tools were becoming able to analyse nanoscale behaviour. The peak (2020-2022) is the most active period, where the world communities, strategic funding schemes, and the concern of society about the toxicity, stability, and environmental impact of nanomaterials came together to generate the most publication output in the past. Nevertheless, the latest time span, 2023-2025, indicates the decrease of the intensity of publications. This decline implies a saturation of underlying themes and a transition in research interests in the world to next-generation materials, including two-dimensional structures, and green nanotechnology, as well as to digital strategies, including AI-assisted material discovery.

Figure 1. Publication trend on corrosion in MNPs (1996-2025) (Source: Lens Database) (Lens, 2025).

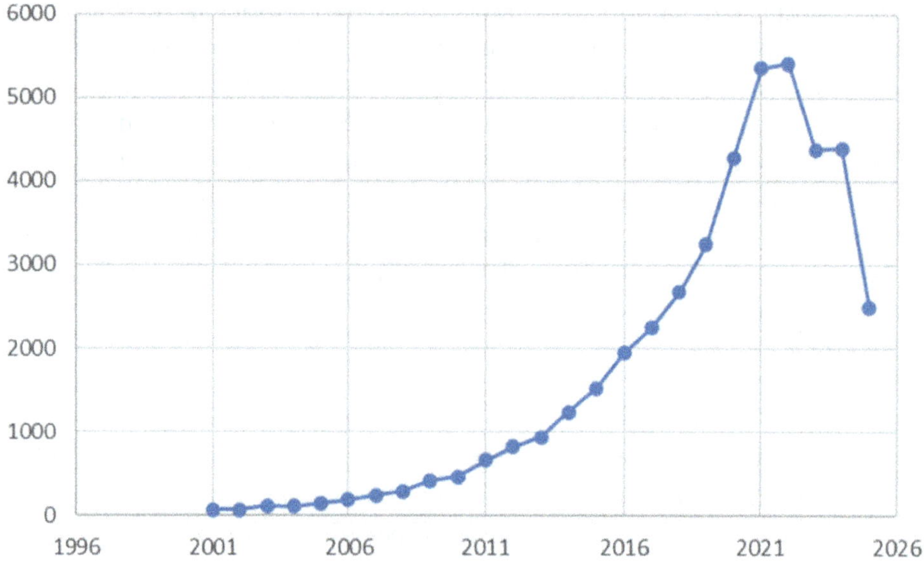

The bibliometric network visualization (Figure 2) will give additional data on the intellectual organization of this area. There are clear bodies of research activity that manifest interdisciplinary approaches to incorporating corrosion research into broad scientific and engineering perspectives. Among the key areas that relate corrosion in MNPs to catalysis, energy technologies, and material performance, electrocatalysis, photocatalysis, graphene, nanocomposites, and mechanical properties will be connected in one of the largest clusters. The other cluster is substituted by the presence of electrochemical processes, including oxidation, reduction, processes, and iron-based chemistry, which indicates the strong linkage of this discipline with the classical corrosion science. The third cluster focuses on the biomedical aspects, such as cytotoxicity, drug delivery, and biocompatibility, as the issue of nanoparticle degradation in physiological conditions is gaining momentum. Other clusters represent the applied and engineering viewpoints, and provide connections between corrosion and biomaterials, coating systems, dental implants, and tissue scaffolds, and mechanical performance concerns of welding, microstructure, and fuel cells. All of these clusters affirm the interdisciplinary nature of the field by its nature but corrosion is at the border of materials science, chemistry, biomedical engineering, and energy research. The key most central words are nanoparticles, corrosion, nanotechnology and humans, which demonstrate the dual scientific and societal topicality of the subject.

The temporal overlay representation (Figure 3) adds value to this image through plotting the development of the thematic focus over time. The initial research focus until around 2017-2018 was on toxicological and chemical background, and the key search terms include oxidation -reduction, toxicity, and electrochemistry. These were the initial anxieties about the reactivity of nanoparticles or environmental or biological hazards. By 2019, greater focus was extended to the nanostructured systems and functional approaches to corrosion mitigation where graphene oxide, carbon nanotubes, coating, and nanocomposites appeared. The 2020-2021 saw a new energy and sustainability-related application diversification and the research in the field started to concentrate on electrocatalysis, photocatalysis, hydrogen evolution reaction, and energy storage. At the same time, biomedical growth was also obvious, and such words as 3D printing, dental implants, and bone tissue engineering became popular. This development shows how corrosion studies have shifted to limited research on the study of chemical degradation to an overall understanding of how corrosion is a performance-defining aspect in a range of technologies.

Figure 2. Bibliometric network map of keywords related to corrosion in MNPs (Source: Lens Database; created using VOSviewer) (Lens, 2025; Van Eck & Waltman, 2022).

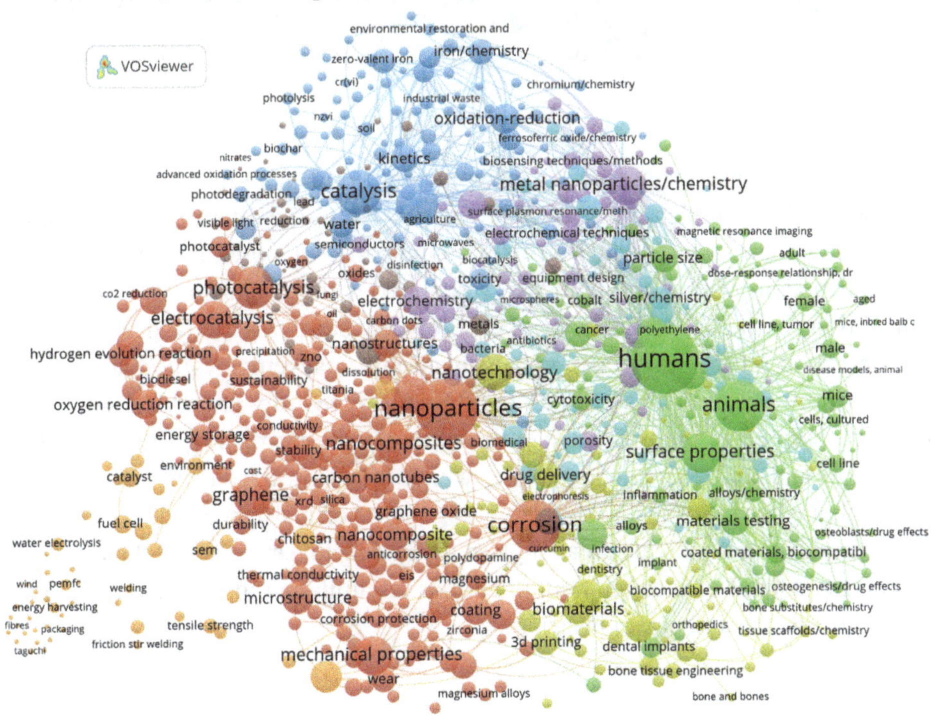

Figure 3. Overlay visualization of temporal keyword evolution (2017–2021) (Source: Lens Database; created using VOSviewer) (Lens, 2025; Van Eck & Waltman, 2022).

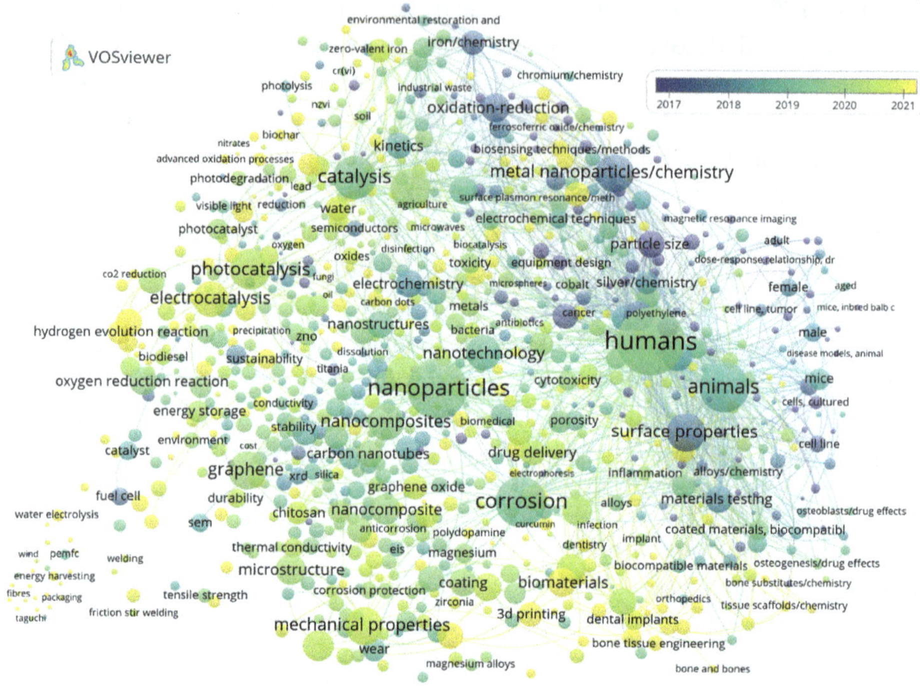

The density map (Figure 4) is a complement to the analysis of centrality, also supporting the dominant presence of key terms, but also outlining new directions. The most common and related terms are nanoparticles, corrosion, surface properties, nanocomposites, and humans, which proves their status as the conceptual foundations of the field. The close association between cytotoxicity, drug delivery, inflammation, biomaterials, coatings, and materials assays employs the dual meaning of corrosion in as much as stability and biological security are addressed. These may be observed in the periphery in the form of such words as graphene, carbon nanotubes and chitosan nanocomposites which show how dependent on composite materials are becoming in terms of providing corrosion resistance and functionality. This peripheral but growing areas suggest the intellectual structure of the field is in a constant state of expansion in all directions, assimilating the notions in the new materials science and bioengineering.

Figure 4. Density visualization of co-occurring terms in corrosion research (Source: Lens Database; created using VOSviewer) (Lens, 2025; Van Eck & Waltman, 2022).

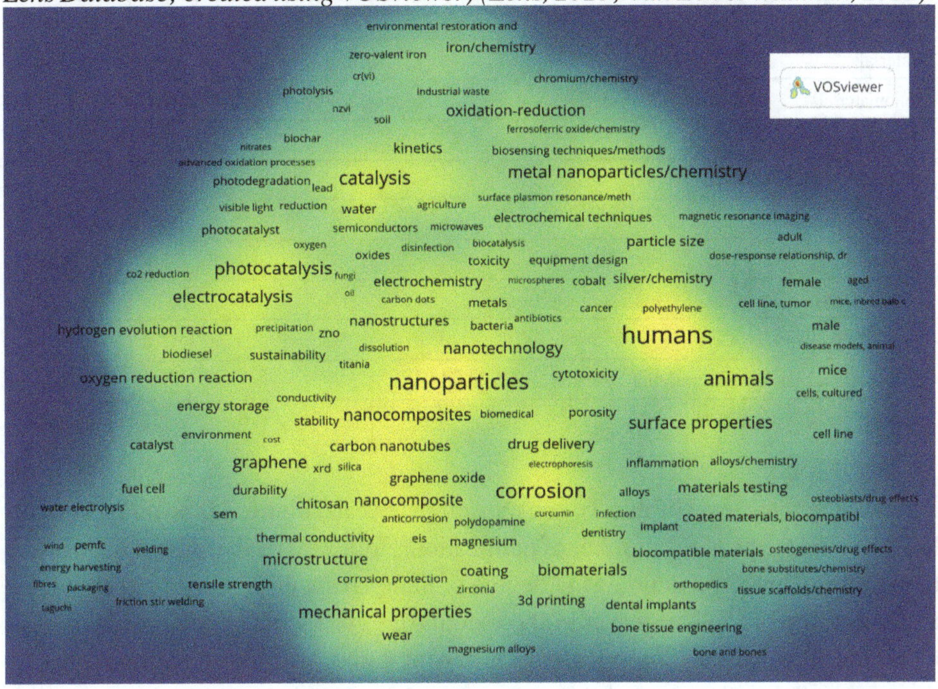

The fact that the field is multidisciplinary is also revealed in Table 2 on the tendency of publications in journals. Dominated sources are Materials, Nanomaterials, Journal of Materials Science and Coatings, which is connected with the great inclination to materials science and engineering. Chemistry journals such as RSC Advances and Molecules are also significant components of showing the applicability of chemical synthesis and mechanistic understanding. The other notable dimension that can be provided with the assistance of such environmental oriented informational sources as Environmental Science and Pollution Research would illustrate that the problems of environmental impact and environmental sustainability have taken the central place in the MNPs corrosion research. The publications list indicates how interdisciplinary the nature of communications that characterize this specialist field is and the results of the research are presented not only in the field of materials science but also in chemistry, the environmental sciences and even in biomedical research.

Table 2. Leading journals publishing research on corrosion in MNPs (Source: Lens Database) (Lens, 2025).

Publisher	Count
Materials	1466
Nanomaterials	1019
Scientific reports	795
Journal of Materials Science	772
RSC advances	753
Polymers	739
Journal of Materials Science: Materials in Electronics	731
Molecules	544
Environmental science and pollution research international	530
Coatings	506

At the country level (Table 3), one may observe that China holds an unquestioned leadership in terms of the number of publications, followed by the United States, and then, India. This trend denotes the degree of national spending on nanotechnology and focus in materials research in these countries. The great participation of UK, Germany, Iran, Russia, South Korea, Saudi Arabia and Australia speaks volumes of the very globalist character of this research project. The Asian countries are taking the lead in terms of volume, but the western countries perform the interdisciplinary and application-oriented innovations in excess. It is this difference that denotes a highly complementary world system where the huge quantity of research productions is compensated by conceptual and practical innovations elsewhere in the world. The co-authorship data also indicate that the degree of international collaboration is elevated, and some of the publications can be viewed as the fruit of the collaboration between the disciplinary and the geographic levels. All these bibliometric results centrally show that corrosion in MNPs is an emerging, cross-disciplinary and globally diverse research problem. The field was initially developed in the framework of biological safety and electrochemical degradation but has evolved gradually to incorporate nanocomposites, graphene-based materials, and coatings, as well as a wide range of surface engineering. It has since expanded more recently into energy and sustainability applications, regenerative medicine and biomedical engineering. This change brings about a basic redefinition: Corrosion is no longer considered as a degradation phenomenon but as a key performance, safety and reliability determinant in advanced nanomaterials.

Table 3. Country-level publication output on corrosion in MNPs

Country	Count
China	10648
United States	4595
India	3356
United Kingdom	1918
Iran	1698
Russia	1568
Germany	1546
Korea, Republic of	1281
Saudi Arabia	1114
Australia	1078

(Source: Lens Database) (Lens, 2025).

In the future, the decline in publications after 2023 cannot be understood as the decrease in the level of significance but can be viewed as the sign of the maturity of the field and redirection. The new fields in corrosion modelling (guided by machine learning), prediction of nanoparticle stability using data, and sustainable development of corrosion-resistant nanomaterials are likely to transform the discipline. Simultaneously, overlaps with the principles of the circular economy, green deterrents, and environmentally-friendly nanocomposites indicate the new directions of socially responsible innovation. With the global scientific agenda on transitioning to clean energy, biomedical safety, and environmental stewardship, the field of corrosion research in MNPs will probably not be left out as it will keep on redefining itself at the threshold of these issues.

6. CORROSION CONTROL AND APPLICATIONS OF METAL NANOPARTICLES

Anti-MNPs corrosion and its mitigation has turned into a highly critical issue in the light of the frequent applications of this technology in biomedical, catalytic, environmental, and structural pipelines. Coating and functionalizing the surface is one of the most commonly employed methods to delay nanoparticle degradation whereby protective layers are either inorganic or organic and they can act as a physical or chemical barrier to corrosive species (Muresan, 2023). Such modifications drastically slow down the rate of dissolution and oxidation by avoiding direct contact of the metal core with water, oxygen, or aggressive electrolytes, and maintain the stability of the particle. Polymeric films and silane layers as well as oxide passivation have been demonstrated to reduce ion diffusion as well as block reactive surface sites thus prolonging the life of nanoparticles in hostile conditions such as saltwater, acidic conditions or physiological fluids. Functionalization is also known to increase the corrosion resistance, by fine-tuning interfacial chemistry to enhance adhesion, hydrophobicity or biocompatibility without affecting intrinsic metallic core properties (Muresan, 2023; Kumar & Sharma, 2022). In addition to coatings, core-shell nanostructures can be very effective since a metallic core is incorporated with an outer shell, which serves as a passive barrier, or as an active corrosion inhibitor. Core-shell systems, including ZnO@SiO_2 and CeO_2@polymer systems, offer two types of protection, namely, physical shielding, as well as the release of corrosion-inhibitory ions. It has been considered essential to optimize the shell thickness, interfacial bonding, and dispersion of composite matrices to provide greater durability, and the electrochemical impedance spectroscopy and salt spray test have shown a consistent enhancement in the resistance of the material over bare or single-phase nanoparticles (Muresan, 2023).

Along with synthetic modifications, more research focus has been on environmentally friendly processes by developing green inhibitors and polymer composites. Such approaches use naturally available extracts, microorganisms or non-toxic building blocks in order to produce corrosion-resistant nanomaterials with better biocompatibility and less ecological footprint (Yadav et al., 2024). Given into polymer matrices, green nano inhibitors serve not only physical protection to the nanoparticles but also active passivation through the prevention of corrosive routes at the nanoparticle surface. The nature of such nanocomposites is synergistic because of the ability to couple mechanical reinforcement of polymer and electrochemical inhibition efficiency of nanostructures which results in a significant reduction of rust formation, a longer service life, and low toxicity of the surrounding environments (Bairagi et al., 2024). The other crucial corrosion prevention axis is the internal environment that must be kept at the correct level since the pH, oxygen supply and electrolyte make a huge impact on the rate of corrosion. Acidic conditions are most likely to act in favour of solubility of metal ions and chloride rich environments are likely to destabilize passive films, thus encourage pitting. Conversely, there are anions that can lead to passivation through the formation of protective precipitates and the oxygen concentration directly influences cathodic reactions. This adjustment of solution chemistry, dissolved oxygen restraint, or pH buffering is, therefore, an important step to stabilizing nanoparticles in their storage or use (Peng et al., 2017).

The effects of corrosion can go beyond the protection measures and have a direct effect on the reliability and safety of MNP-based technologies. Uncontrolled corrosion may also impair biocompatibility in biomedical applications by the use of nanoparticles, such as imaging, targeted drug delivery, and implant coatings, which release cytotoxic ions, leading to inflammation, immune reactions, or allergic responses (Eliaz, 2019). A notably hostile environment is the physiological one, which is characterized by the presence of chloride ions, proteins, and changing pH, and it also requires the use of advanced surface treatments to ensure the integrity of the implants and avoid systemic toxicity (Kim et al., 2024; Manam et al., 2017). The stability problem in catalytic and energy uses is a two-sided sword: active sites can be inactivated neither by partial oxidation nor thin films of passivation, but in other situations, they are antithetically active to catalyst activity and can be harnessed to control the surface reactivity. A major issue in the design of fuel cell catalysts, hydrogen evolution electrocatalysts, and other energy-related nanomaterials is the appropriate balance between the structural integrity and the retention of open catalytic surfaces (Astruc, 2020). As far as the environmental and structural factor is concerned, nanoparticle corrosion contributes to the leaching of ions in soil and water body and makes individuals worried about ecological toxicity and the interest of human health. At the same time, the structural elements of construction, transportation, and electronics may be weakened by corrosion-based degradation in the nanoparticle-reinforced composites. It is essential to know the environmental fate of corroded nanoparticles such as aggregation, dissolution, and contact with organic matter, which contributes to risk assessment and mitigation (Meeran et al., 2025).

Nevertheless, the problem of stability of nanoparticles against functionality still exists and needs to be resolved. Protective coating or shells tend to restrict the availability of catalytically active or biomedical sites of activity, and the corrosion resistance and activity provide a trade-off. A likely direction is the quest of so-called smart materials with selective permeability, self-healing reactions, or stimuli-triggered reaction, allowing nanoparticles to maintain functionality under normal conditions and instigate protective responses in the presence of corrosive attack (Kumar & Sharma, 2022). Beyond that, the scalability, reproducibility and cost-effectiveness issues of this technology pose an industrial level challenge to the deployment of state-of-the-art corrosion protection measures. How both core-shell nanoparticles and hybrid nanocomposites, as well as green inhibitors, need to be synthesized should offer uniform dispersion and morphology and predictable surface chemistry at high concentrations. Automated manufacturing, artificial approaches and optimizing the processes with the help of AI will also be central drivers in overcoming such barriers (Ndukwe et al., 2025). Finally, the most important gaps in the research which remain critical are the long-term behaviour of nanoparticles in the real world, the nature of multi-component corrosion process, and the environmental effects of the degradation products of nanoparticles. Recent technologies, such as atomic-level synthesis, in situ nanoscale measurements, and predictive modelling by artificial intelligence have new opportunities to create these gaps in knowledge (Kumar & Sharma, 2022). Interdisciplinary integration of material science, electrochemistry, biology and environmental science will then be done so as to fully enjoy the benefits of mitigation strategies of corrosion without jeopardizing the wellbeing of the application of metal nanoparticles in high-level technological systems.

7. CONCLUSION

MNPs corrosion may be researched to explain a dynamic research area in the interface between the materials science, electrochemistry, and nanotechnology disciplines too, where the smaller nanoscale size multiplies possibilities and hazards of these substances. As indicated in the discussion on the transformative use of MNPs in the electronics, energy, medicine and environmental technologies, it is evident that its high reactivity as a result of high surface energies and quantum effects is the major force behind innovation and at the same time, it is a challenge in that it degrades very fast. The fundamentals' part is a description of the interaction of such factors as particle size, shape, surface area, and environmental conditions to determine the rate of corrosion, which is often to the extent that the dissolution and structural rearrangement proceeds faster than in bulk metals. Again it is decomposed in chapter on mechanisms where oxidation, galvanic corrosion, pitting and stress induced cracking are established to have their origin in defects, grain boundaries and high-resolution imaging refutes the claim of Kirkendall voids dominance and selective leaching in the bimetallic systems.

Such metals as gold and platinum have been found to be naturally resistant to corrosion as analysed in corrosion behaviour, and can be used as biosensing and catalysts, whereas transition metals such as iron and copper are naturally unstable in use in remediation or biomedical applications. Bimetallic and alloyed nanoparticles that are hybridized will take advantage of synergies to gain a longer shelf life yet not reducing the activity as in core shell structures that shield reactive cores. The bibliometric analysis is a macro-level perspective, showing the maturation of the field in five phases of rise and fall in publications, peaked in the world nanotechnology booms, and revealing the thematic shifts between global toxicity questions and sustainable applications in energy storage and regenerative medicine. The journals like Materials and Nanomaterials focus on spreading on the multidisciplinary frontiers, but the national output is a suggestion of a united effort on the global level, with the largest amount of contribution being Asia and the largest innovation being West.

Speaking about the corrosion reduction, the chapter points at the practical methods involving polymeric and oxide coating, green production of the inhibitors, and the environmental modulation, which does not only enhance the lifetime of MNPs but is also adapted to the environmental-friendly paradigm. These techniques play a major role in practice in which corrosion would render viability in which, in implant applications, toxic ions are released, in fuel cell catalysts, or environmental pollution. Trade-offs are however, protecting catalytic sites or drugs release can inhibit this which is why a smart nanomaterial that is self-healing or stimuli-responsive must be utilized. Scalability issues, long-term real-world performance issues, and opportunities exist in the fields of AI-based modelling, atomic-precision synthesis and interdisciplinary integration. In conclusion, it is not only that corrosion in MNPs is a degradation phenomenon, but that it is, by far, a design consideration, and that determines the safety, efficacy and viability of nanomaterial technologies. As the current paper is progressing into the green nanotechnology, information-based forecasts, the future will see the creation of the strong, multi-purpose MNPs that will be able to reduce environmental hazards and bring about the prospects of energy conversion, personalized medicine, and closed economies. This changing perception of corrosion control has seen it play a major role in the realization of the overall benefits of nanotechnology where the tiny particles will be used in the macroscopic formation without the undesirable occurrences.

REFERENCES

Ahmad, T., Ganguly, A., Ahmed, J., Ganguli, A. K., & Alhartomy, O. A. (2010). Nanorods of transition metal oxalates: A versatile route to the oxide nanoparticles. *Arabian Journal of Chemistry*, *4*(2), 125–134. DOI: 10.1016/j.arabjc.2010.06.041

Astruc, D. (2020). Introduction: Nanoparticles in catalysis. *Chemical Reviews*, *120*(2), 461–463. DOI: 10.1021/acs.chemrev.8b00696 PMID: 31964144

Bairagi, H., Vashishth, P., Ji, G., Shukla, S. K., Ebenso, E. E., & Mangla, B. (2024). Polymers and their composites for corrosion inhibition application: Development, advancement, and future scope–A critical review. *Corrosion Communications*, *15*, 79–94. DOI: 10.1016/j.corcom.2023.10.006

Bayat, M., Nasernejad, B., & Falamaki, C. (2021). Preparation and characterization of nano-galvanic bimetallic Fe/Sn nanoparticles deposited on talc and its enhanced performance in Cr(VI) removal. *Scientific Reports*, *11*(1), 7715. Advance online publication. DOI: 10.1038/s41598-021-87106-0 PMID: 33833296

Beura, V. K., Karanth, Y., Darling, K., & Solanki, K. (2022). Role of gradient nanograined surface layer on corrosion behavior of aluminum 7075 alloy. *NPJ Materials Degradation*, *6*(1), 62. Advance online publication. DOI: 10.1038/s41529-022-00271-z

Boruah, J. S., Devi, C., Hazarika, U., Reddy, P. V. B., Chowdhury, D., Barthakur, M., & Kalita, P. (2021). Green synthesis of gold nanoparticles using an antiepileptic plant extract: In vitro biological and photo-catalytic activities. *RSC Advances*, *11*(45), 28029–28041. DOI: 10.1039/D1RA02669K PMID: 35480751

Chen, X., Liu, J., Yuan, T., Zhang, Z., Song, C., Yang, S., Gao, X., Wang, N., & Cui, L. (2022). Recent advances in earth-abundant first-row transition metal (Fe, Co and Ni)-based electrocatalysts for the oxygen evolution reaction. *Energy Materials*, *2*(4), 28. DOI: 10.20517/energymater.2022.30

Du, J. S., He, K., Xu, Y., Wahl, C. B., Xu, D. D., Dravid, V. P., & Mirkin, C. A. (2021). Galvanic transformation dynamics in heterostructured nanoparticles. *Advanced Functional Materials*, *31*(46), 2105866. Advance online publication. DOI: 10.1002/adfm.202105866

Dwivedi, D., Lepková, K., & Becker, T. (2017). Carbon steel corrosion: A review of key surface properties and characterization methods. *RSC Advances*, *7*(8), 4580–4610. DOI: 10.1039/C6RA25094G

Eliaz, N. (2019). Corrosion of Metallic Biomaterials: A Review. *Materials (Basel)*, *12*(3), 407. DOI: 10.3390/ma12030407 PMID: 30696087

Goswami, M., Pal, D., Ghosh, M. K., Dwivedi, S., & Chandraker, S. K. (2025). Bimetallic Nanoparticles: A Comprehensive review of synthesis methods and their Potential application. *ChemistrySelect*, *10*(29), e01493. Advance online publication. DOI: 10.1002/slct.202501493

Hiromoto, S. (2010). Corrosion of metallic biomaterials. In *Elsevier eBooks* (pp. 99–121). https://doi.org/DOI: 10.1533/9781845699246.2.99

Huynh, K., Pham, X., Kim, J., Lee, S. H., Chang, H., Rho, W., & Jun, B. (2020). Synthesis, properties, and biological applications of metallic alloy nanoparticles. *International Journal of Molecular Sciences*, *21*(14), 5174. DOI: 10.3390/ijms21145174 PMID: 32708351

Jeong, H., Alarcón-Correa, M., Mark, A. G., Son, K., Lee, T., & Fischer, P. (2017). Corrosion-Protected hybrid nanoparticles. *Advancement of Science*, *4*(12), 1700234. Advance online publication. DOI: 10.1002/advs.201700234 PMID: 29270338

Kadyrzhanov, K. K., Kozlovskiy, A. L., Egizbek, K., Kenzhina, I. E., Abdinov, R. S., & Zdorovets, M. V. (2022). Study of corrosion mechanisms in corrosive media and their influence on the absorption capacity of Fe_2O_3/$NDFeO_3$ nanocomposites. *Nanomaterials (Basel, Switzerland)*, *12*(13), 2302. DOI: 10.3390/nano12132302 PMID: 35808138

Kim, S., Kim, C., Lee, S., & Park, S. (2024). Carbon materials and their metal composites for biomedical applications: A short review. *Nanoscale*, *16*(35), 16313–16328. DOI: 10.1039/D4NR02059F PMID: 39110002

Kumar, N., & Sharma, A. (2022). Surface coatings and functionalization strategies for corrosion mitigation. In *ACS symposium series* (pp. 291–316). https://doi.org/DOI: 10.1021/bk-2022-1418.ch014

Lasserus, M., Knez, D., Schnedlitz, M., Hauser, A. W., Hofer, F., & Ernst, W. E. (2019). On the passivation of iron particles at the nanoscale. *Nanoscale Advances*, *1*(6), 2276–2283. DOI: 10.1039/C9NA00161A PMID: 36131962

Lens, (2025). *Search results for "Corrosion in Metal Nanoparticles (MNPs)"*. The Lens. Retrieved September 19, 2025, from https://www.lens.org/lens/search/scholar/list?q=Corrosion%20of%20Metal%20Nanoparticles&n=10

Liu, M., & Zhao, Z. (2025). Corrosion and mechanical Micro-Interaction behavior of metal materials. *Materials (Basel)*, *18*(17), 4114. DOI: 10.3390/ma18174114 PMID: 40942539

Liu, W., Herrmann, A., Bigall, N. C., Rodriguez, P., Wen, D., Oezaslan, M., Schmidt, T. J., Gaponik, N., & Eychmüller, A. (2015). Noble Metal Aerogels—Synthesis, Characterization, and Application as Electrocatalysts. *Accounts of Chemical Research, 48*(2), 154–162. DOI: 10.1021/ar500237c PMID: 25611348

Luciano, K., Wang, X., Liu, Y., Eyler, G., Qin, Z., & Xia, X. (2022). Noble Metal nanoparticles for Point-of-Care Testing: Recent advancements and social impacts. *Bioengineering (Basel, Switzerland), 9*(11), 666. DOI: 10.3390/bioengineering9110666 PMID: 36354576

Manam, N., Harun, W., Shri, D., Ghani, S., Kurniawan, T., Ismail, M., & Ibrahim, M. (2017). Study of corrosion in biocompatible metals for implants: A review. *Journal of Alloys and Compounds, 701*, 698–715. DOI: 10.1016/j.jallcom.2017.01.196

Meeran, S., Begum, A. S., Sheit, H. K., Mohamed, M., & Mohan, K. (2025). The deep investigation of structural and electrochemical analysis on the anti – corrosion behavior of MGCO2O4 and ZNCO2O4 spinel nanoparticles in acid medium. *Results in Surfaces and Interfaces, 100416*, 100416. Advance online publication. DOI: 10.1016/j.rsurfi.2025.100416

Muresan, L. M. (2023). Nanocomposite coatings for Anti-Corrosion properties of metallic substrates. *Materials (Basel), 16*(14), 5092. DOI: 10.3390/ma16145092 PMID: 37512373

Ndukwe, A. I., Nwadirichi, B., Okolo, C., Tom-Okoro, M., Medupin, R., Uche, R., Arukalam, I., Onuoha, C., Egole, C., Okorafor, O., & Nwakuba, N. (2025). *Corrosion Control in Metals: A review on Sustainable approach using Nanotechnology*. Zastita Materijala., DOI: 10.62638/ZasMat1187

Peng, C., Zhang, W., Gao, H., Li, Y., Tong, X., Li, K., Zhu, X., Wang, Y., & Chen, Y. (2017). Behavior and potential impacts of Metal-Based Engineered nanoparticles in aquatic environments. *Nanomaterials (Basel, Switzerland), 7*(1), 21. DOI: 10.3390/nano7010021 PMID: 28336855

Shahalaei, M., Azad, A. K., Sulaiman, W. M. W., Derakhshani, A., Mofakham, E. B., Mallandrich, M., Kumarasamy, V., & Subramaniyan, V. (2024). A review of metallic nanoparticles: Present issues and prospects focused on the preparation methods, characterization techniques, and their theranostic applications. *Frontiers in Chemistry, 12*, 1398979. Advance online publication. DOI: 10.3389/fchem.2024.1398979 PMID: 39206442

Shaheen, K., Shah, Z., Asad, A., Arshad, T., Khan, S. B., & Suo, H. (2020). Synthesis, characterization, and multifunctional applications of CU-FE and NI-Fe nanomaterials. *ACS Omega*, *5*(26), 15992–16002. DOI: 10.1021/acsomega.0c01259 PMID: 32656420

Shi, F., Tieu, P., Hu, H., Peng, J., Zhang, W., Li, F., Tao, P., Song, C., Shang, W., Deng, T., Gao, W., Pan, X., & Wu, J. (2024). Direct in-situ imaging of electrochemical corrosion of Pd-Pt core-shell electrocatalysts. *Nature Communications*, *15*(1), 5084. Advance online publication. DOI: 10.1038/s41467-024-49434-3 PMID: 38877007

Van Eck, N. J., & Waltman, L. (2022). *VOSviewer* manual (Version 1.6.18). Leiden University, CWTS. Retrieved from https://www.vosviewer.com/documentation/Manual_VOSviewer_1.6.18.pdf

Yadav, S., Raman, A. P., Singh, M. B., Massey, I., Singh, P., Verma, C., & AlFantazi, A. (2024). Green nanoparticles for advanced corrosion protection: Current perspectives and future prospects. *Applied Surface Science Advances*, *21*, 100605. DOI: 10.1016/j.apsadv.2024.100605

Yanardag, T. (2019). Corrosion behavior of noble metals in different oral media. *International Journal of Scientific and Technological Research*. https://doi.org/DOI: 10.7176/JSTR/5-4-14

Yaraki, M. T., Nasab, S. Z., Zare, I., Dahri, M., Sadeghi, M. M., Koohi, M., & Tan, Y. N. (2022). Biomimetic metallic nanostructures for biomedical applications, catalysis, and beyond. *Industrial & Engineering Chemistry Research*, *61*(22), 7547–7593. DOI: 10.1021/acs.iecr.2c00285

Zeng, Q., Wang, B., Lai, X., Li, H., Chen, Z., Zeng, X., & Zhang, L. (2024). A multifunctional flame-retardant TA-MXene based nanocoating for cotton fabric. *Progress in Organic Coatings*, *189*, 108333. DOI: 10.1016/j.porgcoat.2024.108333

ADDITIONAL READING

Anjum, M. J., Ali, H., Khan, W. Q., Zhao, J., & Yasin, G. (2020). Metal/metal oxide nanoparticles as corrosion inhibitors. In *Elsevier eBooks* (pp. 181–201). https://doi.org/DOI: 10.1016/B978-0-12-819359-4.00011-8

Ejileugha, C., Ezealisiji, K. M., Ezejiofor, A. N., & Orisakwe, O. E. (2021). Microbiologically influenced corrosion: Uncovering mechanisms and discovering Inhibitor—Metal and metal oxide nanoparticles as promising biocorrosion inhibitors. *Journal of Bio- and Tribo-Corrosion*, *7*(3), 109. Advance online publication. DOI: 10.1007/s40735-021-00545-0

Elechiguerra, J. L., Larios-Lopez, L., Liu, C., Garcia-Gutierrez, D., Camacho-Bragado, A., & Yacaman, M. J. (2005). Corrosion at the nanoscale: The case of silver nanowires and nanoparticles. *Chemistry of Materials*, *17*(24), 6042–6052. DOI: 10.1021/cm051532n

Jain, P., Patidar, B., & Bhawsar, J. (2020). Potential of nanoparticles as a corrosion inhibitor: A review. *Journal of Bio- and Tribo-Corrosion*, *6*(2), 43. Advance online publication. DOI: 10.1007/s40735-020-00335-0

Rodriguez, P., Tichelaar, F. D., Koper, M. T. M., & Yanson, A. I. (2011). Cathodic corrosion as a facile and effective method to prepare clean metal alloy nanoparticles. *Journal of the American Chemical Society*, *133*(44), 17626–17629. DOI: 10.1021/ja208264e PMID: 21981303

Yanson, A. I., Rodriguez, P., Garcia-Araez, N., Mom, R. V., Tichelaar, F. D., & Koper, M. T. M. (2011). Cathodic Corrosion: A quick, clean, and versatile method for the synthesis of metallic nanoparticles. *Angewandte Chemie International Edition*, *50*(28), 6346–6350. DOI: 10.1002/anie.201100471 PMID: 21626623

Yu, H., Zhang, Z., & Han, M. (2012). Metal corrosion for nanofabrication. *Small*, *8*(17), 2621–2635. DOI: 10.1002/smll.201200475 PMID: 22707341

KEY TERMS AND DEFINITIONS

Corrosion: The gradual degradation of a metal through chemical or electrochemical reactions with its environment.

Galvanic Corrosion: Corrosion occurring when two dissimilar metals are electrically connected in an electrolyte, causing one to corrode preferentially.

Metal Nanoparticles (MNPs): Nanoscale metallic particles (1-100 nm) exhibiting unique physical and chemical properties due to high surface area and quantum effects.

Nanocoating: Ultra-thin protective or functional layers applied at the nanoscale to enhance surface properties.

Passivation: Formation of a thin protective oxide layer on a metal surface that reduces its corrosion rate.

Surface Plasmon Resonance (SPR): Collective oscillation of surface electrons in metal nanoparticles induced by incident light.

Chapter 2

Corrosion Inhibition in Automobile Industry Using Nano-Bio Composites

Yadnyesh Khapekar
Ajeenkya D.Y. Patil University, India

Aditya Shrivastav
https://orcid.org/0009-0000-8461-8171
Ajeenkya D.Y. Patil University, India

Sunil Sankathala
https://orcid.org/0009-0004-7809-7647
Ajeenkya D.Y. Patil University, India

Susanta Das
https://orcid.org/0000-0002-9314-3988
Ajeenkya D.Y. Patil University, India

ABSTRACT

Corrosion is a major problem affects many different businesses and causes significant structural damage and financial strain. Efficient corrosion control is essential for practical and financial reasons as well as for aesthetic, intellectual, and environmental grounds. One of the most effective and economical ways to preserve metal surfaces among the several techniques available is the use of corrosion inhibitors. Because of their enormous surface area, distinct physicochemical characteristics, and improved protective function, inhibitors based on nanomaterials have drawn interest. The production of these nanomaterials by green synthesis techniques, which

DOI: 10.4018/979-8-3373-2617-7.ch002

reduce dependency on hazardous chemicals and promote more general sustainability objectives, has been the focus of recent study. This review highlights the various materials that are used in automobile industry that face the problem of corrosion, the nano composites that are used to inhibit corrosion in them and the challenges that are yet to be addressed by the industry in the implementation of these composites.

1. INTRODUCTION

Corrosion is a prevalent and persistent issue in the automotive industry, with far-reaching ramifications for vehicle longevity, safety, and performance (Chami et al., 2021; Sawyer-Beaulieu et al., 2022). Environmental agents such as humidity, road salts, and pollutants contribute to the erosion of automotive metals, resulting in the loss of structural integrity and functionality. The imperative for effective corrosion protection methods is critical for the safety of the automotive parts and to meet the expectations of environmental sustainability. Traditional anticorrosion methods have relied on toxic substances for historical precedent, so there has been an immediate shift to green, environmentally friendly methods. Recent advances in nano and bio-based material studies have expanded the range of corrosion control methods available to the automotive industry. Development of nano-bio and hybrid nanocomposite coating properties has shown to be particularly viable through their enhanced protective performance, improved mechanical strength, and life cycle (durability) in harsh environments typically encountered in automotive service. In addition to reducing corrosion rates, advanced and new coatings also improve scratch and wear resistance, decrease maintenance frequency, and ultimately lead to a longer total service life and therefore greater reliability in the vehicle (Shafique & Luo, 2019). The economic impact of corrosion on motor vehicles is also significant as it encompasses increased maintenance costs, increased vehicle downtime and decreased depreciation. The main materials that are especially vulnerable to corrosion are critical materials that are used in car manufacturing, such as steel, aluminum, and several types of alloys. In addition to the obvious economic effects, corrosion can also pose an environmental issue through toxic discharge and actual material waste, meaning manufacturers must find a balance between efficiency, cost-effectiveness, and strict environmental standards. The most recent trends in industry are showing an increase in awareness about the importance of corrosion prevention as a way to promote safe performance of individual components, particularly in safety-critical subsystems in vehicles, such as brakes and suspension systems (Mazumder, 2020).

Corrosion inhibitors are chemicals that slow the degradation of metallic substrates by forming passivate surfaces or changing the electrochemical mechanisms that facilitate corrosion. In the past, automotive paint formulations frequently included chromates, phosphates, & amines as inhibitors; however, this application has generally been limited by acute toxicity as well as restrictions on those compounds. As a result, bio-composites & nano-bio composites have been developed as alternatives that generally demonstrate better inhibition efficiencies, biodegradability & negligible effect on the environment. Inhibition demonstrated through bio-based composites is typically based on the generation of strong passivation layers, & can be designed to be released in controlled amounts, resulting in significant life-cycle enhancement of automotive components, as shown by studies done by Km et al. (2024). The non-toxic commercial inhibitors industry has also been developed in bio-composite space & uses naturally available sources such as ionic liquids, chitosan, cellulose derivatives and furan derivatives. The inhibitors are effective corrosion suppressors, non-toxic and environmentally compatible, which is synergistic, and contemporarily regulatory change demonstrative of the trend toward green chemistry in industrial practice. Demonstrative of this, recent studies of choline-based ionic liquids and furan-carboxylic acids, have shown excellent inhibition efficiencies that support future biomaterials based green automotive anticorrosive systems (Aslam et al., 2024).

Nano-bio composites consist of an incorporation of nanoscale entities such as nanoparticles, nanofillers, and nano-coatings with bio-derived polymers, or biomolecules such as chitosan, cellulose, and graphene oxide. These composites can deliver many functional benefits, like improvements in mechanical integrity, self-healing abilities, and responsiveness to dynamic environmental conditions. It has been demonstratively shown that using these in automotive products will enhance the service-life of an automotive part and address ever-expanding environmental requirements (Idumah et al., 2020). This review seeks to investigate the potential of nano-bio composites as a means to prevent corrosion in the automotive sector, discuss the associated challenges pertaining to scaling, durability and commercial viability, and will discuss the future directions and opportunities for development of sustainable corrosion prevention based on the specificities of the sector.

2. LITERATURE REVIEW

2.1. Fundamentals of Corrosion in the Automobile Industry

Most of the corrosion of automotive parts is due to an electrochemical process whereby the metals corrode by a sequence of chemical reactions with the chemical components in the environment. This form of corrosion typically entails the formation of corrosion cells on the surface of the metal in a series of anodes (the site of oxidation), cathodes (the site of reduction), and an electrolyte (water with a myriad of ionic levels dissolved in it) that may act as a base media within which the ions and electrons can move between these areas (Sagar et al., 2024). In this case as an example, rusting of steel elements in water and some added, dissolved salt, generates corrosion cells where iron is used as the anode & is oxidized, and at the cathodic areas where the major reduction of oxygen takes place in the presence of water. Besides, it results in a localized response that generates rust & other corrosion products in the oxidation area. These general electrochemical reactions normally create localized pitting & cracking of the surface & eventually loss of material. Atmospheric corrosion is in general less degradable in aluminum, but localized corrosion may also be present, particularly when the surface oxide layer is damaged by wear, or when chloride ions are introduced into it to both corrode the layer or corrode in a galvanic interaction with a more noble metal, such as steel or copper (Sagar et al., 2024). Furthermore, an assembly of modern vehicles can be reinforced by two different metals, and a galvanic corrosion may occur provided that the two metals are somehow in contact with each other as well in the presence of an electrolyte. Bimetallic alloys, including steel/aluminum and steel/magnesium, may exhibit a high rate of corrosion in the less noble metal (typically aluminum or magnesium) as polarization experiments and actual deicing salt exposure prove (Hagarová et al., 2021).

Environmental factors play a major role in determining how quickly and how severely automotive corrosion occurs. Important factors include:

I. Temperature & Humidity: Higher temperatures and higher humidity, especially in the 60% range or higher, increase both corrosion rates and the mobility of corrosion agents.
II. Road Salts: Vehicles are subjected to excesses of chloride ions due to widespread use of sodium & calcium chloride as deicing salts in cold weather climates. Chloride ions will attack the protective layers on metals and cause pitting corrosion in aluminum & steel alloys.
III. Acid rain & pollutants: Acidic precipitation, sulfur dioxide and nitrogen oxide in the atmosphere increase the acidity of the surface film (both the metal alloy and the paint) and weaken the coatings, rapidly leading to rust formation and failure of the paint.
IV. Ultraviolet (UV) Radiation: The long-term UV radiation may cause the breaking of paints, sealants, and polymer-based protective layers that expose the underlying metal or composite to the susceptibility of moisture and ion intrusion (Cai et al., 2018).

The salty air constantly exposes coastal areas to a highly conductive layer of electrolyte on the surface of cars & intensively enhances corrosion risk. Cities are open to a blend of chemicals & pollutants of industry & traffic, which may cause complicated types of corrosion such as stress corrosion cracking of metal elements. Extreme winter conditions are associated with large amounts of deicer application and cyclic wet-dry environments, promoting both uniform and localized corrosion on the bodies of car and the undercarriage (Xie et al., 2025; Liu et al., 2018).

The effects of corrosion have a detrimental effect on various areas related to the operation and value of vehicles. Critical load bearing components such as the chassis, frame, brake lines, and suspension components can be weakened by corrosion directly impacting on the safety of the vehicle. Minor corrosion in these areas decreases the overall design bearing capacity and can result in disastrous failures when exposed to the load. The corrosion stains, the blisters of paint and surface rust also weaken the visual attractiveness of cars. This damage particularly to areas of the car that are visible such as doors, hood and under body panels negatively impacts resale value and consumer satisfaction. During the process of corrosion of metal substrates, the roughened surfaces, the product accumulation can cause the disruption of aerodynamic profiles, interference with moving components or fidelity of electrical connections. The failures of paint & sealants admit more water and debris into areas that are supposed to be covered which contributes to further corrosion & could also affect the fuel efficiency by increasing weight through rusting or decreasing mechanical performance. Repair due to corrosion also makes up a large portion of the total vehicle maintenance costs. The replacement of corroded components e.g. exhaust systems, braking pipes, wheel arches, fastening reduces the working life of vehicles and may also increase the downtime. According to the results of industrial surveys, corrosion has always been one of the main reasons of warranty claims and unplanned visits (Golubović-Bugarski et al., 2021).

Many different materials are commonly used in the construction of automobiles and each material has different corrosion susceptibility. The steel that is used easily oxidizes when its paint or galvanization is broken especially in wet or salty conditions. Welded joints are especially susceptible to heat-affected areas where grain growth & pitting are notorious in promoting the development of cracks and structural failure. Galvanic corrosion occurs commonly in assemblies made of steel & more or less noble metallic inserts. Lightweight panels, wheels, structural, and heat exchangers are widely made of aluminum. The passive aluminum oxide coating offers the protection on baselines, but the exposure to the chloride ions or the development of galvanic couples with steel or magnesium may lead to the rapid pitting and localized attack as it was observed by the recent research on the aluminum alloys during the deicing conditions (Sagar et al., 2024). Zinc is mostly used as galvanic coating on steel to serve as a sacrificial anode, which prevents the corrosion of the underlying steel. Zinc is prone to white rust & may be exhausted on aggressive cyclic exposure to salt sprays undermining long-term protection. Climatic stress is also susceptible to other materials such as plastics, rubber, paint coatings, organic coatings (paints, rubber gaskets & polymeric seals). Cracking, peeling and loss of adhesion exposes the underlying metals and therefore corrosion occurs rapidly. Although these layers do not corrode, the disintegration of these layers is a significant corrosion initiator route on metallic surfaces (Liu et al., 2018; Hagarová et al., 2021). For example, steel body panels that have paint damage (scratches, chips) often rust at these exposed locations by showing surface staining that progresses to pitting & cracking over several weeks to months in wet environments or on salted roads. Aluminum alloy wheels often show pitting and surface flaking for the same reason once the oxide layer is broken, with the initiation by chloride ions noted in long-term exposure studies and microstructural studies. Mixed assemblies (for example, steel/copper wiring harnesses & steel/aluminum frames) are especially vulnerable to galvanic corrosion, where lab studies and studies of real vehicles have consistently demonstrated rapid dissolution of the less noble metal in the mixed assembly.

Below are the equations that explain the chemical reaction that occurs in iron which converts it to rust. Oxygen and moisture are the key elements which convert iron to rust.

An equation showing the rusting of iron is shown below in Equation 1.

1. Overall iron corrosion reaction forming rust:

$$4Fe + 3O_2 + 6H_2O \rightarrow 4Fe(OH)_3 \tag{1}$$

Eq. (1) shows chemical reaction between iron, oxygen & water to form rust.

This represents iron reacting with oxygen & water to form hydrated iron (III) oxide, or rust.

2. Electrochemical half-reactions of corrosion:
   - Anodic (oxidation of iron):

$$Fe \rightarrow Fe^{2+} + 2e^- \tag{2}$$

   - Cathodic (reduction of oxygen):

$$O_2 + 2H_2O + 4e^- \rightarrow 4OH^- \tag{3}$$

3. Subsequent rust formation:

$$Fe^{2+} + 2OH^- \rightarrow Fe(OH)_2 \tag{4}$$

$$4Fe(OH)_2 + O_2 + 2H_2O \rightarrow 4Fe(OH)_3 \quad (5)$$

These reactions explain the electrochemical corrosion mechanism where Eq. (2) shows iron oxidizes releasing electrons, Eq. (3) shows oxygen is reduced and then Eq. (4) & Eq. (5) show us how iron hydroxides are formed which leads to rusting of Iron (Klenam et al., 2021).

2.2. Corrosion Inhibition: Theory & Mechanisms

Corrosion inhibition is a basic approach where the rate of global or deterioration of metal is controlled or avoided by the electrochemical reactions, which are of particular importance in the severe conditions that are presented to the automotive components. Essentially, corrosion inhibition is achieved by interfering with redox reactions, i.e. anodic oxidation of metals and cathodic reduction of environmental species, by the addition of chemical compounds called inhibitors. Such inhibitors act in a number of critical ways. One of such methods is adsorption, during which the molecules of the inhibitor are attached to the metal surface & a protective layer is formed that is compact & blocks the corrosive substances like water, oxygen & ions. It is possible to include organic molecules or inorganic particles in this layer, which can be stabilized by a physical or chemical interaction. The other mechanism entails the inhibitor creating a barrier film, where the inhibitor encourages the growth of an uninterrupted, pore-less film that impedes the movement of corrosive elements to the metal. This film lowers the electron transfer rate as well as ion exchange which reduces the corrosiveness of the metal. Also, there are certain inhibitors that encourage passivation by forming a stable, insoluble oxide or salt film that does not allow further oxidation, especially with metals such as aluminum & stainless steel. There are also other specific inhibitors that act by selectively inhibiting either the anodic or cathodic reactions, slows down one or both of these reactions to minimize the total corrosion current (Sawyer-Beaulieu et al., 2022).

These principles find extensive application in the automotive sector by means of protective coating, surface films & additive systems in lubricants & coolants, and these are all designed to suit a wide range of metallic substrates & the harsh environments that vehicles are used in. Conventionally, it has been customary in the automotive industry to make use of corrosion inhibitors that contained compounds such as chromates, which are efficient in passivating steel and aluminum but are gradually being replaced by their extreme level of toxicity and carcinogenicity. Other traditional inhibitors include phosphates and amines, which provide anodic or cathodic inhibition but are usually poorly biodegradable & environmentally persistent and so are being distanced in favor of safer alternatives (Al-Amiery et al., 2023). The recent years have seen the emergence of new strategies based on nano-bio and bio-based inhibitors which use sustainable and biocompatible raw materials. There are bio-like inhibitors, which are natural polymers such as chitosan & cellulose and plant extracts, which are renewable, biodegradable, and have high adsorption capacity on metal surfaces due to such functional groups as hydroxyl, amino, and carboxyl. Their performance is further increased upon incorporation of nanoparticles of zinc oxide, titanium dioxide, graphene oxide or nano-silica into the biopolymer matrix to form nano-bio composites. These nanoparticles enhance the surface coverage, enhance the barrier properties, and can also play a role in passivation or self-healing (Idumah et al., 2020). Studies have revealed that nano-bio inhibitors are more promising as far as automotive use is concerned in the future as they have improved corrosion resistance, reduced ecological footprint and increased biocompatibility over the existing systems. The processes of such sophisticated systems are complex & interdependent. Polar or aromatic rich biopolymers adsorb onto the metal surfaces effectively inhibiting reactive sites, decreasing electron and ion transfer with the effect of reducing both the anodic dissolution of the metal & cathodic depolarization. Nanoparticles in the composite form thick & convoluted diffusion routes that inhibit the intrusion of the corrosive agents, as well as, occupying the micro voids & cracks, to provide a self-healing aspect to the protective layer. Additionally, these nanoscale additives can modify the surface's electrochemical traits, can modify the distribution of surface charge & improve passivation, which naturally shifts the corrosion potentials to more noble values, with improved resistance, as established by electrochemistry. Overall, this combination of bio-based formulations & nanotechnology is a sustainable high-performance form of corrosion inhibition that can adapt to the changing demands of the automotive industry (Sagar et al., 2024).

2.3. Nano-Bio Composites: Composition & Properties

A nano-bio composite is a composite material that combines bio-based materials based on natural polymers and composite materials either comprised of metal nanoparticles or carbon-based nanomaterials, typically under the nano-scale. Such a combination utilizes the high surface area and reactivity of nanoparticles along with the barrier properties of biopolymer materials, such as chitosan, cellulose, and gelatin, alongside their biodegradable, biocompatible, and film-forming traits. The composites are multifunctional in features to prevent corrosion, especially with respect to all types of dynamic automotive environments (Muresan, 2023). Corrosion inhibition with the use of nano-bio materials can take on various forms. The role of the nanoparticles is important because of the large surface energy & activity. ZnO, TiO_2 & CeO_2 are metal and metal oxide nanoparticles that are extensively utilized because of their high degree of chemical stability and ability to develop passivating oxide layers that prevent corrosive elements. Carbon based nanomaterials (Graphene oxide (GO), carbon nanotubes (CNTs), & fullerenes) add value, enhance mechanical strength, impermeability, and electrical conductivity, which could reduce charge transfer & offer effective mechanical barriers. In addition, the silica and clay-based nanostructures provide mechanical stability, form labyrinthine structures, which improve the stability and protection of the coatings (A & Xavier, 2025). There are biodegradable and environmentally friendly corrosion-inhibiting composite matrices using natural polymers. The amino and hydroxyl groups of chitosan is a cationic polysaccharide and has been identified to be very adhesive as well as metal binding and is used extensively due to its antimicrobial & film making qualities. The polysaccharides, such as cellulose and starch, can easily form hydrogen bonds and dense film, which inhibit corrosion, polysaccharides contain hydroxyl groups, and gelatin & lignin possess antioxidant properties and increased barrier properties (Kumar et al., 2025, Rahman et al., 2023). The biopolymer matrix nanoparticles are hybrid nano-bio composites, and they are formed by the combination of nano-particles to create composite materials of higher mechanical strength, and they are flexible to withstand operation pressure in the automotive industry. Corrosion performance in these composites is also better due to synergistic action of inorganic fillers & organic coating, which lead to a higher corrosion resistance due to not only the action of chemical passivation processes, but also the physical shielding mechanisms (Randis et al., 2023).

Nano-bio composites are very effective as corrosion inhibitors because of several critical properties. The surface area & reactionary capability of the nanomaterials result in uniform dispersal and proximity with the substrate. Biopolymers are biodegradable & biocompatible which implies that they are safe to the environment and do not compromise on their performance. In addition, its film forming ability, good hydrophobicity, and adhesion are excellent, and the metals are covered by continuous and imperfect layers (A & Xavier, 2025). Some composites even are stimuli-reactive or self-healing and hence, permitting the coating to be fixed or the discharging of inhibitors in answer to external factors such as pH changes or mechanical damages. Nano & bio components also synergistically interact to give improvement in corrosion resistance. Nanoparticles form tortuous diffusion paths which obstruct the availability of the ions & the moisture and the biopolymers form adherent films which are cohesive & isolate the active sites of corrosion and increase the resistance to charge transfer. Electrochemical modifications through the composite stabilize the growth of passive film on automotive metals & decrease the electron mobility. Anti-corrosion effect of the bio-component based antibacterial systems, and also of biofouling caused corrosion reduction by biocidal nanomaterials like ZnO is of special significance in the car cooling systems. One of the automotive steel corrosion inhibition studies conducted on chitosan- graphene oxide with ZnO nanoparticle demonstrated over 95 percent corrosion inhibition efficiencies due to the presence of saline, which validates the practice utility of the nano-bio hybrid material (Kumar et al., 2025).

2.4. Nano-Bio Composites and their use and mechanism in automobile corrosion inhibition

Some of the common metals & alloys that are used in automotive parts include mild steel, galvanized steel, aluminum & magnesium alloys. Mild steel and galvanized steel are also used because of their mechanical strength and low cost of production. However, they easily corrode in harsh environment. The alloys of aluminum & magnesium are also lightweight alloys that do not require any additional work, but, unfortunately, are prone to corrosion, which is formed by natural oxide covering. The corrosive degradation of these materials is primarily due to the environmental conditions, which include salt deposition (mostly road salts), moisture due to rain, condensed water and change in temperatures, which encourage recurring stress corrosion and oxidation. Chassis, underbody, fasteners & fuel systems, and exhaust systems are some of the most frequent and most exposed areas with the exposure to corrosive agents being typical and frequently magnified by mechanical damage or coating failures (Km et al., 2024). It is important in the study of the weakness of these materials to the usual automotive working environment to dictate the right corrosion inhibition practice.

2.4.1. Nano-Bio Composites Use in Automotive Corrosion Inhibition.

Nano- bio composites have also been found to be a good material to be used as coating on the automotive surfaces therefore enhancing the corrosion resistance by offering protective surfaces on metallic surfaces. Even the deposition of good thickness & bonding properties of coating can be deposited evenly through the using of several approaches, like dip-coating, spray-coating, electrodeposition, and layer-by-layer assembly. These coating exploit the synergetic characteristics of the bio-based polymers with nanomaterials - e.g. graphene oxide, TiO_2 or ZnO nanoparticles - to increase the integrity & longevity of the film. The recent developments in improvements have encompassed enhanced adhesion, reduced porosity, and enhanced resistance to corrosive species. Epoxy-based nanocomposite coating, which continuously increases the mechanical strength & corrosion prevention when compared to traditional coating, is reported to be due to the increase in crosslinking density, and nanofiller dispersion (Zaghloul et al., 2023; Thakur et al., 2025). Nano-bio composites have also been added to the surface coating and in the automotive lubricants and automotive coolants to reduce corrosion internally. This is generally involving nano-encapsulated bioactive molecules which enable the bioactive molecules to be released in a controlled manner and provide extended protection to the engine components, radiators & fluid pipelines. The encapsulation procedures improve the stability & diffusion of bio-inhibitors in either aqueous or oil based medium to reduce the corrosive contact between the metal interfaces. It has been shown that nano-bio emulsions will be useful in the effective passivation of metal surfaces to minimize the amount of wear & corrosion when subjected to operational pressures as well as thermal variations (Parthipan et al., 2024). This will be by their environmental concerns, which involve the involvements of biodegradable and renewable assets in automotive fluids additives. Introduction of nano-bio composite technologies into the real-life car scenarios has been in a gradual growth pattern. Industrial pilot projects provide an insight into how eco-friendly, waterborne nanocomposite coating, which incorporates polymers of plant origin and nanoparticles can be applied to provide an additional layer of defense against corrosion as well as eliminate the emission of volatile organic compounds. Commercial coating materials have also been developed through academia/industrial relations and have self-healing qualities through the use of microcapsules that release inhibitors when the coating is damaged. These new coating are long-lasting in components that operate in extreme conditions of automobiles leading to less maintenance. One of the most famous examples of such is a report on a set of reports on a zinc oxide reinforced epoxy coating which was applied to protect against corrosion in chassis of cars and was shown to have improved corrosion resistance significantly as justified by electrochemical impedance spectroscopy (EIS) and salt spray tests (Bao et al., 2019).

2.4.2. Nano-Bio Composites Mechanisms for Corrosion Inhibition.

The nano-bio composites have a number of corrosive inhibiting mechanisms that are intertwined. The effect is based on the barrier, that the composite is formed in a thick and non-permeable film that literally prevents the physical accessibility of the corrosive agents to the metal surface. Nanomaterials are incorporated to seal the micro-voids as well as strengthen the coating matrix to reduce its perviousness. Surface passivation occurs when bioactive functional groups (e.g. hydroxyl, amine, carboxyl) in bio polymers react chemically with metal ions and create stable protective layers that prevents metals dissolution. In addition, anodic & cathodic suppression is also achieved using nano-compositions elements to alter the reaction rate in electrochemical reactions on the interface and hence lower corrosion current. Adsorption of active molecules on the metal surface to form chemisorbed protective films increases these effects. This is enhanced by nanomaterials with high surface area which is more electrically conductive & adhesive to substrate hence enhancing hydrophobicity which repels water based corrosive media. The reason why the inhibitors are sustainable is the biodegradability and environmental friendliness offered by the biopolymers (Yadav et al., 2024). The reason is due to the synergies that are formed through nanoscale materials combined with bio-based polymer matrices, in terms of the enhancement of corrosion protection. This kind of synergy increases corrosion resistance of charge transfer through interference with the flow of electrons to form corrosion reactions, increases barrier behavior by creating narrow, defect-free films, & enables self-healing reactions by the release of inhibitors sequestered in microstructures in response to stimuli. Moreover, some nano-bio composites can act as antibacterial & antifoulants that mitigate the risk of biocorrosion in wet or salty automotive conditions. These benefits are supported by the recent experimentations that have been done to establish these benefits through accelerated corrosion experiments such as salt spray tests, electrochemical impedance spectroscopy & scanning electron micro graphics. The example of introducing graphene into polysaccharide resulting in an increase of salt spray resistance greater than 50 percent higher than the conventional coating and the ability to restore its protective layer after mechanical damage is the reason the potential of nano-bio synergy in automotive corrosion inhibition is promising (Ielo et al., 2021; Daoudi et al., 2024).

3. RESEARCH & BIBLIOMETRIC ANALYSIS

The data for the below analysis has been taken from two sources – Len.org and PubMed databases. Figures 1 to 4 use schematics that are generated in Lens.org "analysis tab".

The bibliometric word cloud in Figure 1 offers an extensive view of the research environment that is present in the field of materials science and other related fields. In this distribution, the predominance of Materials Science (6,610) and Composite Materials (4,013) suggests the peripheral role these research areas contribute to the issue of deterioration of material properties (corrosion in this case) and performance optimization through the use of inhibitors. This indicates the communal approach to research the automotive industry has, where areas that save weight, resist wear & tear, and mitigate environmental damage are required components of the design.

Figure 1. Keyword Map of Research Domains Relevant to Nanocomposite-Based Corrosion Inhibition (Source: Lens Database) (Lens, 2025).

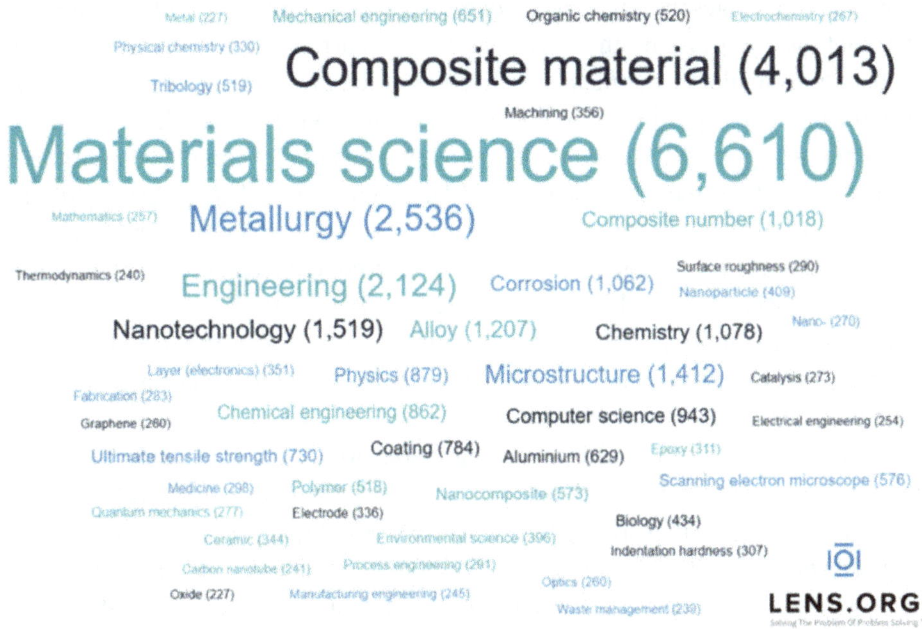

Metallurgy (2,536), Engineering (2,124) & Alloy (1,207) also have high counts that are somewhat connected to the fact that the automotive industry is still using metal systems like steels & aluminum which are prone to corrosion under different conditions in the environment, while protection occurs via sophisticated methods. The fairly ubiquitous appearance of Corrosion (1,062) as a key word further connotes that corrosion is a significant research problem, as it directly affects not only the safety of the vehicle, but also life cycle and maintenance costs. Importantly, the number of Nanotechnology (1,519), Microstructure (1,412), and Nanocomposites (573) points to the movement to solutions at the nanoscale to mitigate corrosion. Nanoparticles & nano-structured coating, as well as carbon-based nanomaterials like Graphene (260) and Carbon Nanotubes (241), have evolved into one of the main methods that is used for protective layers. These materials have unique physicochemical properties. They have large surface area exposure, greater barrier functions and catalytically inhibit corrosion reactions, which has generated great interest in automotive corrosion protection. The bibliometric evidence shows that nanotechnology has ceased to be viewed as another technique, & has developed into almost a scheme for innovation in corrosion science. In another link in figure 1, the word strength of words like Chemistry (1,078), Electrochemistry (267) & Catalysis (273), is a clue that suggests the interdisciplinary skill sets of the corrosion inhibition studies. The chemical processes that underpin oxidation and electrochemical degradation of metals is becoming an increasing area of research related to nanoscale changes. For example, a nanostructured coating is usually designed to serve not just the passive functions of a physical barrier but also to provide active contributions to electrochemical stabilization of the metal surface itself. This kind of multifunctionality is in high demand in the automotive sector where conditions of temperature, road salts & humidity expose the material to complex conditions of corrosion. At the same time, Coating (784), Ultimate Tensile Strength (730), and Tribology (519) are identified as key topics because there is a growing interest in the development of nanostructured protective surfaces for decreased corrosion along with wear resistance & mechanical integrity. The synergizing of surface engineering and corrosion inhibition is particularly relevant to any motor components that are subjected to both chemical & mechanical load like the body, suspension system, & undercarriage.

There is also a clear concern regarding sustainability, as indicated by the words that are present in the word cloud such as Environmental Science (396), Waste Management (239), and Biology (434). These words signify the increasing interest regarding nanomaterials that are environmentally sustainable and the green synthesis of corrosion inhibitors. This direction is significant to the auto-industry where environmental regulations are becoming increasingly stringent. Nanoproducts that derive not only through enhanced corrosion protection, but also biodegradable, low-toxicity, & recyclable nanomaterials will be found to be part of the wider objective of sustainable manufacturing & to reduce ecological footprint.

Mechanical Properties is the most frequent keyword listed, and more than 270 documents were found that emphasize the priority placed on the strength, durability and performance of the advanced materials, as seen in Figure 2. This is of substantial concern particularly to the automotive industry where structural integrity & light weight design are of the utmost importance in terms of safety & efficiency. Second on the list is Microstructure (~160), which illustrates that there is a considerable extent of literature directed to the study of the correlation between nanoscale changes and the operational work of bulk materials. This carries significant weight for the case of corrosion inhibition which microstructural refinement has a substantial impact in the direction of corrosion and the protective behavior. Additive manufacturing (~140) is the third most common subject that indicates the growing intertwining of available fabrication methods with the construction of corrosion resistant materials. Alongside being process suitable to the incorporation of nanostructured materials complicated geometries, additive manufacturing is also a possible process that shapes microstructures for improved corrosion resistance in automotive parts. The similar terms of Wear/Hardness/Powder Metallurgy are used as keywords, which points to the complementation nature of the mechanical degradation & corrosion challenges, meaning that the studies frequently deal with the two simultaneously via the advanced surface engineering & nanocomposites reinforcement.

Figure 2. Bar chart showing the document count associated with top research keywords relevant to nanocomposites (Source: Lens Database) (Lens, 2025).

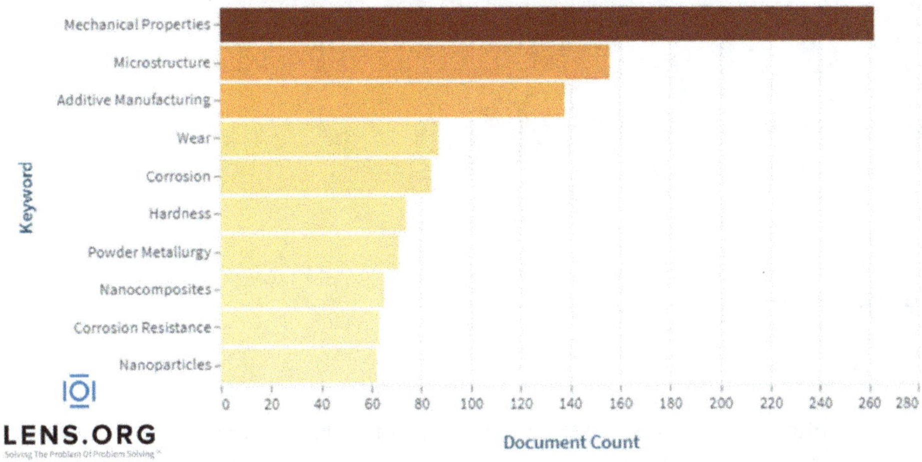

Most importantly, keywords 'Corrosion' & 'Corrosion Resistance' are used together with 'Nanocomposites & Nanoparticles' in figure 2, which proves that nanoscale interventions have become increasingly accepted as effective solutions to increasing material longevity. The frequency of such words is rather high, which reflects the active direction of research to use nanotechnology-like graphene, carbon nanotubes, or metal-oxide nanoparticles as active agents in the field of corrosion inhibition. These materials do not only create shields between them, but also, they alter the electrochemical behavior, thus enhancing the life of the components in harsh automotive service environments.

Figure 3. Bar chart displaying the number of research documents by country/region (Source: Lens Database) (Lens, 2025).

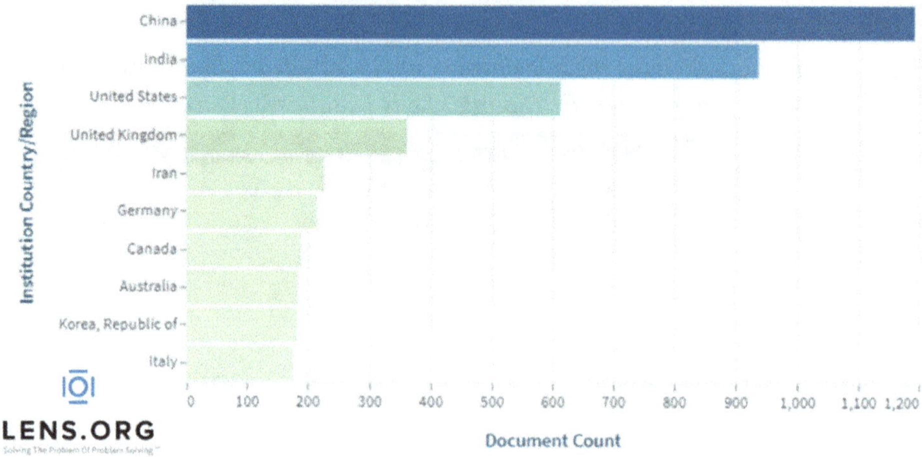

The bibliometric analysis in Figure 3 of the contributions made by the countries shows the different trends of research across the world in the area of nanotechnology and corrosion inhibition with the automotive characteristic. China is the leading country with more than 1,100 publications, next comes India (~950) and United States (~620). This superiority echoes the large governmental and industrial investment in research of advanced materials in these countries, especially in the automotive industry & transport industry where corrosion prevention has a direct effect on vehicle performance, safety, & sustainability. The cohesiveness of China is explained by its extensive research facilities, its rapidly expanding automotive sector, and the focus on the development of nanomaterials, as the surface finishes, and structural alloys. Equally, the large contribution level in India highlights its increasing academic productivity in materials science & nanotechnology and its strategic interest in lightweight, corrosion-resistant automotive materials, with the local automobile industry growing. Although the United States publishes fewer publications than China and India, it has long been a source of high-impact nanotechnology research, especially in the applied fields of nanotechnology & protective materials integration and smart materials. Strong research cluster is also represented by European countries like United Kingdom, Germany and Italy. They typically work on high-performance materials, green inhibitors, and new testing methods for automotive corrosion resistance. It is also significant that the research presented here is from sources in Iran, Canada, Australia, and South Korea, confirming that the topic of corrosion inhibition using nanotechnology is a global interest & consideration in new automotive production/engineering sites or regions, as well as existing regions.

This geographic distribution of the research indicates that while the research is global in nature, there are local priorities that influence the focus of the research. For instance, in some Asian nations (e.g., China and India), the mass industrial requirements and growth in automotive manufacture is driving interest in the potential advantages of nanostructured coatings and inhibitors at potentially lower costs; while North America and Europe are focusing on optimization of performance, green corrosion inhibitors, & advanced nano-compositions. Thus, these variances provide a fair (potentially, a stage) of balance globally: to develop corrosion solutions based on a nanoscale approach that results in a balance between performance, cost, & green sustainability.

The leading institutions contributing to research on nanotechnology & corrosion inhibition are primarily located in Asia, with the Chinese Academy of Sciences, King Saud University, Islamic Azad University, Harbin Institute of Technology, and Tsinghua University producing the highest volume of publications as seen in Figure 4. This reflects the global shift of material innovation hubs towards Asia, where industrial demand for corrosion-resistant automotive materials is growing.

Figure 4. Top Universities Involved in Corrosion Inhibition Research Using Nanocomposites (Source: Lens Database) (Lens, 2025).

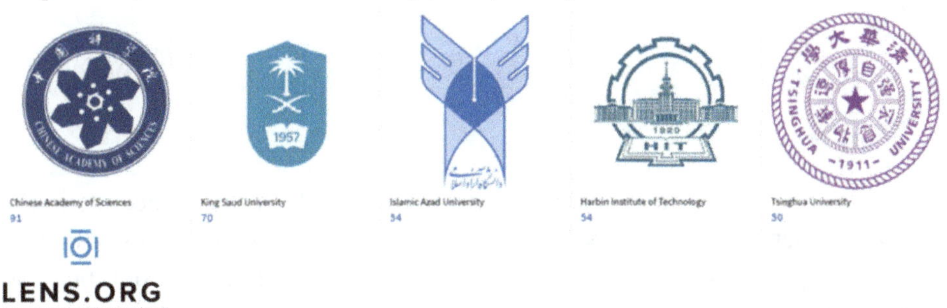

The bibliometric network shown in Figure 5 highlights corrosion as the thematic center of several research themes into which corrosion has been studied, including nanotechnology, composite materials, electrochemical surface treatments and thermal management. Much of the research on corrosion is connected to aluminum alloys & low alloy steels which highlights the importance of lightweight metallic systems in the automotive & aerospace industries. With the potential development of improving the damage tolerance in corrosive environments, there is strong representation by design methodologies to alloys, additive constituents or materials and methods of assembly, such as brazing. The fact that the automotive node has been connected to Al alloys on more than one occasion emphasizes the notion of practicability to this body of research especially in terms of developing lightweight corrosion free vehicle bodies.

Figure 5. Conceptual Map of Corrosion-Related Research Domains in Nanocomposite Based Automotive Applications (Source: Lens Database; created using VOSviewer) (Lens, 2025; Van Eck & Waltman, 2010).

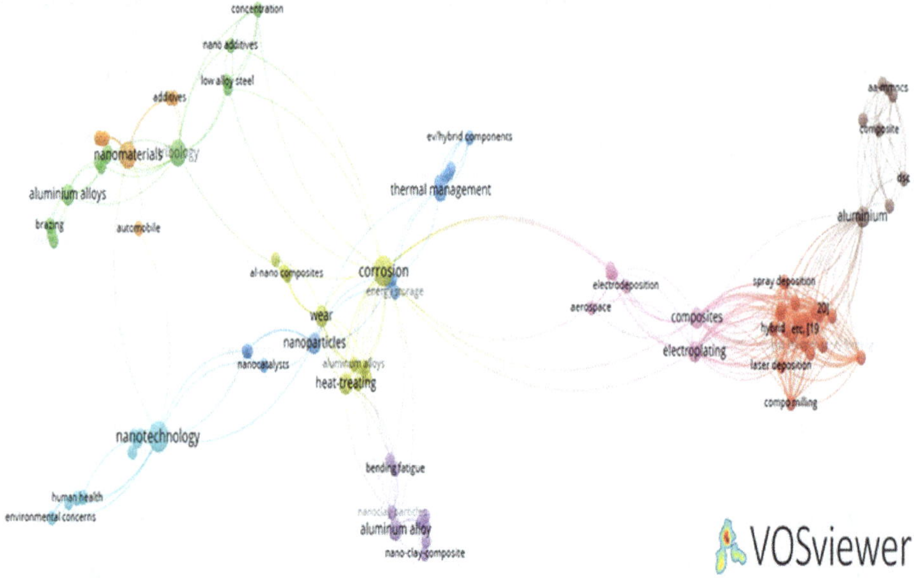

Nanotechnology is strategically situated in the middle of the connection in the networks, where keywords such as nanoparticles, nano sensors, & nano catalysts derive a frequent association; these factors link the science of corrosion but also with broader discussions on energy storage, wear resistance, & environmental sustainability. The linkage of the phrases 'human health' and 'environmental impact' implies that, although nanotechnology has offered a new means to improve corrosion resistance & heat transfer, it has raised consideration of long term ecological & biological impact. A different branch of research is spawned around thermal management, which is being more closely aligned with the research for electrification, hybrid cars, and the next-generation automotive system. In the part of the map with thermal management, aluminum alloys dominate. Aluminum alloys serve a dual purpose as a structural lightweight material as well as a desirable heat conductor that fits the development of electric mobility. This suggests that the intersection of corrosion science and thermal management is now developing as a significant avenue of research to advance high-performance, robust, and green electrified drivetrains. The subdomain of composites-electroplating is depicted as being quite dense around the outside edge of the network. In this subnetwork, all types of processes are applicable to aluminum substrates, which are spray deposition, laser deposition, compo milling, hybrid coating processes. The aerospace & electrodeposition link strengthens the industrial relevance to any studies involving anti-corrosive surface finishes which are essential in mitigating service life in harsh environments. Other peripheral nodes, such as nano clay composites, fatigue behavior, and additive manufacturing highlights the new integrations between mechanically durable, processing-advanced, and corrosion preventive treatments. More specifically, the correlation between 3D printing & nanomaterials indicates that it is becoming more attractive to design specific corrosion resistance properties at microstructure design level using design-driven manufacturing.

Overall, the network illustrates that corrosion studies are becoming more layered and interdisciplinary, it's also based on aluminum & it alloys, but nanotechnology, composite fabrications & thermal management are the common themes that may allow it to expand industrially.

4. CHALLENGES AND FUTURE OUTLOOK

Nano-biological and bio-composites used in preventing corrosion within the automobile sector hold high potential; though, the major challenges that exist with respect to the use of these materials encompass technical, environmental, regulatory and economic which have to be tackled systematically to ensure that the full potential is achieved. Stability of nano-bio coating in the extreme conditions of the auto environment is an outstanding point of challenge. These finishes should be able to withstand thermal cycling, mechanical abrasion, and exposure to chemical elements in the automotive fluids & the contaminants on the atmosphere without losing the protective properties. It has been reported that the nano-bio coatings have problems of adhesion and delamination leading to poor durability & resistance to corrosion in dynamic service environments. Synthesis methods of nano-bio composites are also subjected to significant challenges because of their reproducibility and scalability. Natural biopolymer sources can cause batch-to-batch variation, and nanofiller dispersions due to variations in dispersion can be difficult to control quality in production (Farooq et al., 2022). The environmental effects of nano-bio composites, especially on the release of nanoparticles during service & disposal should be subjected to question. Nanoparticles incorporated in the coats can be emitted through ageing, abrasion or degradation, and the toxicity and ecological impacts are of concern. Moreover, car rules and the environment standards set strict boundaries of hazardous substances and require life-cycle tests, which make it hard to implement new nano-bio materials without much testing & certification. A proper waste management plan should be established in order to manage the end-of life automotive parts that may have nano-bio composites so that they do not cause environmental pollution (Ruggiero et al., 2019). Although the functional advantages, the high price of nanomaterials and special biopolymers makes nano- bio composites still economically difficult to use in large scale in the automotive manufacturing. Setups of production need a lot of capital and the incorporation of these materials to the available automobile production lines is hampered by the compatibility with the available OEM coatings, lubricants & fluids. These reasons prevent the preparedness to mass production and slow down commercialization. There is poor standardization of corrosion tests on nano-bio composites under realistic automotive conditions e.g. salt, humidity, and thermal cycling. Stricter and standardized testing conditions would ease convenient benchmarking and obtain regulatory permits. There are also promising prospects in the development of smart coating systems with self-healing or pH-responsive or stimulus-responsive capabilities to increase coating lifetimes and improve corrosion protection. The introduction of machine learning approaches is also increasing the identification and optimization of corrosion inhibitors & nano-bio composite formulations through predicting performance based on molecular and structural properties, and thus decreasing costs and time of experimentation (Km et al., 2024).

The use of nano-bio composites is destined to become a more significant part of the automotive corrosion protection process because of the necessity of the sustainable and lightweight materials, which are multifunctional. To realize this future, interdisciplinary cooperation between materials scientists, automotive engineers, environmental experts, and regulatory agencies are needed to consider the complex issues. Potential commercial avenues will enjoy pilot projects that prove to be economical, reliable in performance, and environmentally friendly. The choice of material, the production process, and end of life management must be environmentally conscious and must be done considering the changing standards in the automotive industry & social expectations. All these will help the nano-bio composites to leave their potential laboratory applications to automotive mainstreams so that they can be useful to offer lasting & friendly corrosion protection which is environmentally friendly (Lloyd & Lave, 2003).

5. CONCLUSION

Corrosion has also been a major challenge to the performance, safety, and the life of automotive parts especially with the high usage of metals such as mild steel, galvanized steel, aluminum, & magnesium alloys. The conventional methods of corrosion protection, although to some degree efficient, usually imply the use of substances that are harmful to the environment, including chromates & phosphates, and cannot be used in the environment with complicated service conditions. Nano-bio composites in this regard have come out as a revolutionary tool in terms of corrosion inhibition in the automotive sector in the next generation. The nano-bio composites are versatile corrosion inhibitors by preventing corrosion by using different mechanisms, such as physical barrier, release of active inhibitors, control of release, and electrochemical stabilization. Popular biopolymers such as chitosan, cellulose, and starch provide natural adhesion, flexibility, and biodegradability while nanofillers provide enhanced mechanical strength, UV resistance & electrochemical stability. In terms of effective performance, nano-bio composites are also closely aligned with the growing interest of the automotive sector in green production & environmental compliance. Even though nano-bio composites have demonstrated promise in previous studies at laboratory and pilot projects, they are still facing challenges in terms of large-scale processing, costs, and long-term use in field performance. There needs to be more interdisciplinary work & research studies to review their formulations to achieve effective, consistent applications & respect compatibility with current production technologies. Nonetheless, the broad trend in the area of innovation suggests that nano-bio composites are likely to become part of the corrosion resistance toolbox in the automotive industry, as they have great potential to support the future development of more sustainable, durable, high-performance vehicles, as a result of the combination of high efficiency, multifunctionality, & environmental compliance.

REFERENCES

A, P., & Xavier, J. R. (2025). A Review on Functional nanocomposites for enhanced anticorrosion and flame retardant properties. *Surfaces and Interfaces, 70*, 106850. https://doi.org/DOI: 10.1016/j.surfin.2025.106850

Al-Amiery, A. A., Yousif, E., Isahak, W. N. R. W., & Al-Azzawi, W. K. (2023). A review of Inorganic corrosion Inhibitors: Types, Mechanisms, and applications. *Tribology in Industry*, *45*(1), 313–339. DOI: 10.24874/ti.1456.03.23.06

Aslam, R., Zhao, J., Sun, X., Zhou, X., Wang, Q., Aslam, J., & Yan, Z. (2024). Bio-based ionic liquid as a corrosion inhibitor for mild steel in 5% HCl solution: Experimental and theoretical investigation. *Sustainable Chemistry and Pharmacy*, *39*, 101614. DOI: 10.1016/j.scp.2024.101614

Bao, W., Deng, Z., Zhang, S., Ji, Z., & Zhang, H. (2019). Next-Generation composite coating system: Nanocoating. *Frontiers in Materials*, *6*, 72. Advance online publication. DOI: 10.3389/fmats.2019.00072

Cai, Y., Zhao, Y., Ma, X., Zhou, K., & Chen, Y. (2018). Influence of environmental factors on atmospheric corrosion in dynamic environment. *Corrosion Science*, *137*, 163–175. DOI: 10.1016/j.corsci.2018.03.042

Chami, A., Benabbou, R., Taleb, M., Rais, Z., & Haji, M. E. (2021). Analysis of survey data on corrosion in the automotive industry. *Materials Today: Proceedings*, *45*, 7636–7642. DOI: 10.1016/j.matpr.2021.03.113

Daoudi, W., Dagdag, O., Nik, W. B. W., Oussaid, A., & Aatiaoui, A. E. (2024). Introduction to nano-hybrid smart coatings. In *ACS symposium series* (pp. 1–16). https://doi.org/DOI: 10.1021/bk-2024-1469.ch001

Farooq, S. A., Raina, A., Mohan, S., Singh, R. A., Jayalakshmi, S., & Haq, M. I. U. (2022). Nanostructured Coatings: Review on processing techniques, corrosion behaviour and tribological performance. *Nanomaterials (Basel, Switzerland)*, *12*(8), 1323. DOI: 10.3390/nano12081323 PMID: 35458032

Golubović-Bugarski, V., Petković, S., & Globočki-Lakić, G. (2021). The effect of corrosion on a structural integrity and vehicle safety. *Mobility and Vehicle Mechanics*, *47*(3), 49–62. DOI: 10.24874/mvm.2021.47.03.05

Hagarová, M., Brezinová, J., Baranová, G., Viňáš, J., & Maruschak, P. (2021). Degradation of components in cars due to bimetallic corrosion. *Materials (Basel)*, *14*(12), 3323. DOI: 10.3390/ma14123323 PMID: 34208438

Idumah, C. I., Obele, C. M., Emmanuel, E. O., & Hassan, A. (2020). Recently emerging nanotechnological advancements in polymer nanocomposite coatings for anti-corrosion, anti-fouling and self-healing. *Surfaces and Interfaces*, *21*, 100734. DOI: 10.1016/j.surfin.2020.100734 PMID: 34957345

Ielo, I., Giacobello, F., Sfameni, S., Rando, G., Galletta, M., Trovato, V., Rosace, G., & Plutino, M. R. (2021). Nanostructured Surface Finishing and Coatings: Functional properties and applications. *Materials (Basel)*, *14*(11), 2733. DOI: 10.3390/ma14112733 PMID: 34067241

Klenam, D. E. P., Bodunrin, M. O., Akromah, S., Gikunoo, E., Andrews, A., & McBagonluri, F. (2021). Ferrous materials degradation: Characterisation of rust by colour – an overview. *Corrosion Reviews*, *39*(4), 297–311. DOI: 10.1515/corrrev-2021-0005

Km, S., Praveen, B., & Devendra, B. K. (2024). A review on corrosion inhibitors: Types, mechanisms, electrochemical analysis, corrosion rate and efficiency of corrosion inhibitors on mild steel in an acidic environment. *Results in Surfaces and Interfaces*, *16*, 100258. DOI: 10.1016/j.rsurfi.2024.100258

Kumar, S., Arora, A., Mathur, D., Chaudhary, A., Pant, V., Guchhait, S., & Singh, B. K. (2025). A review on chitosan and chitosan-based bionanocomposites: Promising biological macromolecules for sustainable corrosion inhibition. *International Journal of Biological Macromolecules*, *140392*, 140392. Advance online publication. DOI: 10.1016/j.ijbiomac.2025.140392 PMID: 39880227

Lens, (2025). *Search results for "Corrosion AND Nano"*. The Lens. Retrieved September 19, 2025, from https://www.lens.org/lens/search/scholar/analysis?q=(corrosion%20nano)&p=0&n=10&s=_score&d=%2B&f=false&e=false&l=en&authorField=author&dateFilterField=publishedYear&orderBy=%2B_score&presentation=false&preview=true&stemmed=true&useAuthorId=false

Liu, M., Guo, Y., Wang, J., & Yergin, M. (2018). Corrosion avoidance in lightweight materials for automotive applications. *NPJ Materials Degradation*, *2*(1), 24. Advance online publication. DOI: 10.1038/s41529-018-0045-2

Lloyd, S. M., & Lave, L. B. (2003). Life cycle economic and environmental implications of using nanocomposites in automobiles. *Environmental Science & Technology*, *37*(15), 3458–3466. DOI: 10.1021/es026023q PMID: 12966996

Mazumder, M. J. (2020). Global Impact of Corrosion: Occurrence, cost and mitigation. *Global Journal of Engineering Sciences*, *5*(4). Advance online publication. DOI: 10.33552/GJES.2020.05.000618

Muresan, L. M. (2023). Nanocomposite coatings for Anti-Corrosion properties of metallic substrates. *Materials (Basel), 16*(14), 5092. DOI: 10.3390/ma16145092 PMID: 37512373

Parthipan, P., Kumar, A. S., Dhandapani, P., Elumalai, P., & Thirumurugan, D. (2024). Sustainable design and production of nano-hybrid smart coatings. In *ACS symposium series* (pp. 121–138). https://doi.org/DOI: 10.1021/bk-2024-1469.ch006

Rahman, S. a. A., Priyadharsini, P., Deeksha, R. V., & Arun, J. (2023). Grafted Biopolymer Composites and Nanocomposites as Sustainable Corrosion Inhibitors. *Grafted Biopolymers as Corrosion Inhibitors: Safety, Sustainability, and Efficiency*, 397–416. https://doi.org/DOI: 10.1002/9781119881391.ch18

Randis, R., Darmadi, D. B., Gapsari, F., Sonief, A. A., Akpan, E. D., & Ebenso, E. E. (2023). The potential of nanocomposite-based coatings for corrosion protection of metals: A review. *Journal of Molecular Liquids, 390*, 123067. DOI: 10.1016/j.molliq.2023.123067

Ruggiero, E., Vilsmeier, K., Mueller, P., Pulbere, S., & Wohlleben, W. (2019). Environmental release from automotive coatings are similar for different (nano) forms of pigments. *Environmental Science. Nano, 6*(10), 3039–3048. DOI: 10.1039/C9EN00227H

Sagar, K., Vergis, B. R., & Devendra, B. K., & SampthKumaran, P. (. (2024). Electrochemical corrosion influencing the al beryl alloy system subjected to extrusion process. *Results in Surfaces and Interfaces, 100399*. Advance online publication. DOI: 10.1016/j.rsurfi.2024.100399

Sawyer-Beaulieu, S., Tam, E., & Hussein, A. (2022). Measuring corrosion on vehicles, in Real-Time, using digital imaging and analysis techniques. *Materials (Basel), 15*(9), 3053. DOI: 10.3390/ma15093053 PMID: 35591387

Shafique, M., & Luo, X. (2019). Nanotechnology in Transportation Vehicles: An Overview of its applications, environmental, health and safety concerns. *Materials (Basel), 12*(15), 2493. DOI: 10.3390/ma12152493 PMID: 31390752

Thakur, A., Zarrouk, A., Selvaraj, M., Assiri, M. A., Khanna, V., Kumar, A., Berdimurodov, E., & Eliboev, I. (2025). Nanomaterial-based smart coatings for sustainable corrosion protection in harsh marine environments: Advances in environmental management and durability. *Inorganic Chemistry Communications, 114280*, 114280. Advance online publication. DOI: 10.1016/j.inoche.2025.114280

Van Eck, N. J., & Waltman, L. (2010). *VOSviewer [Computer software]*. Centre for Science and Technology Studies, Leiden University. https://www.vosviewer.com

Xie, R., Wei, Y., Luo, K., Lin, N., Yu, Y., Cheng, P., & Zeng, Q. (2025). Effect of environmental factors on the corrosion behavior of CoCrNi and CoCrFeMnNi alloys in 3.5% NaCl solution. *Journal of Materials Research and Technology*, *35*, 2994–3007. Advance online publication. DOI: 10.1016/j.jmrt.2025.01.227

Yadav, S., Raman, A. P., Singh, M. B., Massey, I., Singh, P., Verma, C., & AlFantazi, A. (2024). Green nanoparticles for advanced corrosion protection: Current perspectives and future prospects. *Applied Surface Science Advances*, *21*, 100605. DOI: 10.1016/j.apsadv.2024.100605

Zaghloul, M. M. Y., Zaghloul, M. M. Y., & Fuseini, M. (2023). Recent progress in Epoxy Nanocomposites: Corrosion, structural, flame retardancy and applications — A comprehensive review. *Polymers for Advanced Technologies*, *34*(11), 3438–3472. DOI: 10.1002/pat.6144

ADDITIONAL READING

Jain, P., Patidar, B., & Bhawsar, J. (2020). Potential of nanoparticles as a corrosion inhibitor: A review. *Journal of Bio- and Tribo-Corrosion*, *6*(2), 43. Advance online publication. DOI: 10.1007/s40735-020-00335-0

Kordas, G. (2022). Nanocontainers Against Biofouling and Corrosion Degradation of Materials: A Short Review with Prospects. *Frontiers in Nanotechnology*, *4*, 813908. Advance online publication. DOI: 10.3389/fnano.2022.813908

Kotnarowska, D., Przerwa, M., & Szumiata, T. (2014). Resistance to erosive wear of Epoxy-Polyurethane coating modified with nanofillers. *Journal of Materials Science Research*, *3*(2). Advance online publication. DOI: 10.5539/jmsr.v3n2p52

Ndukwe, A. I., Nwadirichi, B., Okolo, C., Tom-Okoro, M., Medupin, R., Uche, R., Arukalam, I., Onuoha, C., Egole, C., Okorafor, O., & Nwakuba, N. (2025). *Corrosion Control in Metals: A review on Sustainable approach using Nanotechnology*. Zastita Materijala., DOI: 10.62638/ZasMat1187

Negi, P., Chardra, V., Kumar, A., Kumar, P., Priya, N., Yashmin, K., Kumar, N., Joshi, N., Saxena, R., Kumar, A., Dhiman, N., & Rawat, B. (2025). Plant-mediated synthesis of metal oxide nanocomposites: Emerging prospects in electrochemical corrosion inhibition and biosensing. *Sustainable Chemistry for Climate Action*, *100107*, 100107. Advance online publication. DOI: 10.1016/j.scca.2025.100107

Singh, P., Rana, A., Karak, N., Kumar, I., Rana, S., & Kumar, P. (2023). Sustainable smart anti-corrosion coating materials derived from vegetable oil derivatives: A review. *RSC Advances*, *13*(6), 3910–3941. DOI: 10.1039/D2RA07825B PMID: 36756545

Zamindar, S., Mandal, S., Murmu, M., & Banerjee, P. (2024). Unveiling the future of steel corrosion inhibition: A revolutionary sustainable odyssey with a special emphasis on N+-containing ionic liquids through cutting-edge innovations. *Materials Advances*, *5*(11), 4563–4600. DOI: 10.1039/D4MA00156G

KEY TERMS AND DEFINITIONS

Automotive: Related to automobiles.

Bio-Based Corrosion Inhibitor: Inhibitors made from bio-derivatives for corrosion protection.

Biopolymer Nanocomposites: A nanomaterial made from bio-derivatives involving polymer like structure

Chitosan: Natural biopolymer made from chitin.

Corrosion Inhibition: Protection form corrosion.

Graphene Oxide Coatings: A coat of Graphene oxide, mainly used for its corrosion resistive properties.

Nano-Bio Composites: Particles whose measures are in nanometer scale and are made from biotic derivatives.

Nanoparticles: Particles in nanometer scale.

Protective Coatings: An external layer that is meant to protect the internal surface from external damage.

Chapter 3

Role of Soil Microorganisms in Corrosion of Agricultural Implements and Irrigation Systems

K. Gayathri
https://orcid.org/0009-0007-3786-2009
Sacred Heart College, Tirupattur, India

M. Kokila
Sacred Heart College, Tirupattur, India

P. Saranraj
https://orcid.org/0000-0002-5052-6532
Sacred Heart College, Tirupattur, India

K. Kesavardhini
Sacred Heart College, Tirupattur, India

M. Manigandan
https://orcid.org/0000-0002-8115-2713
Sacred Heart College, Tirupattur, India

B. Lokeshwari
Sacred Heart College, Tirupattur, India

S. Monisha
Sacred Heart College, Tirupattur, India

ABSTRACT

Soil microorganisms assumes in the corrosion of agricultural implements and irrigation systems, a significant economic issue for the agricultural sector. Soil corrosion is primarily electrochemical, driven by variations in soil properties that create potential differences on metal surfaces. Microbial Induced Corrosion (MIC), facilitated by bacteria and fungi such as Sulfate-Reducing Bacteria, accelerates this process by producing corrosive metabolites like acids, leading to localized pitting and structural degradation. The problem is exacerbated in agricultural environments by

DOI: 10.4018/979-8-3373-2617-7.ch003

factors including soil salinity, sodicity, and exposure to agro-chemicals like fertilizers, pesticides, and acidic manure slurries. These conditions not only directly corrode metals and concrete but also alter soil microbiology, further influencing corrosion dynamics. This chapter also discusses mitigation strategies, including the use of corrosion-resistant materials like stainless steel, protective coatings, and system maintenance practices such as acid treatment for drip irrigation to prevent clogging.

1. INTRODUCTION

Grounding in the electrical grid is an essential part of the system that ensures safe operation. It stabilizes voltage and provides a controlled path for current flow during events such as lightning strikes or equipment failures. However the grounding conductor is buried underground for several years, it may deteriorate and malfunction according to the soil and operating circumstances. demonstrates the deterioration of the conductor's ground. Since it is a ground conductor defect in the power system, the current cannot spread entirely into the soil. In this case, high voltage travels to the secondary circuit as the ground resistance and ground potential increase. In addition to increasing the risk to humans and electrical equipment, this could result in serious accidents and financial losses (Li et al., 2025).

Irrigation water frequently contains salts, were accumulate in the soil over time and may drag down plant growth, modify the chemical makeup of the soil, and contaminated groundwater, amongst additional problems. Approximately 25% of the world's land (2 billion acres) is excessively salted, rendering it unusable for agriculture. Although cations such as calcium, iron, magnesium, and others are frequently found in soils, sodium salts are the primary cause of salinity. The primary causes of salinity and sodicity's greater effects on arid and semi-arid regions include the deposition of salts from parent rock, ancient drainage basins, and inland seas, as well as an absence of adequate natural drainage. The effects of salinity and sodicity, if any, are usually seasonal in humid regions. Nevertheless, mineral salts that are thought to have already leached away could contaminate the groundwater (Doshvarpassand et al., 2018).

Given that plant requires water to grow, agriculture uses the most freshwater worldwide. Although water is a renewable resource, the Food and Agriculture Organization (FAO) estimates that the world's irrigated area is currently around 300 million square kilometres, and projections for 2050 indicate that agricultural water supplies will become more scarces. This has exacerbated poverty, climate change, and global food insecurity. Providing food and fiber to almost 8 billion people worldwide is the largest challenge facing agriculture as freshwater resources deplete. In farms and agricultural settings full of bio-waste, agro-food effluents, silage juices, manure, and molasses, corrosion forecasts the lifespan of structural elements (Moreira-Filho & Figueiredo, 2021).

Further variables that may increase the likelihood of corrosion include high concentrations of acids (from spilled meal-water mixes), salts, chlorine-containing sterilizing agents, aggressive gasses, Hydrogen Sulfide (H_2S), dust, and warmth. Microbial-induced corrosion is additionally influenced by microorganisms, particularly Bacteria, Yeast, and Fungi. In addition to addressing the effectiveness of high-strength concrete, specialty concretes, stainless steel, and galvanized steel in reducing corrosion and material degradation in the environment, to improve understanding of the durability of concrete structures exposed to severe corrosion caused by acids, moisture, bacteria, and other corrosive agents (Salgar-Chaparro et al., 2020).

Agriculture is the primary thrust of all the major economies in the world. To several big corporations, the farming economy remains essential. Several additional companies rely heavily on agricultural operations. A lot of equipment and tools are needed in agriculture. Therefore, one of the main issues that farmers deal with is that these tools and equipment are prone to corrosion, which initially damages their finances and shortens their lives. Life on Earth would be nearly impossible without soil, which is a crucial component of the planet's ecology. It is a natural habitat for plants since it contains an integrated system for recycling both organic and inorganic materials (Karadoğan et al., 2025).

A living being, soil unites the Hydrosphere (water), Biosphere (living things), Atmosphere (air), and Lithosphere (rock). It depicts a region's interactions between robust rocks, living organisms, and the air and water that flow through, in, and around them. The interactions between the physical, biological, and chemical components of soil are just as important as the actual materials that comprise it. The way soil functions as the foundational element of the ecosystem is determined by its properties. An evaluation of the soil at a specific location and scale, based on a set of criteria that include healthy functioning, is known as soil health. Utilizing various forms of organic manure has a significant impact on soil health, both directly affecting soil microbes and indirectly through modifications to the physical and chemical characteristics of the soil (Procópio, 2020).

2. SOIL MICROORGANISMS INVOLVED IN CORROSION

The soil microorganisms are important in Microbiologically Induced Corrosion (MIC) process in which the microbial activity enhances the corrosion rate of metals buried in soil. This biological corrosion is associated with different types of bacteria which react biochemically and electrochemically with metals resulting in their corrosion. Sulfate-Reducing Bacteria (SRB) are one of the major groups of microbes that are involved in soil corrosion. These aerobic bacteria will lower sulfates to Hydrogen Sulfide (H_2S), which is a corrosive liquid that will result in pits and weakening of metal surfaces. SRB are common in soils and are particularly active in environments with limited oxygen such as layers of corrosion products and under deposits. Their metabolic processes produce acidic and sulfide substances which chemically assault the metal that corrodes underground metal structures and pipelines faster (Su et al., 2019).

Gallionella and *Sideroxydans* (Iron Oxidizing Bacteria (IOB) also play a major role in corrosion. The microaerophilic bacteria reduce soluble ferrous iron (Fe^{2+}) to ferric iron (Fe^{3+}) to form insoluble iron oxides which add to the layers of rust and corrosion. The biofilm on metal surfaces is usually started by IOB, and this process allows the further development of microbial colonies. These bacteria produce iron oxides which form anaerobic microenvironments that enhance the growth of SRB which increases the corrosion process. Bacteria that have been involved in the process of nitrogen cycling like the nitrifying bacteria such as *Nitrospira* and the denitrificing bacteria such as *Hydrogenophaga* are enriched in the soils that are affected due to metal corrosion. The bacteria play a role in availing nitrogen compounds, cycling of nitrogen compounds, which affect redox conditions of soil and corrosion dynamics are determined by electron transfer processes. The interaction of the nitrifying bacteria and the denitrifying bacteria aids in the preservation of the biochemical environment of facilitating the microbial corrosion activities (Kumari & Mulky, 2024).

The corrosion process leads to effects of altering the microbial community structure and functioning of soil microbes. Metal dissolution also releases nutrients and ions that promote growth of bacteria that are related to corrosion and decrease the overall microbial diversity and stability around the metal surfaces. With time, there is a complex co-occurrence network of microbes formed which forms a synergistic environment that encourages more corrosion. The association of microbial groups leads to complex biotic interaction and increasing microbial activity in the environments of corroding metals as a result of this close association. The formation of biofilm by these microorganisms on metal surfaces is essential to MIC. Biofilms present a protective and stable niche to the microbial communities to allow the transfer of electrons between the metal and the microbes. This effect of movement of electrons increases the rate of oxidation of metal, which is one of the basic stages of corrosion. Corrosion-related microbes rely on the extracellular electron transfer processes that are closely connected to the biogeochemical cycles of iron and nitrogen (Javaherdashti, 2016).

Soil Temperature, Soil Moisture, Availability of Oxygen and pH are all environmental factors that have a strong influence on microbial activity and corrosion rate. An increase of moisture content, such as, can boost SRB activity and corrosion, and the oxygen gradient generated by the activity of microbes can further shape the community dynamics on the metal surface. The type of metal and the type of soil influence the corrosion mechanisms and the pattern of microbial colonization. Soil microbial corrosion is a complex process and it is catalyzed by a consortium of bacteria, which interact with each other via iron, sulfur, and nitrogen cycles. Corrosive sulfides are generated by sulfate reducing bacteria; rust is generated by iron oxidizing bacteria and biofilm formation is controlled by nitrogen cycling bacteria. The microbial biofilms promote metal to metal electron transfer and therefore corrosion. These microbial interactions and environmental effects are understood to aid in the development of corrosion control of buried infrastructure and soil-metal. This microbiological view of metal corrosion gives emphasis to the complex interconnections between soil microbiology and material decay (Jones et al., 2024).

In Table – 1 Demonstrates the Soil microorganisms play a major role in accelerating metal deterioration through a process called Microbiologically Influenced Corrosion (MIC). Different groups of microbes, such as Sulphate-Reducing Bacteria, Sulfur-Oxidizing Bacteria, Iron-Oxidizing And Iron-Reducing Bacteria, Nitrate-Reducing Bacteria, Methanogenic Archaea, Acid-Producing Bacteria, Fungi and Actinomycetes, contribute to corrosion through various biochemical activities. These microorganisms produce corrosive metabolites like hydrogen sulfide, sulfuric acid, organic acids and nitrites, or they alter the redox conditions around metal surfaces. Their activities lead to pitting, rust formation, weakening of protective coatings and rapid metal degradation in pipelines, storage tanks and underground structures.

Table 1. Soil Microorganisms Involved in Corrosion

Group of Microorganism	Examples	Type of Corrosion Caused	Mechanism / Role
Sulphate-Reducing Bacteria (SRB)	*Desulfovibrio desulfuricans, Desulfotomaculum*	Anaerobic corrosion (pitting)	Produce H_2S which reacts with metals → metal sulphides → deep pitting corrosion
Sulfur-Oxidizing Bacteria (SOB)	*Thiobacillus thiooxidans, Thiobacillus ferrooxidans*	Acid corrosion	Produce sulfuric acid by oxidizing sulfur → lowers pH → metal dissolution
Iron-Oxidizing Bacteria (IOB)	*Gallionella, Leptothrix, Sphaerotilus*	Rust tubercle formation	Oxidize Fe^{2+} to Fe^{3+} → deposits iron oxides → differential aeration corrosion
Iron-Reducing Bacteria (IRB)	*Geobacter, Shewanella*	Structural weakening of metal	Reduce ferric (Fe^{3+}) oxides → destabilize protective layers → accelerate corrosion

continued on following page

Table 1. Continued

Group of Microorganism	Examples	Type of Corrosion Caused	Mechanism / Role
Nitrate-Reducing Bacteria (NRB)	*Pseudomonas, Bacillus, Paracoccus*	Accelerated corrosion in underground structures	Produce nitrite and nitric acid → metal oxidation
Methanogenic Archaea	*Methanobacterium, Methanococcus*	Hydrogen-induced corrosion	Utilize hydrogen → depolarize cathodic reaction → increase metal dissolution
Acid-Producing Bacteria (APB)	*Acetobacter, Lactobacillus*	Acid attack on metal	Produce organic acids → reduce pH and enhance corrosion
Filamentous Fungi	*Aspergillus niger, Penicillium, Fusarium*	Chemical and physical deterioration	Produce organic acids and penetrate metal coatings, causing corrosion
Actinomycetes	*Streptomyces* species	Surface pitting	Produce acidic metabolites that degrade coatings and metal surfaces

3. SOIL SALINITY AND SODICITY

The natural weathering of rocks or the parent material is the primary cause of salinity and sodicity in soils. Seawater leaking into coastal areas, inadequate drainage and an increase in the groundwater table, limited surface evaporation, seawater sprays (moisture) close to coastal areas, and the consumption of saltwater or brackish water to irrigate plants are additional sources (Li et al., 2021). Poor drainage and prolonged cultivation of shallow-rooted crops (which results in the formation of salt beds in the soil layers) are the main reasons of high salinity in areas where the water table is rising, which is typically the case in drylands that are not watered frequently. In a result, the soil becomes less absorbent, and eventually the groundwater's salts rise to the surface and increase the salinity. When water is scarce, particularly in arid or dry regions with no additional supply of saline or sodic water, repeatedly using this water for irrigation results in the salinization of both surface and subsurface formations (Basera et al., 2019).

Roughly one-third of the world's population lives in dry regions, which make up 41% of the planet's total land area. Rainfall in the arid region occurs sparsely, sporadically, and mostly during a few strong storms. Droughts are common, and it frequently doesn't rain when it should. Rainfall can occasionally be detrimental to the crops in the field. Plants cannot thrive in dry land situations when there is insufficient precipitation for longer than three months. The rainfall is less than 40% of the possible evapotranspiration, and the average temperature is at least 27°C. Dry and semi-dry regions feature a variety of soil types and vegetation cover, intermittent droughts, and low, erratic rainfall of up to 700 mm annually (Lv et al., 2024).

The primary variables influencing soil quality at the field level are variations in agricultural methods, cultivation systems, and fertilizer. Phosphorus (P) is an essential macronutrient that organisms require in large quantities. It contributes to up between 0.2 and 0.8% of their weight. Because it stimulates tillering and root formation, the P is particularly essential for early vegetative growth. Nevertheless, a well-known nutritional issue for salt-affected soils is a deficiency in P. On these soils, using P fertilizer at standard rates may not result in the highest crop production and quality. Poorly soluble calcium phosphate raw materials occur in alkaline and calcareous soils, making it difficult for plants to absorb and use P. This suggests P will adhere to the soil and eventually settle out. P fertilizer is less effective in calcareous alkaline soils due to its limited solubility. Because the soil contains so much calcium, insoluble calcium phosphate complexes temporarily separate from the soil, lowering the amount of P that is available. In response to the lack of phosphorus in certain soils, plants can become stunted, resulting in shorter internodes and weaker roots (Singh & Pokhrel, 2017).

The Table – 2 demonstrated Plant growth is more strongly impacted by salinity during the reproductive stage. Grain yields were reduced in wheat plants under stress from 100–17 m NaCl because they had fewer spikelets per spike, spikes that emerged more slowly, and lower fertility. However, these wheat plants' shoot apex had concentrations of Na^+ and Cl^- that were too low to impede metabolic reactions less than 50 and 30 mM, respectively. The precise way that salt stress controls the cell cycle and differentiation may be the root source of salinity's harmful effects. According to recent research, salinity inhibits enzyme activity, seed germination, and seedling growth, all of which have a negative impact on plant growth and development (Lv et al., 2024).

Table 2. Comparison Between Salinity and Sodicity in Soil

Parameter	Salinity	Sodicity
Main Cause	Accumulation of salts like NaCl, $MgCl_2$, $CaCl_2$	High exchangeable sodium percentage (ESP > 15%)
Effect on Soil Structure	Causes soil to flocculate (particles come together)	Causes soil to disperse (particles separate)
Impact on Plants	Creates osmotic stress; reduces water uptake	Poor aeration, reduced root growth, toxicity
Effect on Infiltration	Moderate reduction	Severely reduces infiltration due to clay swelling

continued on following page

Table 2. Continued

Parameter	Salinity	Sodicity
Impact on Microbes	Reduces enzyme activity and microbial diversity	Reduces microbial biomass and disrupts nutrient cycling
Common in microbes	Coastal areas, poor drainage, saline irrigation water	Arid and semi-arid regions with poor-quality water

4. IMPACT OF FERTILIZERS, PESTICIDES, AND AGRICULTURAL CHEMICALS ON MICROBIAL

The impact of fertilizers, pesticides, and agricultural chemicals on microbial corrosion in soil is a complex and increasingly relevant topic. These chemicals, widely used to enhance crop production, can alter the soil microbial communities responsible for corrosion processes affecting agricultural implements and irrigation systems. Chemical fertilizers, especially nitrogen- and phosphate-based formulations, influence microbial activity and soil chemistry (Sharma et al., 2018). They can modify soil pH, nutrient availability, and redox conditions, directly affecting the growth and metabolic functions of corrosion-related microbes such as Sulfate-Reducing Bacteria (SRB), Iron-Oxidizing Bacteria (IOB), and Nitrate-Reducing Bacteria (NRB). Enhanced nutrient concentrations can stimulate microbial metabolism, potentially accelerating Microbiologically Influenced Corrosion (MIC) by increasing microbial biomass and biofilm formation on metal surfaces. Conversely, some fertilizers may inhibit specific microbial groups through changes in soil chemistry, thereby altering corrosion rates and patterns (Kumar et al., 2025).

Pesticides and herbicides, designed to control pests and weeds, have documented toxic effects on soil microorganisms. These chemicals can reduce microbial diversity, disrupt enzymatic activities critical to nutrient cycling, and interfere with microbial community structure. The reduction or shift in microbial populations can affect the balance between corrosive and non-corrosive microbes (Telegdi et al., 2017). Selective pressure from pesticides may favors resilient MIC-relevant microbes that can thrive despite chemical stress, potentially leading to localized intensification of corrosion near affected implements and pipelines. Additionally, pesticide residues might change soil organic matter composition and bioavailability, indirectly influencing microbial corrosion activity by altering microbial ecology (Jia et al., 2018).

Agricultural chemicals also affect extracellular electron transfer mechanisms and biofilm dynamics on metal surfaces. Changes in microbial community function can affect the production of corrosive metabolites like sulfides and organic acids, which directly participate in metal degradation. Some chemicals may promote biofilm resilience or alter its composition, facilitating microbial electron uptake from metals and accelerating MIC. Others may disrupt these processes by inhibiting key microbial groups or enzymatic pathways (Eid et al., 2018). The complex interaction between applied agricultural chemicals and corrosion-related soil microbiomes also impacts the biogeochemical cycling of iron, sulfur, and nitrogen central elements in MIC. Fertilizers raising nitrate levels can enhance nitrate-reducing bacteria activity, changing redox potential and affecting corrosion chemistry. Sulfur-containing agrochemicals might alter sulfur-oxidizing or reducing bacteria populations, influencing acid production and metal sulfide formation. These chemical-microbe interactions create dynamic and often site-specific corrosion environments in agricultural soils (Huang et al., 2021).

Furthermore, the long-term use of fertilizers and pesticides leads to cumulative effects on soil microbial health and corrosion potential. Persistent chemical residues may cause chronic microbial community shifts, some possibly reducing microbial diversity and destabilizing soil ecosystems. This ecological disruption can lead to unpredictable corrosion behavior, sometimes exacerbating metal degradation in irrigation systems and farming tools.In managing microbial corrosion in agriculture, understanding the impacts of fertilizers and pesticides is crucial. Practices such as optimized fertilizer use, pesticide selection with minimal microbial toxicity, and soil health monitoring can help mitigate adverse corrosion effects. Developing corrosion-resistant materials and applying biocontrol agents that modulate microbial communities are emerging strategies targeting chemical-induced MIC problems (Hua et al., 2023).

Overall, fertilizers, pesticides, and other agricultural chemicals profoundly influence soil microorganisms involved in corrosion processes. Their effects on microbial abundance, diversity, metabolic activity, and community interactions determine the extent and nature of microbiologically influenced corrosion on agricultural implements and irrigation infrastructure. Effective management requires integrated approaches considering both chemical use and microbial ecology to reduce corrosion-related losses and maintain sustainable agricultural operations (Atalah et al., 2022).

5. CORROSION AND DEGRADATION OF STRUCTURAL MATERIALS IN FARM ENVIRONMENT

Farm structures are created employing concrete, steel reinforcement, timber, and concrete replacements. The durability of concrete, however, is the most important factor to think about. Due to its susceptibility to rot, swell, shrink, color changes, and fungal and insect damage, natural wood is only utilized in a limited number of circumstances. In addition, metal constructions are more prone to rust than concrete ones. Concrete buildings on farms are subjected to a range of corrosive substances, including as acids, bacteria, and meal-water mixes. These substances cause the pH to drop, lactic acid and other organic acids to accumulate, soluble calcium salts to develop, and leaching to occur. The loss of structural integrity of the material is the cumulative effect. In addition to the weather, metal reinforcements such as galvanized Z-girders, which are frequently utilized in agricultural storage silos, have an impact on the buildings' strength. There are other factors besides concrete and metal that affect how long timber support structures last. The weather, the wood's treatment, and the joint reinforcements all have a role. Researchers investigate the effects of bacteria on corrosion induced by bases, acids, and concrete in metal and concrete buildings. The first sections look at how bacteria and fungus contribute to the deterioration of metal and cement structures (Xu et al., 2024).

5.1. Functions of Bacteria Plant-Engineered Root Zone Communities

It relates to the widely acknowledged complexity of soil agro-ecosystems. However, the plant root system acts as a guiding force that can bring order to the natural disorder found in productive agricultural soils. Numerous microorganisms, including pathogens, mutualistic symbionts, endophytes, transient epiphytic saprophytes, and epiphytic commensals, reside on the vast surface area of roots (Chen et al., 2024). In order to support the growth of particular, plant-beneficial microbial communities, plants can "load up" the root zone environment with substrates. People mostly do this by either selectively exuding specific sugars, carboxylic acids, and amino acids, or by releasing carbon-rich material in the form of root border cells (Tripathi et al., 2021).

Latest research indicates that the consequences of fungus and bacteria on metal and concrete constructions differ. The presence of non-aggressive microbes is linked to positive effects, whilst the establishment of unwanted bacteria is linked to negative effects. demonstrates how fungus and bacteria work together to generate corrosion. It illustrates that while mechanical and chemical corrosion is occurring, heterotrophic bacteria and fungi function at their peak efficiency (Tran et al., 2021). Cement (C-S-H-phases, Ca (OH)), silicate/carbonate aggregates, and sulfate salts (such as Gp, Anh, Bas, and Ett) are necessary for the process. *Acidithiobacillus thiooxidans, Acidithiobacillus ferrooxidans, Thiobacillus intermedius, Halothiobacillus neopolitanus, Starkeya novella* and *Thiobacillus tioparus*are the most active bacterial species at this phase. Importantly, corrosion is unlikely to happen until certain parameters are met, such as high relative humidity, average temperature, and elevated CO_2 and H_2S levels. Everyone will talk about the durability of concrete and the effects of acid corrosion on it below (Frah et al., 2023). Allelopathy can be used to describe all these root-microbe interactions. They comprise microorganisms or communities of endophytically competent bacteria that can aid in plant growth and prevent the spread of illnesses, as well as biochemical interactions between and within the same species (Custódio et al., 2021).

5.2. Corrosion due to Slurries/Manures and its Solution

Since animals initially were hunted in the wild, animal husbandry has gone through significant change. Manure and waste removal issues have gotten worse due to grasscutters, rabbits, the growing demand for cow's milk, goats, and other associated species. In order to create materials that can endure the sharp increase in waste production, manufacturers are taking special notice of the equipment and the supplementary tools needed to handle it. It combines dung and urine with litter, which is typically straw or wood shavings, to create farm yard manure (Frah et al., 2023).

Fermentation of Farm Yard Manure (FYM) produces Carbon dioxide, Ammonia, and Moisture. These gasses are the most corrosive component of slurries and farm yard manures when combined with naturally occurring animal chlorides. Neglected agricultural equipment and materials, as well as steel structures, are very corrosive. Manufacturers continue to utilize concrete or galvanized steel for order floors *via* livestock barns. According to tests, piglets fed copper compounds lose roughly 0.1 mm of galvanized metal thickness annually. The local deposit/galvanic corrosion effect could be the source of this. Researchers have discovered that combining slurry and silage effluent can produce a hazardous mixture of hydrogen sulfide gas, which may partially cause corrosive acids to build beneath deposits (Garrido et al., 2019).

6. THERMAL REMOTE SENSING AND ITS APPLICATION IN AGRICULTURAL SECTOR

In agriculture, thermal RS can be used in a variety of techniques to monitor crops and soil. It can be used to map soil texture, identify residue cover, recognize tiles in fields, continuously monitor crop maturity for harvesting, map crop yield, recognize diseases and pathogen-infected crops, and determine the amount of water stress crops and soil are under for irrigation scheduling. This section discusses studies conducted to determine the potential utility of thermal imaging in agriculture. highlights a few studies on the efficient application of thermal imaging in agriculture (Knisz et al., 2023).

6.1. Soil Moisture Detection

Soil moisture is crucial for plant growth because it acts as a solvent and carrier of nutrients, regulates soil temperature, supports microbial life, and influences agricultural practices such as fertilization, irrigation, and planting. It also serves as a nutrient on its own right. To determine when to irrigate the soil, researchers have employed thermal RS to measure the soil's moisture content. Two popular methods for measuring soil moisture using thermal imaging are the triangle method and the soil moisture index. The triangle approach is based on the distribution of pixels in the fractional plant cover image of surface radiant temperature. As the vegetation cover increases, the range of surface radiative temperatures decreases, resulting in the formation of a triangle. The narrow range of surface radiative temperature throughout the dense vegetation is depicted by a vertex (Thomas et al., 2024).

Plants that are under water stress usually have an enhanced radiant temperature than plants that are well watered. The Perpendicular Soil Moisture Index (PSMI) using raw digital count data from Landsat satellite images in the red, NIR, and thermal bands. They showed that PSMI is strongly related to soil moisture that was measured. Measurement of the temperature of shaded leaves as a dependable indicator of leaf temperature and, consequently, stomatal conductance. It also showed that the temperature differences between stressed and non-stressed plants could be very useful for finding out if plants need water, which could help with scheduling irrigation. Utilizing thermal images obtained from UAVs, the CWSI was estimated to pinpoint water-stressed regions within a commercial orchard, thereby validating this method for precision irrigation management (Tirandaz et al., 2024).

6.2. Plant Disease Detection

In agriculture, plant disease resulted to substantial losses, especially in terms of production and financial losses. Particularly, plant diseases contribute to crop losses in the USA that cost over $33 billion in lost income annual. According to some estimates, farmers might cut this loss by almost $11 million if they eradicate 20% of the widespread disease known as soybean rust. In Alabama's Coastal Plain and Appalachian Plateau physiographic regions, studies have shown that it is feasible to assess and track wheat (*Triticum aestivum L.*) straw cover rates (0, 10, 20, 50, and 80%) using thermal remote sensing. The residue cover rate affected the heat capacity of the soil and residue surfaces, assuming all other factors remained constant. According to their findings, visible and near-IR photos could only explain 77% of the variations in crop residue cover, however thermal images could explain almost 95% of the variations (Zhang et al., 2018).

6.3. Soil Texture Mapping

Other chemical and physical properties of the soil, like its structure, porosity, hydraulic attributes, and holding of nutrients capacity, can be inferred from some soil properties, specifically texture. These factors all have an impact on soil quality, which in turn has an impact on crop growth efficiency. Therefore, being able to rapidly and properly determine the texture of soil is crucial when making farming decisions. Soil texture and land surface temperature are closely related. The soil's texture influences how much water it can retain, which in turn influences the land surface's temperature. For example, dry, sandy soil that is less able to retain water is likely to lose water more quickly and have less water in it overall (Shentu et al., 2023).

The soil's surface will be chilled if the clay soil, which can hold more water, loses it more slowly and contains more water. Because soil pH fluctuates constantly due to "internal" and "external" influences, it is a crucial component of soil fertility. It impacts the quality of organic matter, the capacity to buffer and filter, the ease of nutrient uptake by plants, and the biomass production of most crops. Soils with a pH of 6 to 7.5, which is slightly acidic to slightly alkaline, are ideal for growing most vegetables that can be cultivated on land. A pH of less than 5.5 is undesirable and requires lime treatment to improve. The soil maintains its shape more effectively because of to the organic materials (Chatterjee et al., 2021).

7. CHALLENGES AND FUTURE PERSPECTIVES ON THE USE OF THERMAL IMAGES FOR PRECISION AGRICULTURE

Thermal RS provides information regarding crop temperature and soil surface temperature during time and space, however there are a few considerations when using thermal RS. These include the impact of the atmosphere's quality and functionality, the spatial and temporal resolutions of the captured, and the fact that the majority of thermal cameras only have a 640 x 480 pixel resolution. Thermal cameras are still more costly and have lower resolution than commercial off-the-shelf optical sensors used in UAVs, despite the fact that these represent significant advancements in thermal technology. The resolutions of the optical cameras that are sold in stores are more than 4200 by 2800 pixels (Ali, 2019).

7.1. Spatial and Temporal Resolution

Scientists around the world have tracked the temperatures of crops and soil surfaces using satellite thermal. Except for most farming locations, the pixels in these images are larger than a single field, so they cannot be used for site-specific farming applications. For example, some satellite-based thermal sensors that continuously monitor the entire world, like the Advanced Very High-Resolution Radiometer (AVHRR), the Geostationary Environmental Satellite (GOES), and the Moderate-Resolution Imaging Spectroradiometer (MODIS), can record thermal images with resolutions ranging from 1000 to 4400 meters. On the other hand, only images with a resolution of 250 to 1000 meters may be captured using visible, NIR, and SWIR images. Some satellites have improved the spatial resolution of thermal sensors, such as ASTER (90m), Landsat 7ETM (60m), and Landsat 8 (100m). But these satellites only come once every sixteen days (Pozzer et al., 2020).

The low spatial resolution and temporal frequency present major barriers for farmers who also want to use images from satellites to monitor their crops and soil during growth. Based on the surface emissivity determined by the object, the precise manner in which that radiation is transmitted by the environment, and the speeds at which it rises and falls, a thermal sensor monitors radiation. The ratio of radiation released by an object's surface to that emitted by a black body, or perfect emitter of radiation, at the same temperature is known as emissivity. Other air quality parameters, such as humidity and temperature, can also be measured. Some of you can obtain an estimate of how far away the object is from the sensor due to glancing at the flight altitude (Milovanović & Pečur, 2016).

7.2. Future Use of Unmanned Aircraft in PA

Since UAVs are far more adaptable and less expensive for taking pictures from, they are anticipated to be employed far more for agricultural and soil monitoring in the future than traditional satellite and manned aircraft platforms. UAV services are probably going to increase the fastest, per a poll conducted in early 2015 among US PA service retailers. It is anticipated that 38% of individuals will use UAV services, up from 19% in 2015. UAV-generated imagery could be a good alternative to aerial and satellite imagery, but there are many technological and regulatory obstacles from the time of acquisition to the distribution of the final photos. UAV has a lot of potential for soil and crop monitoring with less stringent flying regulations and improved algorithms for image processing, geo-referencing, mosaicing, and classification. (Pinck et al., 2020).

8. TYPES OF IRRIGATION SYSTEM

Surface irrigation, sprinkler irrigation, and micro irrigationwhich includes drip, trickle, and spray are the three primary categories of irrigation systems. Each has advantages and disadvantages in specific circumstances – Figure - 1.

Figure 1. Types of Irrigation system in Agricultural Corrosion

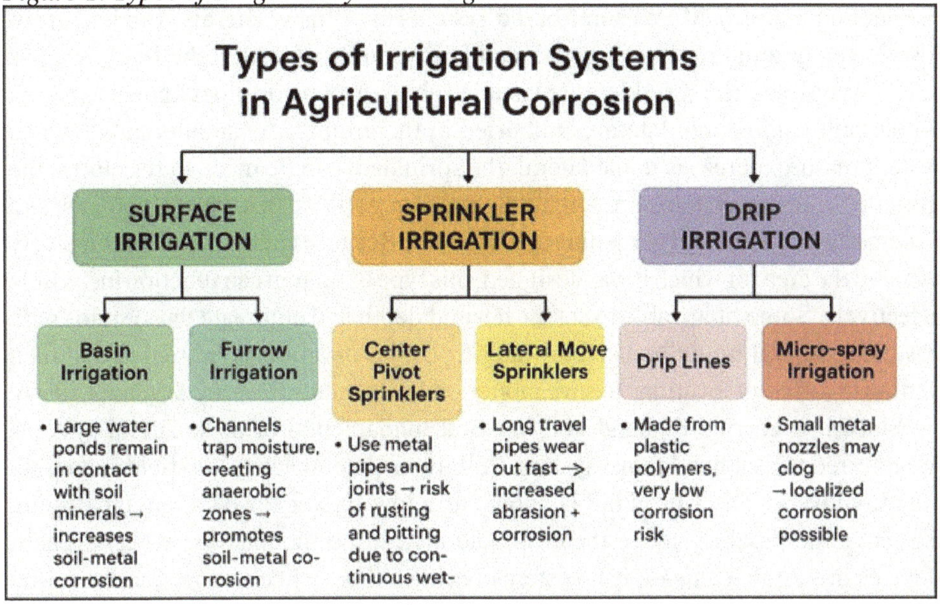

8.1. Sprinkler Irrigation

Sprinkler irrigation systems are utilized by locals for gardening, farming, and landscaping. The principles of operation are the same for all applications, despite the possibility of varied management objectives. Sprinkler systems can be divided into four primary categories: Solid-Set, Moved Lateral, Moving Lateral, and Single-Sprinkler. There are two types of sprinkler systems that D uses. With single-sprinkler systems, a vast area is watered by a single sprinkler. The sprinkler can move around the space automatically or manually with the fingertips. A single sprinkler positioned around the yard and automatic systems with a large water-firing sprinkler that may spray water hundreds of feet are two examples (traveler irrigation system). The location and duration of sprinkler placement determine how well a particular sprinkler system performs (Pousti et al., 2018).

A disadvantage of these constructions is that, in order to ensure that the targeted field receives adequate water they typically irrigate regions outside of the ones that are being irrigated. A further advantage, however, is that the single sprinkler system is incredibly adaptable and functions well on non-square terrain. A system with several sprinklers connected by a lateral pipe is more complicated than one with just one sprinkler. A lateral-based sprinkler system's primary components are the mainline and one or more laterals. Water is transported from the source to the laterals using a system of pipes called the mainline. Sprinkler devices are stored in the lateral pipelines. Most lateral-based systems have many laterals. The laterals of a solid-set system are fixedly positioned in just one spot within the field. In most solid-set systems, the sprinklers are installed above ground on pipes known as risers, while the mainline and laterals are buried in the ground. As an alternative, when water pressure increases in the lateral, the sprinklers can be made to rise above the ground. Solid-set systems are frequently used on golf courses, lawns, gardens, and in some agricultural and gardening applications. Because each sprinkler exclusively watered the area for which it was designed, this type of system can function incredible effectively. Some of are able to set the desired depth and automate the systems with ease. One lateral might be used to water a portion of an irrigated region and then shifted to different locations to save money on investments (Geesey & White, 1990).

Since it is carried by hand, the first and fundamental of these lateral systems are referred to as a hand move system. Pulling the lateral across the field is another way to move it. The latter kind is known as a tow line or towed sprinkler system. To move the pipeline above the crop, laterals can be mounted on wheels. These devices are known as side roll systems because their wheels move the lateral by moving across the field. The laterals are typically left in one spot for eight, twelve, or even twenty-four hours because repositioning them requires a tremendous amount of effort. As a result, each irrigation in the systems often uses a lot of water. The lateral has been moved across the field using automated systems. Examples of moving lateral systems include center pivots and linear or lateral motion systems. One lateral is used by all of these systems to irrigate a sizable region. However, because the lateral travels at a fixed speed, the water's depth can be changed over a large range (Rathour & Pathania, 2025).

8.2. Surface Irrigation

Several forms of surface irrigation are utilized, depending on the crops cultivated, soil texture, and property type. Furrows, boundaries, and basins are instances of this. On farms, orchards, and additional locations with moderate slopes, people utilize surface irrigation systems. Once surface irrigation is implemented, the water distributes out across the field as it travels over the land. Surface irrigation methods usually more affordable for operation per unit of water utilized since they require less pressure than sprinkler irrigation. Installing surface systems could be less expensive than installing sprinklers if it’s don't need to level the ground. In addition, surface irrigation frequently encounters three issues. To water the area uniformly, water must flow across it quickly. A portion of the water will therefore run off the field. Rules in several states prohibit irrigation water from leaving the field (Gaur et al., 2022).

The runoff problem is substantially solved if the surface system incorporates a return flow or runoff recovery system. The second concern is that surface irrigation can be exceptionally physical demanding. Much of the time, irrigators are either unable or unwilling to dedicate themselves to the time required to carry out it correctly. Too many applications result from this, which causes water to be wasted through deep percolation or runoff. Deep percolation, which occurs when infiltration is not evenly distributed, is another common problem with surface irrigation. Similar to a sprinkler system's mainline, a surface irrigation system has a means of delivering water to it. This supply system may include of subterranean pipelines with surface-mounted valves, a gated pipe, or a "head" ditch. Employing siphon tubes for transferring water from a supply canal is one method for doing this. Water will flow over a tiny portion of the field regardless of the type of water supply equipment employed. This area of the field may be bounded by furrows in furrow irrigation or by small dikes in a border irrigated field. Sometimes a flat area has little dikes surrounding it. This type technology is known as basin irrigation. If the field is nearly level in both the transverse and flow directions, the water that would typically run off it may be obstructed and compelled to remain on the field (Deland et al., 2021).

8.3. Microirrigation

Laterals with emitters (drip irrigation), micro sprinklers, or laterals that allow water to flow continuously along their lengths (soaker hose) have all been used in micro irrigation systems. It demonstrates dropping irrigation, likewise referred to as drip irrigation, on the soil's surface. Micro irrigation differs in that the discharge devices are designed to irrigate a single plant or a collection of plants rather than the entire soil surface. In landscape applications, each emitter's flow rate might be extremely low, while in orchard applications, multiple devices might be required to supply the proper amount of water. Microsystems can be expensive and are frequently installed permanently. Maintenance can be costly if the water needs to be purified, but there isn't much effort involved. High-value crops frequently use micro irrigation systems in areas with limited, costly, or subpar water supplies. The holes or openings of emitters and micro sprinklers are extremely tiny (Custódio et al., 2021).

It is necessary to prevent soil particles and microorganisms such as germs from becoming trapped in the tiny holes. Because microsystems require numerous costly pipelines and filters, they are among the priciest methods of watering plants. Because they are too expensive and must be removed each season, they are typically not utilized to produce row crops. By burying the laterals beneath the tillage zone, subsurface drip irrigation addresses the second issue. Furrow irrigation using a gated pipe is one type of surface watering. The image was provided by the Nebraska Extension and Steve Melvin. Micro irrigation is widely used in landscaping, particularly for gardens, trees, and plants. Numerous advantages come with these systems, including: (1) their high efficiency due to not wetting the entire plant area, which initially results in low evaporation loss; (2) their extremely low water application rates, which result in virtually nonexistent runoff, even on steep slopes; and (3) their ease of automation to save labor (Deland et al., 2021).

9. DRIP IRRIGATION AND ITS IMPORTANCE

Drip irrigation is an efficient and sustainable irrigation practice that makes water flow through the root zone of plant in controlled and slow way. This technique saves a lot of water as opposed to the conventional irrigation methods such as sprinklers or flood irrigation because it reduces water loss which are evaporation, runoff, and deep percolation. An effective drip irrigation system can save on the consumption of water by 30-60 percent and is therefore especially useful in regions where there is a water shortage or droughts. Water is delivered at the point where it is required, and this saves water besides delivering nutrients to the plants since fertilizers can be directly fed to plants and fertilization efficiency has increased by approximately 30 percent and environmental pollution has been reduced. Drip irrigating is not only important to the conservation of water. It delivers water and nutrients directly to the root zone which makes plants grow healthier and stronger, thus likely to yield more crops- up to 230 percent. The accurate application of water also lowers the growth of the weed as the water is not scattered all over the land where the weeds commonly grow and it helps in reducing the diseases related to the wet growth of the foliage hence the use of pesticides is reduced (Noreen et al., 2024).

Drip irrigation minimizes soil erosion and nutrient leaching which are prevalent in traditional irrigation systems in terms of soil health. It offers a smooth and uniform distribution of water that sustains soil structure and fertility which play an important role in sustainable farming. The system is also versatile and adaptable to other field conditions because it can be used on any type of soil and field shape without field leveling. Drip systems may be low pressure thus saving on energy consumption as well as reducing operation costs.The extra advantages include labor and cost efficiency. Drip irrigation systems are highly automated using timers and do not need as much manual control meaning that less labor will be required. Drips irrigation will also pay off within a relatively short time period, as an initial investment will be recovered via the savings on water, fertilizer, and energy bills, as well as, enhanced production. The system is easy to expand and can apply it to many crops and landscapes, such as gardens, orchards, vineyards, greenhouses, and row crops (Coenye et al., 2024).

Drip irrigation is an important technology that can enhance sustainable agriculture through saving water, enhancing crop health and yields, lowering the cost of labor and inputs, and safeguarding soil health. Its accuracy and efficiency do not only contribute to the conservation of the environment but also make farmers economically viable, and thus it assists in serving the rising global food demands with lower resources. The wide range of application as well as the long-term costs make the drip irrigation an intelligent investment to both the small-scale gardeners and large agricultural ventures (Giangeri, 2024).

10. CALCULATION TEST FOR DETERMINING THE DURATION OF ACID TREATMENT

By measuring how long it takes for a color dye to reach the ultimate lateral and the last emitter, this test is carried out only once to determine how long the acid treatment will persist. It has been employ the throttle valve variance in pressure and run the drip irrigation system while trying to do this. The injectable liquid is created by combining water with a color dye. A stopwatch is started to record the time when suction begins. The ideal duration for acid treatment is the time it takes for the colored solution in order to reach final lateral and emitter. The end caps are opened to release the color solution after the time has proved recorded. The same procedure has been employed if the field has other sections. In order for the benefit of as a reference, if there are multiple sections, the section number and the amount of time required for acid treatment are noted (Geesey & White, 1990).

10.1. Acid Treatment for Drip Irrigation System

If drip irrigation systems fail to be properly maintained, they might get clogged or blocked. This can occur for several reasons, including microbes present in the system (algal growth and bacterial slime). Large particles including suspended silt and clay particles from the water supply enter the drip system and obstruct it. Deposits may form in the lateral pipes as a result of certain dissolved salts, such as carbonates and bicarbonates, calcium and magnesium, and iron. They may also occasionally be able to block the emitters. Chlorine treatment addresses issues caused by bacteria developing in the drip system, whereas acid treatment addresses issues caused by dissolved or precipitated salts. A Fertigation tank, a Venturi System, or a Fertijet Pump can be used to inject these chemicals. A common issue that arises when salts dissolve or settle is clogging (Jia et al., 2018).

11. REGULATION OF AIR AND WATER IN SOIL

In the root zone, plants require both oxygen and water to thrive as best they can xisting gaps between soil aggregates and particles that allow water, air, and other substances to enter. The flow of water and gasses into and out of the soil is determined by the size of the pores inside and between soil particles. Macropores are typically visible to the naked eye, while micropores inside and between microaggregates are invisible. The soil must have both macro- and micropores to allow water and air to reach the plants and maintain healthy growth. Soil pores are created when soil particles adhere to one another to form a hierarchy of aggregates. In addition to improving aggregate stability. Clay, organic debris, root hairs, bacterial and fungal organic substances, and fungal hyphae are the elements that hold soil aggregates together. Aggregate stability is the capacity of soil aggregates to withstand the erosive pressures of water. Maintaining macropores in the soil, preventing surface crusting, enhancing aeration, preventing rain from running off, and preventing soil erosion are all made possible by good aggregate stability. Because the organic matter particles contained in aggregates are safe for bacteria to consume, aggregates also aid in maintaining organic matter in the soil. This is possible for many large soil creatures. Both nitrogen and phosphorus inputs can be balanced by introducing small amounts of manure or compost to meet phosphorus needs and then increasing nitrogen from legume cover or forage crops in a crop rotation (Giangeri, 2024).

12. CONCLUSION

This research concludes by highlighting the fact that soil microorganism activity is a major driver and accelerator of agricultural infrastructure corrosion, which poses a serious operational and economic problem. Although Microbiologically Influenced Corrosion (MIC) is a significant exacerbating factor, soil corrosion is essentially an electrochemical process. Through the development of biofilms and the generation of corrosive metabolites like acids and hydrogen sulfide, microorganisms such as sulfate-reducing bacteria and acid-producing fungi promote corrosion, resulting in localized pitting, structural deterioration, and the failure of metal and concrete components. The particular circumstances of agricultural settings make the issue even worse. In addition to physically stressing materials, variables like soil salinity and sodicity can change the microbial ecology in the soil, frequently favoring more caustic species. Additionally, the widespread use of agro-chemicals, such as fertilizers, pesticides, herbicides, and acidic manure slurries, produces a highly corrosive environment that targets irrigation systems and implements, especially drip systems that are vulnerable to clogging from microbial slime and chemical precipitates.

To tackle this complex problem, a comprehensive mitigating approach is needed. A first line of defense is the application of corrosion-resistant materials such as galvanization, protective coatings, and stainless steel. To avoid emitter blockage, proactive maintenance using acid and chlorine treatments is crucial for irrigation infrastructure, particularly drip systems. Additionally, using cutting-edge monitoring methods like thermal remote sensing can aid in the early identification of problems linked to corrosion and stress. In the end, improving the robustness and lifespan of essential agricultural infrastructure depends on a better comprehension of the intricate interactions among soil microbiology, environmental factors, and farming methods, which will direct the creation of more efficient, focused maintenance and protection procedures.

REFERENCES

Ali, R. (2019). Deep learning- and infrared thermography-based subsurface damage detection in a steel bridge.

Atalah, J., Blamey, L., Amenábar, M. J., Kelley-Loughnane, N., & Blamey, J. M. (2022). Biochemical and microbiological characterization of a thermotolerant bacterial consortium involved in the corrosion of Aluminum Alloy 7075. *Research Square*. DOI: 10.21203/rs.3.rs-2371670/v1

Basera, P., Lavania, M., & Lal, B. (2019). Potential of dynamic bacterial communities in the bio-corrosion process: A proof study with surface morphology of metal coupons. *RSC Advances*, *9*(30), 17040–17050. DOI: 10.1039/C9RA01959F PMID: 35519851

Chatterjee, M., Fan, Y., Cao, F., Jones, A. A., Pilloni, G., & Zhang, X. (2021). Proteomic study of *Desulfovibrio ferrophilus* IS5 reveals overexpressed extracellular multi-heme cytochrome associated with severe microbiologically influenced corrosion. *Scientific Reports*, *11*(1), 15458. Advance online publication. DOI: 10.1038/s41598-021-95060-0 PMID: 34326431

Chen, H., Kimyon, Ö., Gunawan, C., Ramandi, H. L., Craig, P., Chen, R., et al. (2024). An effective barrier coating technology against premature bolt failures in underground mines. Rock Mechanics and Rock Engineering. https://doi.org/10.1007/s00603-024-03909-3

Coenye, T., Ahonen, M., Anderson, S., Cámara, M., Chundi, P., Fields, M. W., Foidl, I., Gnimpieba, E. Z., Griffin, K., Hinks, J., Loka, A. R., Lushbough, C., MacPhee, C., Nater, N., Raval, R., Slater-Jefferies, J., Teo, P., Wilks, S., Yung, M., & Webb, J. S. (2024). Global challenges and microbial biofilms: Identification of priority questions in biofilm research, innovation and policy. *Biofilm*, *8*, 100210. DOI: 10.1016/j.bioflm.2024.100210 PMID: 39221168

Custódio, V., Gonin, M., Stabl, G., Bakhoum, N., Oliveira, M. M., Gutjahr, C., & Castrillo, G. (2021). Sculpting the soil microbiota. *The Plant Journal*, *109*(3), 508–522. DOI: 10.1111/tpj.15568 PMID: 34743401

Deland, E., An, B. A., Yao, J., Sobol, O., Skovhus, T. L., & Koerdt, A. (2021). *Environmental conditions impact the corrosion layer composition of methanogen induced microbiologically influenced corrosion (MI-MIC)*. Research Portal Denmark.

Doshvarpassand, S., Wu, C., & Wang, X. (2018). An overview of corrosion defect characterization using active infrared thermography. *Infrared Physics & Technology*, *96*, 366–375. DOI: 10.1016/j.infrared.2018.12.006

Eid, M. M., Duncan, K. E., & Tanner, R. S. (2018). A semi-continuous system for monitoring microbially influenced corrosion. *Journal of Microbiological Methods*, *150*, 55–62. DOI: 10.1016/j.mimet.2018.05.018 PMID: 29803719

Garrido, I., Lagüela, S., Cфappa, C., Saavedra, F. J. M., & Arias, P. (2019). Automatic detection of moistures in different construction materials from thermographic images. *Journal of Thermal Analysis and Calorimetry*, *138*(2), 1649–1660. DOI: 10.1007/s10973-019-08264-y

Gaur, V. K., Gupta, S., Sharma, P., Gupta, P., Varjani, S., Srivastava, J. K., Chang, J.-S., & Bui, X.-T. (2022). Metabolic cascade for remediation of plastic waste: A case study on microplastic degradation. *Current Pollution Reports*, *8*(1), 30–50. DOI: 10.1007/s40726-021-00210-7

Geesey, G., & White, D. C. (1990). Determination of bacterial growth and activity at solid-liquid interfaces. *Annual Review of Microbiology*, *44*(1), 579–597. DOI: 10.1146/annurev.mi.44.100190.003051 PMID: 2252395

Giangeri, G. (2024). Novel approaches for steering microbial interactions in anaerobic digestion: Focusing on the role of direct interspecies electron transfer. *Research Portal Denmark*, 165.

Hua, W., Sun, R., Wang, X., Zhang, Y., Li, J., Qiu, R., & Gao, Y. (2023). Corrosion of Q235 carbon steel induced by sulfate-reducing bacteria in groundwater: Corrosion behavior, corrosion product, and microbial community structure. *Environmental Science and Pollution Research International*, *31*(3), 4269–4283. DOI: 10.1007/s11356-023-31422-7 PMID: 38097840

Huang, Y., Xu, D., Huang, L., Lou, Y., Muhadesi, J.-B., Qian, H., et al. (2021). Responses of soil microbiome to steel corrosion. *npj Biofilms and Microbiomes, 7*, 6. **https://doi.org/**DOI: 10.1038/s41522-020-00175-3

Javaherdashti, R. (2016). Microbiologically influenced corrosion (MIC). In *Engineering materials and processes* (p. 29). Springer Nature. DOI: 10.1007/978-3-319-44306-5_4

Jia, R., Ünsal, T., Xu, D., Lekbach, Y., & Gu, T. (2018). Microbiologically influenced corrosion and current mitigation strategies: A state of the art review. *International Biodeterioration & Biodegradation*, *137*, 42–58. DOI: 10.1016/j.ibiod.2018.11.007

Jones, L., Salta, M., Skovhus, T. L., Thomas, K., Illson, T., & Wharton, J. A.. (2024). Dual anaerobic reactor model to study biofilm and microbiologically influenced corrosion interactions on UNS G10180 carbon steel. *Research Square*. **https://doi.org/**DOI: 10.21203/rs.3.rs-5216435/v1

Karadoğan, A., Azgin, F., Yiğittekin, E. S., & Dinçer, S. (2025). Biofilms and their multidimensional impacts: From combating industrial and clinical risks to unlocking opportunities in sustainable agriculture. In *IntechOpen eBooks*. IntechOpen., DOI: 10.5772/intechopen.1010579

Knisz, J., Eckert, R. B., Gieg, L. M., Koerdt, A., Lee, J., Silva, E. R., Skovhus, T. L., An Stepec, B. A., & Wade, S. A. (2023). Microbiologically influenced corrosion—More than just microorganisms. *FEMS Microbiology Reviews*, *47*(5), fuad041. Advance online publication. DOI: 10.1093/femsre/fuad041 PMID: 37437902

Kumar, A. S., Sivakumar, L., Rajadesingu, S., Sathish, S., Malik, T., & Parthipan, P. (2025). Sustainable corrosion inhibition approaches for the mitigation of microbiologically influenced corrosion - a systematic review. *Frontiers in Materials*, *12*, 1545245. Advance online publication. DOI: 10.3389/fmats.2025.1545245

Kumari, P. P., & Mulky, L. (2024). Optimization strategies for corrosion management in industries with artificial neural network and response surface technology: A comprehensive review. *Journal of Bio- and Tribo-Corrosion*, *10*(3), 59. Advance online publication. DOI: 10.1007/s40735-024-00863-z

Li, Q., Gong, L., Chen, X.-G., Gadd, G. M., & Liu, D. (2025). Dual role of microorganisms in metal corrosion: A review of mechanisms of corrosion promotion and inhibition. *Frontiers in Microbiology*, *16*, 1552103. Advance online publication. DOI: 10.3389/fmicb.2025.1552103 PMID: 40270819

Li, Z., Chang, W., Cui, T., Xu, D., Zhang, D., Lou, Y., Qian, H., Song, H., Mol, A., Cao, F., Gu, T., & Li, X. (2021). Adaptive bidirectional extracellular electron transfer during accelerated microbiologically influenced corrosion of stainless steel. *Communications Materials*, *2*(1), 67. Advance online publication. DOI: 10.1038/s43246-021-00173-8

Lv, X., Wang, C., Liu, J., Sand, W., Etim, I. N., & Zhang, Y.. (2024). The microbiologically influenced corrosion and protection of pipelines: A detailed review. *Materials (Basel)*, *17*(20), 4996. DOI: 10.3390/ma17204996 PMID: 39459699

Milovanović, B., & Pečur, I. B. (2016). Review of active IR thermography for detection and characterization of defects in reinforced concrete. *Journal of Imaging*, *2*(2), 11. DOI: 10.3390/jimaging2020011

Moreira-Filho, P., & Figueiredo, P. (2021). The influence of the marine *Bacillus cereus* over carbon steel, stainless corrosion, and copper coupons. *Archives of Microbiology*, *204*(1), 9. Advance online publication. DOI: 10.1007/s00203-021-02607-w PMID: 34873663

Noreen, S., Saleem, M. H., Ali, B., Khan, K. A., Hafeez, A., & Javed, M. A. (2024). Advancing plant resilience against microplastics and metals through nanotechnology. *BioNanoScience, 14*(2), 2065–2079. DOI: 10.1007/s12668-024-01301-7

Pinck, S., Ostormujof, L. M., Teychené, S., & Erable, B. (2020). Microfluidic microbial bioelectrochemical systems: An integrated investigation platform for a more fundamental understanding of electroactive bacterial biofilms. *Microorganisms, 8*(11), 1841. DOI: 10.3390/microorganisms8111841 PMID: 33238493

Pousti, M., Zarabadi, M. P., Amirdehi, M. A., Paquet-Mercier, F., & Greener, J. (2018). Microfluidic bioanalytical flow cells for biofilm studies: A review. *Analyst, 144*(1), 68–87. DOI: 10.1039/C8AN01526K PMID: 30394455

Pozzer, S., Azar, E. R., Rosa, F. D., & Pravia, Z. M. C. (2020). Semantic segmentation of defects in infrared thermographic images of highly damaged concrete structures. *Journal of Performance of Constructed Facilities, 35*(1), 04020131. Advance online publication. DOI: 10.1061/(ASCE)CF.1943-5509.0001541

Procópio, L. (2020). Microbial community profiles grown on 1020 carbon steel surfaces in seawater-isolated microcosm. *Annals of Microbiology, 70*(1), 13. Advance online publication. DOI: 10.1186/s13213-020-01547-y

Rathour, A., & Pathania, S. (2025). Sensing of microplastics using advanced materials: A comprehensive review. *Sensing and Imaging, 26*(1), 80. Advance online publication. DOI: 10.1007/s11220-025-00613-6

Salgar-Chaparro, S. J., Darwin, A., Kaksonen, A. H., & Machuca, L. L. (2020). Carbon steel corrosion by bacteria from failed seal rings at an offshore facility. *Scientific Reports, 10*(1), 69292. DOI: 10.1038/s41598-020-69292-5 PMID: 32703991

Sharma, M., Liu, H., Chen, S., Cheng, Y. F., Voordouw, G., & Gieg, L. M. (2018). Effect of selected biocides on microbiologically influenced corrosion caused by *Desulfovibrio ferrophilus* IS5. *Scientific Reports, 8*(1), 1747. DOI: 10.1038/s41598-018-34789-7 PMID: 30413730

Shentu, H., Wei, J., Lin, X., Zhou, H., Wang, S., & Liu, J. (2023). Effect of microbial community on iron release under low velocity in drinking water supply pipes. *BIO Web of Conferences, 60*, 1008. DOI: 10.1051/bioconf/20236001008

Singh, M., & Pokhrel, M. (2017). A fuzzy logic-possibilistic methodology for risk-based inspection (RBI) planning of oil and gas piping subjected to microbiologically influenced corrosion (MIC). *International Journal of Pressure Vessels and Piping, 159*, 45–55. DOI: 10.1016/j.ijpvp.2017.11.005

Su, H., Mi, S., Peng, X., & Han, Y. (2019). The mutual influence between corrosion and the surrounding soil microbial communities of buried petroleum pipelines. *RSC Advances*, *9*(33), 18930–18937. DOI: 10.1039/C9RA03386F PMID: 35516885

Telegdi, J., Shaban, A., & Trif, L. (2017). Microbiologically influenced corrosion (MIC). In *Elsevier eBooks* (p. 191). Elsevier BV. DOI: 10.1016/B978-0-08-101105-8.00008-5

Thomas, P., Sahoo, B. N., Thomas, P. J., & Greve, M. (2024). Recent advances in emerging integrated anticorrosion and antifouling nanomaterial-based coating solutions. *Environmental Science and Pollution Research International*, *31*(60), 67550–67576. Advance online publication. DOI: 10.1007/s11356-024-33825-6 PMID: 38831147

Tirandaz, H., Dastgheib, S. M. M., & Hamedi, J. (2024). Sub-minimum inhibitory concentration of tetrakis(hydroxymethyl)phosphonium sulfate enhances biocorrosion of carbon steel by *Pseudomonas aeruginosa. Scientific Reports*, *14*(1), 28918. Advance online publication. DOI: 10.1038/s41598-024-70157-4 PMID: 39572582

Tran, T. T. T., Kannoorpatti, K., Padovan, A., & Thennadil, S. N. (2021). Sulphate-reducing bacteria's response to extreme pH environments and the effect of their activities on microbial corrosion. *Applied Sciences (Basel, Switzerland)*, *11*(5), 2201. DOI: 10.3390/app11052201

Tripathi, A. K., Thakur, P., Saxena, P., Rauniyar, S., Gopalakrishnan, V., Singh, R. N., Gadhamshetty, V., Gnimpieba, E. Z., Jasthi, B. K., & Sani, R. K. (2021). Gene sets and mechanisms of sulfate-reducing bacteria biofilm formation and quorum sensing with impact on corrosion. *Frontiers in Microbiology*, *12*, 754140. DOI: 10.3389/fmicb.2021.754140 PMID: 34777309

Xu, L., Khan, A., Kijkla, P., Kumseranee, S., Punpruk, S., & Gu, T. (2024). Prevention of severe pitting corrosion of 13Cr pipeline steel by a sulfate reducing bacterium using a green biocide cocktail. *Frontiers in Materials*, *11*, 1407655. Advance online publication. DOI: 10.3389/fmats.2024.1407655

Zhang, G., Li, B., Liu, J., Luan, M., Yue, L., Jiang, X., Yu, K., & Guan, Y. (2018). The bacterial community significantly promotes cast iron corrosion in reclaimed wastewater distribution systems. *Microbiome*, *6*(1), 222. Advance online publication. DOI: 10.1186/s40168-018-0610-5 PMID: 30545419

ADDITIONAL READING

El Kayal, W., Al-Kafaween, M., & Al-Maadeed, M. (2024). Agricultural mitigation strategies to reduce bacterial contamination in irrigation systems. *Frontiers in Microbiology, 15*, 1099234. DOI: 10.3389/fmicb.2024.1099234

Emerson, D., Fleming, E. J., & McBeth, J. M. (2023). Aerobic iron-oxidizing bacteria secrete metabolites that markedly impede abiotic iron oxidation. *Proceedings of the National Academy of Sciences of the United States of America, 120*(50), e2309893120. DOI: 10.1073/pnas.2309893120

Huang, Y., Xu, D., Huang, L., Lou, Y., Muhadesi, J.-B., Qian, H., et al. (2021). Responses of soil microbiome to steel corrosion. *npj Biofilms and Microbiomes, 7*, 6. https://doi.org/DOI: 10.1038/s41522-020-00175-3

Knisz, J., Eckert, R. B., Gieg, L. M., Koerdt, A., Lee, J., Silva, E. R., Skovhus, T. L., An Stepec, B. A., & Wade, S. A. (2023). Microbiologically influenced corrosion—More than just microorganisms. *FEMS Microbiology Reviews, 47*(5), fuad041. Advance online publication. DOI: 10.1093/femsre/fuad041 PMID: 37437902

Liu, F., Zhang, Y., Li, X., & Wang, N. (2021). Soil microbial communities response to long-term heavy metal contamination in electroplating contaminated site. *Frontiers in Microbiology, 12*, 644493. DOI: 10.3389/fmicb.2021.644493

Sulfate-Reducing Bacteria (SRB). (n.d.). In *Corrosionpedia*. Retrieved December 7, 2025, from https://www.corrosionpedia.com/definition/1046/sulfate-reducing-bacteria-srb

Thermal Remote Sensing in Precision Agriculture. (2025, April 18). *Vastraholding*. Retrieved December 7, 2025, from https://vastraholding.com/en/thermal-remote-sensing-agriculture/

Wang, X., Ma, L., Zhang, D., & Gu, T. (2022). Accelerated sulfate reducing bacteria corrosion of X80 steel welded joint under carbon starvation. *npj Materials Degradation, 6*, 34. https://doi.org/DOI: 10.1038/s41529-022-00291-9

What is an Iron Oxidizing Bacteria Test Tube? (2025, March 3). *Siny Medical*. Retrieved December 7, 2025, from https://www.sinymedical.com/what-is-an-iron-oxidizing-bacteria-test-tube/

KEY TERMS AND DEFINITIONS

Acid-Producing Bacteria (APB): Acetobacter, Lactobacillus generate organic acids, reducing pH and attacking metal surfaces.

Allelopathy: Plant root exudates (sugars, acids) shape beneficial vs. corrosive microbial communities.

Biofilm: Microbial communities on metal surfaces bound by extracellular polymeric substances (EPS), creating anaerobic microenvironments for electron transfer and corrosion.

Extracellular Polymeric Substances (EPS): Microbial matrix protecting communities and facilitating metal-microbe electron transfer.

Farm Yard Manure (FYM): Fermentation produces CO_2, NH_3, moisture, and chlorides, corroding steel structures and equipment.

Fertilizers/Pesticides: Alter pH, nutrients, favoring MIC microbes (SRB, IOB) or reducing diversity.

Iron-Oxidizing Bacteria (IOB): Microaerophilic bacteria (Gallionella, Leptothrix) oxidize Fe^{2+} to Fe^{3+}, forming rust tubercles and differential aeration cells.

Iron-Reducing Bacteria (IRB): Geobacter, Shewanella reduce Fe^{3+} oxides, destabilizing protective layers and accelerating corrosion.

Irrigation Systems: Drip/sprinkler systems prone to microbial clogging from slime, salts, requiring acid/chlorine maintenance.

Methanogenic Archaea: Methanobacterium consume H_2, depolarizing cathodic reactions and increasing metal dissolution.

Microbiologically Influenced Corrosion (MIC): Microbial activity accelerates metal corrosion via biofilms, corrosive metabolites (H_2S, acids), and electron transfer, causing pitting in buried infrastructure.

Nitrate-Reducing Bacteria (NRB): Pseudomonas, Bacillus produce nitrites/nitric acid, promoting metal oxidation in underground structures.

Slurry/Manure Corrosion: H_2S from animal waste mixtures attacks galvanized steel (0.1 mm/year loss), causing galvanic corrosion.

Soil Moisture: Boosts SRB activity, creates oxygen gradients shaping microbial communities on metal surfaces.

Soil pH: Acidic conditions (from SOB/APB) enhance metal dissolution; optimal 6-7.5 for crops.

Soil Salinity: Soluble salt accumulation (NaCl, $MgCl_2$) from irrigation/poor drainage, causing osmotic stress and altered microbial corrosion activity.

Soil Sodicity: High ESP (>15%) causes clay dispersion, poor aeration, and microbial disruption exacerbating corrosion.

Soil Temperature: Influences microbial metabolism rates and corrosion acceleration.

Sulfate-Reducing Bacteria (SRB): Anaerobic bacteria (Desulfovibrio desulfuricans) reduce sulfates to H_2S, forming metal sulfides and deep pitting corrosion.

Sulfur-Oxidizing Bacteria (SOB): Thiobacillus thiooxidans oxidize sulfur to sulfuric acid, lowering pH and causing acid corrosion.

Thermal Remote Sensing: Infrared imaging detects soil moisture (triangle method, PSMI), plant stress (CWSI), and corrosion anomalies.

Chapter 4
Biotechnological Approaches for the Control of Microbial Corrosion

B. Lokeshwari
Sacred Heart College, Tirupattur, India

P. Saranraj
https://orcid.org/0000-0002-5052-6532
Sacred Heart College, Tirupattur, India

U. Subhalakshmi
Sacred Heart College, Tirupattur, India

K. Gayathri
https://orcid.org/0009-0007-3786-2009
Sacred Heart College, Tirupattur, India

R.S. Roohina Tabassum
https://orcid.org/0009-0008-8250-7206
Sacred Heart College, Tirupattur, India

A. Mohanasri
Sacred Heart College, Tirupattur, India

ABSTRACT

Biotechnological solutions represent an environmental friendly and viable alternative for managing Microbiologically Influenced Corrosion (MIC), which is a ubiquitous issue in the industrial infrastructure, such as pipelines, marine ships, and water systems. The conventional procedures of corrosion control that use toxic compounds and synthetic finishes are harmful to the environment and have limited long-term efficacy. In contrast, biotechnology offers self-regulating, bio-inspired solutions. Key biotechnological strategies include the use of beneficial microorganisms as probiotic corrosion inhibitors, enzyme systems to degrade corrosive metabolites and biofilm disruption mechanisms to limit biofouling. The biosurfactant and biopolymers usage can give natural corrosion shield, while biogenic nanoparticles produced via green synthesis exhibit promising antimicrobial and anticorrosive properties.

DOI: 10.4018/979-8-3373-2617-7.ch004

1. INTRODUCTION

Microbial Corrosion or Microbiologically Influenced Corrosion (MIC) is one of the most serious threats to the industrial and environmental infrastructure. Unlike conventional corrosion models, which primarily involve electrochemical and physicochemical reactions, MIC is a complex process involving interactions between microbial communities, their metabolic activities, and material surfaces (Gonzalez et al., 2020). Localized corrosion, and biofilm-based electrochemical changes are caused by a diverse range of microorganisms, such as Sulfate-Reducing Bacteria (SRB), Sulfur-Oxidizing Bacteria (SOB), Iron-Oxidizing Bacteria (IOB), etc. MIC is mostly found in oil and gas pipes, marine structures, power plants, and water distribution systems and failures caused by MIC can result in significant economic losses, environmental pollution, and safety hazards (Dinh et al., 2004). The traditional methods of corrosion control that involve the use of chemical biocides, cathodic protection, coating of materials and physical cleaning has limited against Microbial Industrial Corrosion. These are microbial resistance to biocides, ineffective penetration into biofilms, microbial consortia complexity, and environmental toxicity of their chemical use.

The environmental impact of chemical pollutants has driven industries to seek more effective and sustainable solutions. Biotechnology has become a primary focus for next-generation Microbial Industrial Corrosion control. Biotechnology is a paradigm shift, since it has the biological causes of Microbial Industrial Corrosion instead of the electrochemical symptoms of the conditions. Molecular microbiology, metagenomics, proteomics, and metabolomics have significantly enhanced the identification, characterization, and modeling of the microbial communities in corrosion (Banerjee et al., 2017). These early MIC diagnosis, real time monitoring and identification of major functional genes and metabolic pathways that contribute to corrosive activity. Synthetic biology has novel opportunities to develop microorganisms resistant to the growth of biofilms, the ability to degrade corrosive metabolites, or overcome harmful microorganisms. Other environmental friendly alternatives to conventional chemical treatments are now being employed, including biotechnological enzymes, quorum-quenching molecules, biosurfactants, bacteriophages, probiotics, and microbial inhibitors (Binnig et al., 1986). Moreover, biofilm engineering provides the choice of regulating the organization of microbial communities, disrupting the pathways in cell-cell interaction, or creating protective and beneficial biofilms that inhibit the corrosive species.

Nanobiotechnology offers advanced capabilities, including the targeted delivery of antimicrobial agents, surface functionalization with enzyme-based coatings, and the creation of smart materials. Although significant advancements have been made, challenges remain in validating the long-term stability of biological solutions at field-scale and complex interactions within MIC biofilms (Parkes et al., 2011). The convergence of biotechnology, materials science, and digital monitoring has created novel opportunities to develop sustainability-focused, efficient, and precision-based MIC management approaches. Biotechnological methods represent a rapidly emerging frontier capable of transforming the field of corrosion control. These biotechnological innovations are defining the future of MIC mitigation, including its underlying mechanisms, benefits, constraints, and industrial applications (Enning & Garrelfs, 2014).

2. PRINCIPLES OF BIOTECHNOLOGY IN CORROSION MANAGEMENT

Biotechnology offers several approaches to the anticipation, and intervention of Microbiologically Influenced Corrosion (MIC), a phenomenon in which microorganisms have substantial metabolic and bio electrochemical remodelling of metal-environmental interacions. MIC has been estimated to contribute between 20-40% of the corrosion failures in the world necessitating the adoption of control measures in high-risk sectors like oil and gas, marine transport, wastewater, and power generation (Videla & Herrera, 2005). Biotechnological approaches integrate microbial ecology, enzymology, genomics and synthetic biology to either prevent corrosive microbes or to make microbiota capable of forming protective surface coatings instead of using toxic biocides and long-duration soluble chemical inhibitors. Biotechnological MIC management development is based on biofilm organised communities of microorganisms within an extracellular polymeric substance (EPS) matrix. These biofilms create physicochemical gradients of oxygen, pH, sulfide and redox potential which have a potent impact on the corrosion kinetics.

In such biofilms, the microbes promote corrosion either by the formation of corrosive metabolites (hydrogen sulfide, organic acids and siderophores) or by catalyzing Extracellular Electron Transfer (EET) among metals and electron acceptors (Marsili et al., 2008). Protective effects are observed in other microorganisms which feed on the corrosive metabolites, form stable surface-associated films or generate EPS which inhibits the flow of electrons and dissolution of the metal. Recent progress in omics technologies and metabolic models has clarified biochemical processes of MIC that including direct electron uptake for the Extracellular Electron Transfer systems. These tools can be used to create biotechnological applications that can specifically and precisely regulate the activity of microbes and regulate the behaviour of biofilms to stabilize industrial materials in complicated microbial habitats. Hence, microbial ecology and biofilm engineering are the foundations of future-generation biotechnological interventions to control MIC.

2.1. Microbial Ecology and Biofilm Engineering

Microbial ecology is also fundamental in explaining the mechanisms of MIC because the corrosive microorganisms are found in multi species biofilms, which are metabolically cooperative, competitive, and niche specific. These interactions determine biofilm structure, EPS composition, and the local electrochemical environment. There are other significant MIC organisms like SRB, methanogens, IOB, IRB, and SOB, which cause corrosion through their individual metabolic and environmental preferences (Hamilton, 2003). Sulfide produced by Desulfovibrio spp. is reacted with iron to form conductive films of iron sulfide, and methanogens are able to obtain electrons directly out of steel, perhaps through extracellular complexes of enzymes, which enhances cathodic depolarization (Uchiyama et al., 2010). These interactions highlight the critical importance of ecological knowledge for developing effective corrosion-control strategies.

Biofilm engineering builds upon the principles of microbial ecology to avoid the formation of corrosive biofilms and favouring protective ones through the control of QS, EPS composition, and colonization behaviour (Tan et al., 2016). Biofilm maturation and the production of corrosive metabolites can be minimized by interfering with Quorum-Sensing (QS) signals, such as Acyl-Homoserine Lactones (AHLs), using quorum-quenching enzymes or synthetic inhibitors (Chen et al., 2011). EPS plays a dual role in MIC by regulating nutrient retention, binding metals, and facilitating electron transfer. While conductive EPS networks can accelerate corrosion, modified or protective EPS can serve as diffusion barriers (Reguera et al., 2005).

The development of biotechnological methodologies must account for the ecological stability of MIC biofilms, which exhibit high resistance to biocides due to EPS diffusion barriers and efflux pumps (Stewart and Costerton, 2001). Metagenomics and metatranscriptomics of MIC biofilms, which reveal genes linked to sulfate reduction, methanogenesis, EPS synthesis, and extracellular transfer of electrons. Such tools have identified multiheme c-type cytochromes in SRB and methanogens that mediate the uptake of electrons in steel explaining the underlying MIC processes. Another alternative is the ecological approach incorporating useful microorganisms that compete with corrosive taxa or generate protective metabolites (probiotic corrosion control). Surface engineering also helps mitigate MIC through altering hydrophobicity, surface charge, or roughness with enzyme-functionalized surfaces, bioinspired polymers, or bioengineered biofilms which are self-assembling to form self-healing protective layers. Manipulation of extracellular electron transfer pathways is the focus of biotechnology-based MIC control since SRB and methanogens, as well as metal-respiring bacteria, use nanowires, cytochromes, and soluble electron shuttles as electron exchange with the metals.

Strategies like redox-reactive polymers, non-toxic electron mediators and modified microorganisms can interfere or divert electrons flow to inhibit corrosive metabolism. Bio electrochemical systems (BES) also extend the concept by controlling microbial electron transfer by applying potentials, preventing colonization or metabolic pathways associated with MIC. MIC microbiomes are highly influenced by environmental factors in terms of temperature, salinity, nutrient availability, and hydrodynamics. The biotechnology allows predictive modelling, which involves microbial kinetics, gene regulation, and EPS dynamics to predict the risk of corrosion and direct the interventions (Kato et al., 2012). Early detection of MIC activity using complementary monitoring tools such as biosensors, molecular probes, and bioelectrochemical signatures enables proactive management. Microbial ecology and biofilm engineering establish the basis of biotechnology-based MIC control. Through the combination of omics-based analysis, engineered biofilms, surface modification, and electron-transfer control, biotechnology facilitates the development of accurate, sustainable, and adaptive corrosion mitigation strategies for the efficacy of conventional chemical methods.

3. KEY BIOTECHNOLOGICAL STRATEGIES FOR CORROSION CONTROL

Biotechnology provides an ecologically sustainable framework for managing MIC by targeting the metabolic and ecological pathways that drive microbial corrosion. This approach contrasts with conventional methods, such as chemical biocides and cathodic protection, which are often ineffective against biofilms, generate toxic by-products, and can promote antimicrobial resistance (Macdonald, 2006). Biotechnological methods use enzymatic treatments, microbial community engineering, bio-based coatings, green nano-particles and genetic methods to attain accurate and Eco compatible control of corrosion. The modern technologies in metagenomics, synthetic biology, and materials science are used to develop biologically based corrosion inhibitors that supplement or substitute traditional techniques. The mechanisms of these strategies including the degradation of corrosive metabolites, disruption of biofilm structure, interference of microbial signaling, and alteration of microbe-metal interactions are paving the way for a novel generation of biologically driven corrosion protection technologies.

3.1. Enzyme-Based Corrosion Inhibition

Enzyme-based strategies represent a green technology that utilizes catalytic biomolecules to mitigate MIC. These enzymes degrade corrosive compounds, disrupt extracellular matrices, or modify metal surfaces. Enzymes offer several advantages over chemical inhibitors, including high substrate specificity, biodegradability, low toxicity, and operational efficacy under moderate environmental conditions (Chakraborty et al., 2020). The build-up of corrosive metabolites produced by Sulfate-Reducing Bacteria (SRB), methanogens, and Sulfur-Oxidizing Bacteria (SOB) is neutralized through a tailored selection of enzymes that convert the corrosive metabolites into non-corrosive or less aggressive forms. In addition, the localized and long-term protective benefits of immobilization of enzymes on metal surfaces are through sustained degradation of corrosive molecules at the metal-biofilm interfaces. Enzymatic methods are also useful in applications where chemical biocides cannot be applied i.e., in drinking water piping, food processing systems.

3.1.1. Enzymatic Detoxification of Corrosive Metabolites

Corrosive microorganisms produce a wide variety of metabolites that accelerate metal degradation, including Hydrogen Sulfide (H_2S), organic acids, ammonia, ferrous iron, and extracellular electron mediators. Enzymatic detoxification aims to neutralize these metabolites before they reach concentrations capable of triggering or sustaining MIC. For example, sulfide oxidase and sulfide:quinone reductase can convert H_2S produced by SRB into elemental sulfur or sulfate, thereby reducing its corrosion (Guo et al., 2018). Similarly, lactate oxidase and formate dehydrogenase degrade organic acids that lower pH and destabilize metal layers. Ammonia-oxidizing enzymes have the potential to reduce nitrifying and ammonifying bacteria corrosion in wastewater treatment plants. Enzymes attacking the extracellular electron donors, e.g. flavins, can also inhibit Extracellular Electron Transfer (EET) is an essential process in MIC powered by methanogens and iron-reducing bacteria. Because these enzymes catalyses transform corrosive metabolites, even low concentrations can provide substantial long-term corrosion protection.

3.1.2. Enzyme Immobilization on Metal Surfaces

The immobilization of enzymes on metal surfaces improves corrosion resistance by concentrating catalytic functions to the interface where reactants of corrosion and materials. Immobilization stabilizes enzymes and increases their functional half-life, as well as decreases the necessity of repeated dosage (Vigneron et al., 2016). Common immobilization strategies employed in corrosion applications include covalent bonding, physical adsorption, sol-gel matrices, enzyme entrapment in polymers, and cross-linked enzyme aggregates. These immobilized oxidoreductase enzymes are capable of breaking down biofilm EPS structures, inhibiting the adhesion of microbial cells and breaking down established biofilms. Proteinases, proteases,

DNases, and glycoside hydrolases have the ability to decompose structural parts of EPS, weakens the biofilm and making it more vulnerable to other inhibitors. Enzyme-functionalized surfaces represent an alternative approach for anticorrosion materials to microbial activity.

3.2. Biofilm Engineering and Control

Biofilm engineering targets the structural, regulatory, and ecological foundations of microbial biofilms, which are crucial to MIC development. Biofilms create micro-environments characterized by oxygen gradients, localized acidification, and enhanced electron transfer conditions that intensify corrosion compared to planktonic cells (Flemming & Wuertz, 2019). Controlling biofilm formation is therefore a primary goal in MIC mitigation. Biotechnology provides novel tools to inhibit biofilm development, disrupt established biofilms, or promote beneficial biofilms that protect metal surfaces. Strategies include quorum quenching, enzymatic degradation of EPS, surface modification, and deployment of anti-biofilm compounds derived from microorganisms or plants. These biological approaches mitigate the drawbacks of chemical biocides and show the high specificity toward corrosion-causing microbes.

3.2.1. Quorum Quenching Strategies

Quorum sensing (QS) has been shown to regulate biofilm formation, virulence, adhesion, and metabolic expression in a wide range of microorganisms in MIC, such as SRB, Pseudomonas, and iron-oxidizing bacteria (Dang and Lovell, 2016). Quorum Quenching (QQ) disrupts QS signalling by degrading Acyl-Homoserine Lactones (AHLs), Auto-Inducing Peptides (AIPs), or AI-2 molecules, thereby inhibiting biofilm development and metabolic activity. Enzymes, such as the AHL lactonases, acylases, and oxidoreductases have the ability to break down QS signals and prevent coordinated behaviour of microbes. These purified QQ enzymes into the industrial systems greatly lower the biofilm density of the SRB and H_2S generation. The stability and specificity of QQ enzymes can be enhanced through genetic engineering to enable tailored disruption of corrosive consortia. QQ enzyme immobilization on surfaces is a combination of QS interference and local protection of metal substrates.

3.2.2. Application of Anti-Biofilm Agents

Biological-derived anti-biofilm agents such as antimicrobial peptides, phyto-based secondary metabolites, bacteriocins and microbial metabolites offer environmental friendly alternatives to chemical biofilm dispersants. The rhamnolipids produced by *Pseudomonas aeruginosa* interfere with biofilm matrix organization and stop SRB adhesion (Moura et al., 2018). Coatings incorporating anti-biofilm compounds derived from marine organisms can effectively disrupt bacterial adhesion proteins. Enzyme-based anti- biofilm agents like DNase, proteinases and polysaccharide hydrolases destroy key EPS parts, which facilitates biofilm dissociation and inhibits the reforming process. Extracellular electron transfer is also inhibited by some biopolymers and peptides that decrease the catalysis of redox reactions by biofilm-forming microbes on metal surfaces. The use of anti-biofilm solutions with the technology for coating and microbial interference methods for a total control of MIC.

3.3. Application of Biosurfactants and Biopolymers

Biopolymers and biosurfactants represents a sustainable category of non-toxic, highly biodegradable corrosion inhibitors with significant industrial potential. Biosurfactants cause a change in the surface hydrophobicity, prevent adhesion of the microbes, and disrupt the pre-existing biofilms. Corrosion resistant coatings could be biopolymers like polysaccharides, Poly-Hydroxy-Alkanoates (PHAs) and microbial exopolymers. These are biological molecules that are synthesized by microorganisms under controlled fermentation with the metabolic engineering to increase stability and anticorrosive performance (Cao et al., 2015). They are compatible with metal surfaces and are eco-friendly to the environment, so they are suitable in infrastructure that is exposed to marine, wastewater or petroleum environments.

3.3.1. Role of Biosurfactants in Corrosion Inhibition

The Biosurfactants such as rhamnolipids, sophorolipids, lipopeptides and glycolipids lower the interfacial tension and modify physicochemical characteristics of the metal surface and therefore prevent adhesion of microorganisms and the initial formation of biofilm. Biosurfactants blocks the attachment sites by forming protective monolayers, producing an unfavourable environment for planktonic cells to colonize surfaces (Nitschke & Costa, 2007). Numerous biosurfactants have been found to possess antimicrobial properties against SRB, methanogens, and biofilm-forming bacteria, which has a direct impact on MIC mitigation. Rhamnolipids have shown great impacts in the dispersal of SRB biofilm and the inhibition of H_2S formation, whereas surfactin, a lipopeptide secreted by *Bacillus subtilis*, has the ability to permeabilise microbial membrane and destabilise EPS. Nanoparticles and enzymes are also being used in synergy with biosurfactants to enhance penetration of the biofilm matrix.

3.3.2. Biopolymer Coatings as Green Corrosion Barriers

Biopolymers that are of microbial or plant origin offer environmental friendly coats that form physical and chemical resistant barriers to corrosive compounds. These hydrophilic films made of polysaccharides can be used to inhibit the diffusion of oxygen, chloride ions, and corrosive metabolites, and include alginate, chitosan, dextran, xanthan gum, and others (Kiran et al., 2020). Chitosan, especially, has antimicrobial activity, which is inherent to it, and therefore it is effective against MIC-associated bacteria. The PHAs and other biodegradable polyesters are stable, corrosion-resistant coating that firmly attaches to metal surfaces. Enzymes, nanoparticles, or anti- biofilm agents can be encapsulated in some biopolymer coatings to be delivered at a sustained rates and provide greater protection. The modification of biopolymers to exhibit enhanced viscosity, metal-binding ability or hydrophobicity through engineering of EPS-producing strains has become a novel field in corrosion control.

3.4. Biogenic Nanoparticles for Corrosion Protection

Biogenic nanoparticles, synthesized using microorganisms or plant extracts, present an environmental benign alternative to chemically synthesized nanoparticles. They are highly biocompatible, tunable, and possess inherent antimicrobial characteristics which make them excellent corrosion protection in MIC-prone conditions (Iravani, 2011). Intracellular enzymes, extracellular reductants, or metabolic by-products can be used to reduce metal ions into stable nanoparticles by microorganisms like fungi, bacteria, and algae. The biogenic nanoparticles as inhibitors of corrosion can be used through the formation of protective coating or suppression of corrosive microbes. Biogenic nanoparticles incorporated into the coating or biopolymers or hybrid materials have increased durability and anticorrosive behaviours.

3.4.1. Green Synthesis of Metal and Metal-Oxide Nanoparticles

The nanoparticles of metals (Ag, Cu, Au) and metal oxides (ZnO, TiO_2, $Fe3O_4$) are synthesized with the help of biological reducing agents (flavonoids, polyphenols, peptides, or microbial enzymes) (Ahmed et al., 2016). These techniques avoid toxic solvents and high-energy reactions, yielding nanoparticles with high stability and a narrow size distribution. The efficient producers of silver and zinc oxide nanoparticles are fungi and bacteria, such as *Fusarium oxysporum* and *Bacillus subtilis*. The biological capping agents that are generated when synthesizing include proteinases, polysaccharides, or lipids, which increase nanoparticle adherence to metal surfaces and are corrosion inhibitors. Biogenic nanoparticles have the potential of reducing the microbial load, inhibiting biofilm growth and enhancing the barrier properties.

3.4.2. Antimicrobial and Anti-Corrosive Mechanisms

Biogenic nanoparticles exhibit dual functionality: an antimicrobial effect that inhibits MIC-inducing microorganisms, and a physicochemical effect that enhances corrosion resistance. Nanoparticles of silver and copper damage cell membranes, cause oxidative stress, disrupting enzyme activity, and preventing biofilm formation (Lee et al., 2019). The photocatalytic effect of metal-oxide nanoparticles includes ZnO and TiO_2, which produce reactive oxygen species, making microbes less viable. Another application of nanoparticles is the creation of thick adherent protective layers which hinder diffusion of chloride ions, oxygen and corrosive metabolites. It is used to functionalize magnetic nanoparticles like $Fe3O_4$ with enzymes, bio-surfactants, inhibitors, so that they can be delivered to MIC-prone regions with the help of external magnetic fields.

3.5. Genetic and Molecular Engineering Approaches

Genetic engineering and synthetic biology are emerging technologies in MIC control as they provide the possibility to accurately manipulate the behaviour, metabolic pathways, and corrosion behaviour of microbes. These methods enable the creation of engineered microbes, which detect corrosive environments, synthesize anti-MIC molecules, or modify the local environment to prevent corrosion (Mahadevan et al., 2016). Molecular Tools like CRISPR/Cas systems, promote engineering and gene editing to manipulate the corrosive taxa or beneficial microorganisms so as to improve their functions in the corrosive conditions. Genetic tools can also be used for the MIC pathways by deleting, regulating or inserting genes in sulfate reduction, methanogenesis, and in biofilm formation.

3.5.1. Synthetic Biology for Corrosion Monitoring and Mitigation

Synthetic biology enables the creation of engineered microbial biosensors to monitor the early MIC activity in response to the presence or absence of corrosive metabolites, pH, or redox changes by expressing genes. Fluorescent proteinases, luciferase, or electroactive biosensors can be used to conduct real-time measurements of corrosion in pipes, storage tanks, and sea systems (Zhang et al., 2018). Furthermore, microbes can be engineered to produce anti-corrosive molecules (e.g., biosurfactants, QQ enzymes, biopolymers) directly at metal surfaces. Artificial consortia that are engineered to SRB or methanogens can be used to control community composition to non-corrosive profiles. Additionally, self-healing biocoatings composed of engineered biofilms that cracks and secrete protective biopolymers represent an innovation in corrosion resistance.

3.5.2. CRISPR/Cas Applications in Microbial Genomics

The CRISPR/Cas technologies offer accurate gene-editing technologies to study and regulate MIC-associated pathways in microorganisms corrosive to metals. The CRISPR interference (CRISPRi) has the capability to silence sulfate reduction-related genes, e.g., dsrAB, or methanogenic-related electron transfer pathways, and thereby inhibits metabolic activities that facilitate corrosion. CRISPR genome editing allows deletion of genes that regulate production of EPS, adhesion, or biofilm formation of corrosive bacteria enabling them to be less able to colonize metal surfaces. Moreover, the genes can be inserted into microbial hosts, including those of QQ enzymes or anti-biofilm peptides, with the aid of CRISPR to express them to generate anti-corrosive agents. Furthermore, portable CRISPR diagnostic tools (e.g., Cas12/13) enable rapid, field detection of MIC-related taxa and functional genes, facilitating improved monitoring and early intervention strategies.

4. BIOELECTROCHEMICAL SYSTEMS FOR CORROSION CONTROL

Bio-Electrochemical Systems (BES) are innovative biotechnological tools for MIC mitigation. They exploit microbial extracellular electron transfer to alter redox conditions at metal surfaces. Unlike traditional strategies, BES mitigate corrosion by targeting electron pathways central to MIC, countering processes such as the generation of corrosive metabolites, iron oxidation, hydrogen evolution, and sulfate reduction. Electron transfer processes of cytochromes and conductive pili allow electroactive microorganisms, such as Geobacter, Shewanella, Sulfate-Reducing Bacteria (SRB), and methanogenic archaea to promote BES functionality. BES can be incorporated into a corrosion prone system by using electrodes to create controllable sinks or sources of electron that would cause the microbial metabolism to shift towards less corrosive pathways and stabilize protective surfaces. Microbial Fuel Cells (MFCs), where the electrons are diverted to metal surfaces, and biocathodes (Muyzer and Stams, 2008). Recent progress, including the development of *in situ* BES for submerged within coatings, underscores the potential of these systems as a sustainable, tunable, and bio-integrated approach to long-term corrosion management.

4.1. Microbial Fuel Cells (MFCs) for Corrosion Control

Microbial Fuel Cells (MFCs) mitigate MIC by diverting electrons from metal surfaces, thereby interfering with electron uptake by corrosive microbes and inhibiting the formation of metabolites. Electroactive microorganisms in MFCs, including *Geobacter sulfurreducens* and *Shewanella oneidensis*, oxidize organic matter to produce electrons at one of the electrodes and conduct them via conductive biofilms to an external circuit to form a favorable electron sink, preventing the growth of Sulfate-Reducing Bacteria (SRB) and subsequent sulfide generation (Lovley, 2006). MFCs minimizes the corrosion and suppress cathodic depolarization by SRBs by restricting their access to electron donors, and changing redox conditions on the surfaces.

Other possible contributions of anodic biofilms to corrosion protection are the creation of electrochemically robust, EPS-rich layers which block diffusion of corrosive ions. MFCs have the ability to adjust local pH by cathodic reduction of oxygen to support the formation of protective passivation layers, which oppose MIC processes (Jiang et al., 2016). Although challenges such as electrode durability and biofouling persist, advances in electrode materials and biofilm engineering to improve MFC performance as a sustainable long-term corrosion control strategy.

4.2. Biocathodes for Corrosion Inhibition

Biocathodes also serve as significant bioelectrochemical systems for MIC mitigation. These systems can be reconfigured to manage cathodic electron flow and inhibit corrosive microbial metabolisms (Venkidusamy and Megharaj, 2016). A major advantage is the suppression of cathodic depolarization for the biocathodes that had electrons away from hydrogen evolution, leaving SRB and methanogens without hydrogen to pursue corrosive activity. Biocathodes enhanced with denitrifying bacteria can further inhibit SRB by directing electrons to nitrate reduction sites, thereby reducing sulfide formation. Moreover, biocathodes alter the local pH and redox gradients, which tend to generate protective mineral layers limiting the ability of microbes.

Stable, conductive layers which prevent the colonization of SRB can also be obtained by cathodic biofilms that are formed by beneficial microbes like Geobacter spp. Controlled oxygen reduction biocathodes lead to the reduction of the formation of reactive oxygen species and passive metal films (Zhang et al., 2014). The nitrate reducing biocathodes also have the ability to inhibit sulfate reduction by competition or generation of nitrite, a well-known SrB inhibitor. These novel approaches involve engineered biocathodes that express quorum-quenching or antimicrobial capabilities, and autotrophic systems that are capable of transforming CO_2 into protective biopolymers. Furthermore, fine-tuning the polarization potential of biocathodes allows for precise control over microbial electron uptake and community structure. Biocathodes are a promising, sustainable way of regulating biocorrosion despite the long-term stability issue and the problem of competition with corrosive taxa, as biocathodes provide a biologically directed flow of electrons at the cathode.

5. ANALYTICAL TOOLS FOR BIOCORROSION MONITORING

Effective MIC mitigation relies on analytical instruments capable of identifying microbial communities, characterizing corrosion-impacted surfaces, and monitoring electrochemical alterations that compromise metal stability. Recent findings on the complexity of MIC, which is brought about by synergistic interaction of microbes, formation of biofilms, and effects of metabolites or electron transfer, require integrated biotechnological solutions. Corrosion diagnostics thus integrate molecular, spectroscopic, and electrochemical techniques to generate mechanistic insights into biocorrosion for monitoring and control. Figure 1 summarizes the integration of analytical techniques (electrochemical, microscopic, and chemical analyses) with molecular tools (PCR, sequencing, metagenomics) for biocorrosion monitoring.

Figure 1. Analytical and Molecular Tools for Biocorrosion Monitoring

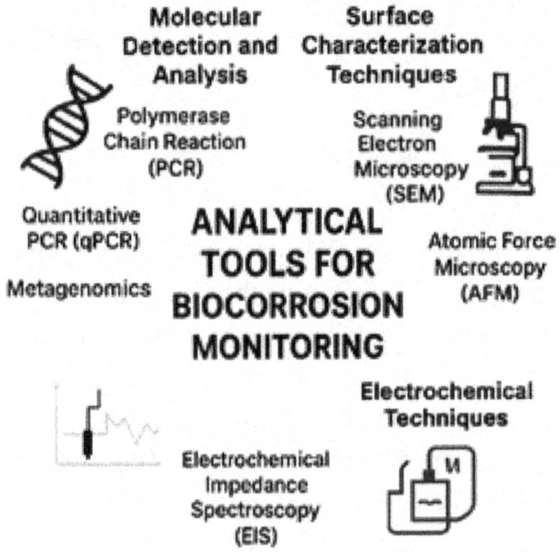

5.1. Molecular Detection and Analysis

Molecular biology can be used to detect, identify and quantify microorganisms causing MIC with accuracy. Conventional culture-based methods would identify less than 1% of the total diversity in the environment of microbial organisms, thus missing significant causes of corrosion (Amann et al., 1995). Therefore, DNA-based investigation of microbial communities made possible by molecular diagnostics can correlate the phylogenetic profiles with the function of corrosion. They can also be used to monitor the dynamics of microbial populations in real-time with their environmental factors, biocide treatments or biotechnological treatment in corrosion reduction.

5.1.1. Polymerase Chain Reaction (PCR)

PCR is a molecular tool to identify microorganisms that are related to corrosion by the amplification of strategic taxonomic genes like the 16S rRNA gene (Mullis et al., 1986). It allows the sensitive detection of major MIC-relevant groups such as sulfate-reducing bacteria (SRB), methanogens, denitrifiers, and acid-producing bacteria directly used for the pipeline fluids, generated water, biofilms, and deposits of corrosion. Amplification of the functional marker as dsrAB is commonly used to confirm the existence of SRB that leads to the release of sulfide and increased corrosion. Despite being the qualitative and susceptible to inhibition in complex samples, PCR remains essential for the microbial fingerprints and identifying taxa associated with corrosions.

5.1.2. Quantitative PCR (qPCR)

Quantitative PCR (qPCR) is an essential molecular technique in MIC diagnostics due to its ability to quantitatively determine the abundance of microbes in traditional PCR (Kubista et al., 2006). qPCR can be directly correlated with the corrosion rates in industrial systems, allows for the quantitative determination of microbial population densities, especially in the Sulfate-Reducing Bacteria (SRB), whose higher concentration has been linked to faster in carbon steel pipelines. In addition to total microbial biomass, qPCR enables specific identification of functional genes, including dsrAB to sulfate reduction, mcrA to methanogenesis, nifH to nitrogen fixation, providing about the metabolic pathways associated with specific corrosion processes. This functional-gene approach is necessary because microbial lineages often contain diverse strains, phylogenetic identification is insufficient to assess corrosion risks. qPCR is also crucial in biotechnological corrosion management, as it allows the assessment of mitigation strategies. A decrease in microbial load treatment with biocides, nutrient control, or Quorum-Quenching agents can be quantified by qPCR, enabling evidence-based optimization of corrosion-control strategies.

5.1.3. Metagenomics

The advanced molecular techniques applicable in the characterization of the microbial communities which are involved in MIC entails the metagenomics that enables the sequencing of complete genomes present in corrosion biofilms. Compared to the PCR-based methods that identify genes that have been pre-selected, metagenomic sequencing allows an entire process of the taxonomic composition and functional properties of the complex biofilms that may change the process of corrosion. Metagenomics also explains key metabolic processes related to corrosion including sulfate reduction, iron oxidation, Extra-Cellular Electron Transfer (EET) and biofilm formation (Doyle et al., 2021). C-type cytochromes are used for the determination of microbes that have the ability to assist in the electron mechanisms directly, which are known to catalyze corrosion in genera such as Desulfovibrio and Geobacter. Metagenomics allows for strain-level differentiation, enabling distinction between corrosive and non-corrosive strains of the same species. This is crucial for the accurate corrosion risk assessment and for developing viable mitigation strategies (Kato, 2016). These Metagenome-Assembled Genomes (MAGs) can also enhance the microbial communities and interactions in corrosion niches and the development of biotechnology-based interventions, such as engineered microbial communities and probiotic biofilms and targeted metabolic inhibitors.

5.2. Surface Characterization Techniques

The methods of surface characterization are the direct visualization and chemical analysis of the surface of metal that is affected by Microbial Industrial Corrosion (MIC). These methodologies analyze the structural, morphological and chemical corrosion signatures unlike the molecular tools which identify the presence of microorganisms. Microbial Industrial Corrosion usually results in the localized basis, heterogeneous deposition and micro scale interactions between the biofilms and the metal surfaces. These current microscopy and spectroscopy equipment can therefore be used to evaluate the corrosion mechanisms, the composition of the layers and the characteristics of biofilm-metal interfaces. The confirmation of biotechnological interventions of corrosion-inhibitions (Nielsen et al., 2005).

5.2.1. Scanning Electron Microscopy (SEM)

Scanning Electron Microscopy (SEM) is also one of the most important tools used to examine the surface morphology of MIC, which allows evaluation of the biofilm structures at the high-resolution level of the metal surfaces. The topographical characteristics and compositional differences are identified with the help of the secondary and electron imaging, while Energy-Dispersive X-ray Spectroscopy (EDS) coupled with SEM enables elemental analysis under conditions of SRB activity (Chavez et al., 2019). Scanning Electron Microscope can also be used to visualize the microbial colonization areas and can be used to distinguish between MIC and abiotic corrosion and to the impacts of biotechnological mitigation measures.

5.2.2. Atomic Force Microscopy (AFM)

Atomic Force Microscopy is a nanoscale microscopic instrument that can be useful in MIC diagnostics at an initial phase because it allows characterization of the metal surfaces and microbial biofilms at the atomic scales. AFM offers three-dimensional area of surface and force measurements which can display the initial biofilm deposition, EPS deposition and nanoscale mechanical degradation of metals (Sreekumari and Ohtsu, 2010). AFM can be used to identify organic and inorganic components of biofilms and measure microbial adhesion forces through phase imaging and force spectroscopy, which allows the measurement of engineered coatings and biofilm-protecting biopolymer barriers. This nano scale facilitates the production of nano-engineered anticorrosion materials. SEM can detect the finer structural changes pertaining to the progression of MIC.

5.2.3. X-Ray Photoelectron Spectroscopy (XPS)

XPS is a spectroscopic technique, which is sensitive to the surface, and is typically used to determine composition of elements on the outer 1-10 nm surfaces (Briggs and Grant, 2003). XPS is used in MIC studies to offer chemical pathways about the products of corrosion, biofilm-derived deposits and binding between the microbial metabolites and the metal surfaces. As an example, the ratio of Fe (II)/ Fe (III), sulfur functional groups (e.g., sulfide, sulfate, thiolate), carbon functional groups, and nitrogen-containing molecules can be identified in EPS by use of XPS. This can be used to determine certain biochemical pathways associated with corrosion, including sulfide deposition in SRB-contaminated habitats or organic acid-mediated complexation in acid-generating bacteria. XPS profiling can be utilized to determine layered structures in corrosion scales and is used to distinguish between biofilm, inorganic deposits and metal oxides. This is critical in assessing the efficacy of anticorrosive coatings, biopolymer shields or microbial inhibitors developed to restrict the interactions of metals biofilms.

5.3. Electrochemical Techniques

Quantitative Corrosion kinetics, electrochemical impedance, and reaction mechanisms are measured real time using electrochemical methods. Microbial Industrial Corrosion (MIC) can either be the transfer of electrons between microbes and the metal surfaces or the production of corrosive metabolites which change the electrochemical parameters (Beech and Sunner, 2004). Electrochemical techniques provide real-time progression on corrosion evolution and enable the evaluation of mitigation strategies in both laboratory and field settings.

5.3.1. Electrochemical Impedance Spectroscopy (EIS) for Corrosion Rate Monitoring

Electrochemical Impedance Spectrophotometry (EIS) is one of the most important analytical instruments to analyze Microbial Industrial Corrosion since it can measure the charge transfer resistance, biofilm impedance at metal-microbe interfaces. EIS is used to determine the dependence of corrosion kinetics on microbial colonization in MIC systems through the fitting of impedance spectra to equivalent circuit models. A decrease in charge transfer resistance that can directly take up electrons, and an increase in biofilm impedance can represent protective or engineered biofilms. EIS can also be used to evaluate the effect of biotechnological inhibitors such as biosurfactants, quorum-quenching agents, bacteriophages and these microbial consortia by measuring a change in resistance, and diffusion before and after treatment. The combination of its non-destructive and highly sensitive features makes EIS very effective in assessing biotechnology-based MIC mitigation in pipelines, storage tanks and marine structures.

6. ENVIRONMENTAL AND INDUSTRIAL APPLICATIONS

Microbial Industrial Corrosion (MIC) is a major concern in the industrial field to cause 20-40% of the corrosion-related expenses in the world and cause significant economic and operational crises. The growing demand for sustainable and precise mitigation has increased the adoption of biotechnological methods for Microbial Industrial Corrosion Control. These technologies are advanced microbial monitoring, engineered microbial systems, biogenic nanomaterials, biosurfactants and quorum- quenching agents which offer specific, ecologically friendly alternatives to standard chemical inhibitors. Their incorporation into industrial infrastructure improves the early detection of the corrosive microorganisms, facilitates mitigation and provides long term corrosion protection under the influence of microbiologically active environment. The Microbial Industrial Corrosion management, which is largely biotechnology-based, has gained special significance in the high-risk areas of oil and gas pipelines, water and wastewater treatment plants, energy and power-plant systems where complex biofilm communities as well as corrosive metabolites require complex interventions (Mand et al., 2014).

6.1. Oil and Gas Pipelines and Storage Tanks

Biotechnological interventions are increasingly central to managing Microbial Industrial Corrosion (MIC) in oil and gas infrastructure with Sulfate-Reducing Bacteria (SRB), Acid-Producing Bacteria (APB) and methanogens prevailing as the dominant corrosive biofilms in pipelines and storage systems. The formation of sulfides with the help of SRB and biofilm production contributes significantly under the deposit corrosion, whereas the Direct Electron Uptake (DEU) corrosion can be brought by methanogenic archaea to increase localized metal degradation (Enning et al., 2012). The recent microbial monitoring tools, such as qPCR, next-generation sequencing, and metagenomics, permits the rapid detection and enumeration of corrosive taxa, including functional gene-based qPCR assays that can be used to estimate these populations and predictive corrosion modelling and optimal biocides usages.

Metagenomic profiling is also used to identify metabolic pathways related to metal dissolution that help to differentiate between highly corrosive and non-harmful pathogens in the microbial community. Biotechnological mitigation strategies increasingly rely on eco-friendly inhibitors like biosurfactants derived by *Pseudomonas aeruginosa*, *Bacillus subtilis* and *Rhodococcus erythropolis*. These compounds interfere with biofilms, decrease surface tension, enhance the penetration of traditional biocides, and reduce environmental toxicity. Such mechanisms as biopolymer-based protective coating based on the Extracellular Polymeric Substances (EPS), which inhibit microbial adhesion and maturation of biofilms on the pipeline surfaces (Oliveira et al., 2017).

A specific and sustainable alternative to chemical treatment is the integration of a microbial community engineering strategy to suppress corrosive taxa, by either ecological competition or antagonistic secretion of potent metabolites. Early identification of metabolic activity and corrosion currents in storage tanks and water-oil interface locations through biotechnology-enabled diagnostics, biosurfactant-assisted cleaning and bioelectrochemical sensors can be used to conduct predictive maintenance (Jia et al., 2019). These biotechnological interventions strengthen the specificity, environmental friendly and long-term usage of MIC management of oil and gas systems.

6.2. Water Treatment and Sewage Systems

Biotechnological intervention has been described as essential step in reducing microbial-based corrosion in water supply systems, wastewater pipes, and sewage treatment systems where mixed aerobic and anaerobic communities (especially Sulfur-Oxidizing Bacteria (SOB), Sulfate-Reducing Bacteria (SRB), nitrifiers, and Iron-Oxidizing Bacteria (IOB)) are involved in corrosion of the steel and concrete infrastructure. SOB mediated oxidation of hydrogen sulfide is the major contributor to biogenic sulfuric acid generation, and the structural damage in sewer environments (Islander et al., 1991). The use of modern molecular diagnostics to identify microbe populations and the detection of metabolic gene targets involved in oxidation of sulfur, has advanced the monitoring of MIC by allowing the detection of functional groups of microbes and major metabolic genes. Such methods ensure the detection of vulnerable areas of corrosion and enhance the evaluations of microbial corrosion.

Manipulation of microbial metabolism and structure is also directed at the biotechnological prevention of corrosion. Nitrate reducing bacteria have been used to inhibit SRB activity where the electron flow is redirected to nitrate reducing instead of sulfide when compared to the use of conventional chemicals (Zhang et al., 2019). Microbially produced biosurfactants and biopolymers also facilitate reduced microbial adhesion of initial biofilms and make them easier to remove during mechanical cleaning, which leads to less MIC initiation. Enzymatic dispersal agents such as proteases and DNases also offer environmental friendly way of disrupting biofilms especially in drinking-water systems where chemical disinfectants are known to have health issues. The use of biogenic nanoparticles, like microbial or plant-produced nanoparticles of silver, zinc oxide, and iron oxide, with proven antimicrobial and anticorrosive activity, controlled release, and lower ecological toxicity, represents an emerging biotechnological development. In addition, Microbial Fuel Cells (MFCs) are used in wastewater treatment to minimize anaerobic niches, suppress SRB populations, and enable energy recovery, highlighting the potential of bioelectrochemical technologies in biofilm prevention.

6.3. Energy and Power Plants

Microbial Industrial Corrosion (MIC) poses a significant threat to energy and power-generation plants such as thermal, nuclear, geothermal and renewable energy plants as cooling-water circuits, heat exchangers, boilers, and condensers with diverse applications. The formation of biofilm in such units reduces the heat-transfer efficiency, hydraulic resistance and enhances the localized corrosion by producing metabolic products and altering the microenvironment. The biotechnological responses towards MIC are currently focused on the identification of microbes, biocontrol approaches that are ecofriendly and on the surface engineering approaches to reduce MIC. The molecular methods of qPCR and high-throughput sequencing allow specific characterization of biofouling communities in cooling-water systems that aid in identifying corrosive microbial shifts for their predictive maintenance. Bio-controlling agents that are ecofriendly such as biosurfactants and biopolymers are also the main innovations in the minimization of microbial attachment on metal surfaces within the nutrient-rich cooling water conditions (Handelsman, 2004). Biosurfactants also reduce surface free energy in preventing the formation of biofilms, and biopolymer surfaces like chitosan or alginates that generate hydrophilic, antimicrobial, and corrosive inhibitory layers on heat-exchange surfaces.

Geothermal plants have unique properties because of thermophilic microbial communities, including Thermococcus and Archaeoglobus that promotes sulfide-mediated corrosion in high temperature, anaerobic conditions. Specific molecular markers for thermophiles also aid in diagnosing microbial activity in geothermal wells and reinjection lines (Stetter et al., 1993). Radiation-resistant Deinococcus-type extremophiles cause biofilm-related stress corrosion cracking in nuclear power plant cooling circuits. Renewable energy systems that are exposed to the sea, especially offshore wind and tidal systems are susceptible to biofouling and MIC mediated by sulfate-reducing bacteria, marine biofilms, and oxidative systems caused by algae. Nanostructured biopolymer coatings and biotechnology-based monitoring solutions can be utilized to minimize the rate of corrosion. Bioelectrochemical sensors have provided real-time on the metabolism, redox activity and localized corrosion within the power-generation industries, which are crucial to the predictive biotechnology-based corrosion control initiatives.

7. CHALLENGES AND LIMITATIONS

Despite these advancements, several inherent challenges arise for the widespread application of biotechnological methods for industrial corrosion management. Microbial consortia and electrochemical processes that are not adequately done by existing predictive models for the MIC control. The structural and functional resistance of microbial biofilms presents a significant obstacle. Mature biofilms exhibit heterogeneous microenvironments and a robust Extracellular Polymeric Substance (EPS) matrix that impedes infiltration and action of biosurfactants, enzymes and quorum-quenching compounds. EPS stability is further enhanced by synergistic interactions among sulfate-reducing bacteria, methanogens, nitrate-reducing bacteria, and iron-oxidizing bacteria, making biofilm disruption. MIC progression remains challenging due to rapid shifts in microbial communities caused by nutrient inflows, hydrodynamics, and materials. Furthermore, challenges remain in the functional interpretation of metagenomic data, metabolic inference, and correlating with corrosion severity (Prosser, 2015).

PCR-based assays and metagenomic analysis are also costly and computational intensive. Biotechnological agents exhibit poor stability under extreme industrial conditions. Enzymes, biosurfactants, and microbial inhibitors can become ineffective under conditions of elevated temperature, high salinity, or in hydrocarbon-rich oil and gas reservoirs, thus reducing their field-level reliability (Muthusamy et al., 2008). The engineered or beneficial microorganisms are designed to compete the corrosive species that are able to survive in an anoxic or high-pressure environment. Major challenges include the biotechnological agents, maintaining uniform dosing, and treating extensive pipeline networks or inaccessible tank bottoms due to the effect of dilution which reduces the effectiveness of the treatment processes.

The regulatory pathways of genetically modified microorganisms and biopolymers emitted to the environment further retards the adoption of technology raised over the ecological effects and gene transfer (Bagwell et al., 2016). Moreover, there are no standardized systems to integrate the electrochemical, microbial and imaging data into the MIC risk evaluation models, which results in inconsistent interpretation and mitigation measures. Economic factors such as the biosurfactants, biopolymers, and advanced molecular diagnostics are more expensive to implement than traditional chemical inhibitors in the short term, and they cannot be quickly transferred to industries. The major constraints are biofilm resistance, environmental factors, technological and analytical limitations, regulatory obstacles, and financial constraints. These will be critical for the scalable and reliable biotechnological based solutions to industrial corrosion management. Table 1 outlines the major technical, environmental, economic, and regulatory challenges associated with applying biotechnological approaches in corrosion management, including issues of microbial instability, scalability, and field adaptability for the large-scale industrial implementation of biocorrosion control strategies.

continued on following page

Table 1. Continued

Table 1. Challenges and Limitations of Biotechnological Approaches in Corrosion Management

S. No	Categories	Challenges
1	Biofilm Resistance	Strong EPS matrix, heterogeneous microenvironment, difficult penetration of enzymes and biosurfactants.
2	Microbial Community Variability	Rapid shifts due to nutrients, hydrodynamics, and material type; unpredictable MIC progression.
3	Analytical & Molecular Limitations	High cost, computational load of PCR/metagenomics; difficulty translating microbial signatures into corrosion severity.
4	Environmental Constraints	Poor stability of biotechnological agents in extreme temperature, salinity, pressure, or hydrocarbon-rich environments; survival issues for engineered microbes.
5	Regulatory Barriers	Concerns about release of engineered microbes/biopolymers and ecological risks.
6	Lack of Standardized Models	No unified framework integrating microbial, electrochemical, and imaging data for MIC risk assessment.
7	Economic Limitations	High cost of biosurfactants, enzymes, biopolymers, and advanced diagnostics compared to conventional inhibitors.

8. FUTURE PROSPECTS

Biotechnology is poised to play a major role in advancing MIC control by providing precise, adaptive, and ecologically sustainable solutions that overcome the limitations of traditional chemical biocides. The development of synthetic biology is an important frontier, where corrosive species and non-corrosive metal-protective biofilms can be engineered, or corrosive by-products can be degraded using this technology. These gene circuits can be designed to enable engineered organisms to detect MIC-related signals (e.g., sulfide concentration, redox variations) with the controlled release of inhibitors or self-destruction to prevent ecological imbalance. MIC prediction and monitoring will also change with the advanced multi-omics technologies. The integration of metagenomics, meta transcriptomics, proteomics, and metabolomics gives high-resolution data about the composition and functional pathways related to corrosion in microbial communities (Brune and Bayer, 2012).

The integration with machine learning and artificial intelligence, these datasets can enable the prediction of corrosion in real time, early warning systems, and mitigation regimes that are applied depending on the community-level metabolic activities (Jia et al., 2019). Nanobiotechnology is a potent, and biogenically synthesized metal and metal-oxide nanoparticles have demonstrated significant corrosion-inhibition and antimicrobial properties. Different applications can be focused on the delivery of nanoparticles to localized surfaces of corrosion, smart nanocoating with the ability to release inhibitors when triggered by microorganisms, and the use of hybrid coatings that incorporate nanomaterials and biopolymers to improve durability and eco-friendly. Biobased green inhibitors, such as biosurfactants, natural polymers, and enzymes, are increasingly becoming prominent as an alternative to chemical treatment, especially as a sustainable solution. Enzyme-based approaches which target Extracellular Polymeric Substances (EPS) or microbial adhesions are promising, especially when enzymes are immobilized or encapsulated to provide greater resilience to the enzyme under industry stress conditions.

Biosurfactants can also be incorporated into self-healing coating in case of microbial colonization or changes in pH. Bioelectrochemical sensors can provide real-time detection of electron transfer signals, microbial metabolism, and corrosion (Logan and Regan, 2006), with future prospects of these sensors being able to provide automated adjustments to treatment due to wireless capabilities and AI-driven analytics. Microbial fuel cell systems also have dual functionality in that they can prevent sulfate-reducing bacteria and produce low-power electricity. Biotechnology also enables environment-specific corrosion control through the use of carefully tailored microbial systems designed for extreme industrial environments characterized by high salinity, temperature, or pressure. This system is capable of safeguarding major infrastructure at geothermal field, pipelines and nuclear cooling systems. Further advances through regulatory development and better techno-economic viability (through advances in fermentation, metabolic engineering and optimization of bioprocesses) will further push biosurfactants, enzymes, nanomaterials and engineered microbes to be widely adopted in industry. The future of MIC mitigation will be based on the coherent structure whereby biotechnology will merge with materials science, electrochemistry, and digital monitoring.

9. CONCLUSION

Biotechnological approaches represent a transformative and sustainable paradigm for managing Microbial-Induced Corrosion (MIC), addressing the significant limitations of conventional chemical-based methods such as environmental toxicity, microbial resistance, and lack of long-term efficacy. Despite existing challenges related to scalability, biofilm complexity, and regulatory hurdles for genetically modified organisms, the convergence of biotechnology with materials science, digital monitoring, and artificial intelligence is paving the way for advanced solutions. Key innovations include the use of enzyme-based systems for detoxifying corrosive metabolites like hydrogen sulfide, the application of biosurfactants and biopolymers to disrupt biofilms and form protective barriers, and the deployment of biogenic nanoparticles with inherent antimicrobial and anticorrosive properties. Furthermore, genetic engineering tools such as CRISPR/Cas enable the design of microbial biosensors for real-time corrosion monitoring and the development of engineered probiotics that outcompete or inhibit corrosive microorganisms. Bio electrochemical systems, including microbial fuel cells and biocathodes, offer dual functionality by mitigating electron-driven corrosion processes for the integration of sustainability with their efficiency. These sophisticated analytical techniques such as metagenomics, electrochemical impedance spectroscopy, and high-resolution microscopy these biotechnological solutions provide a comprehensive overview of MIC mechanisms and enable the corrosion management. The future of corrosion control lies in self-adaptive systems capable of predictive intervention and long-term resilience, enabling environmental friendly and are economical viable corrosion mitigation across these industries including oil and gas, water treatment, and power generation.

REFERENCES

Ahmed, S., Ahmad, M., Swami, B. L., & Ikram, S. (2016). A review on plants extract mediated synthesis of silver nanoparticles for antimicrobial applications: A green expertise. *Journal of Advanced Research, 7*(1), 17–28. DOI: 10.1016/j.jare.2015.02.007 PMID: 26843966

Amann, R. I., Ludwig, W., & Schleifer, K. H. (1995). Phylogenetic identification and in situ detection of individual microbial cells without cultivation. *Microbiological Reviews, 59*(1), 143–169. DOI: 10.1128/mr.59.1.143-169.1995 PMID: 7535888

Bagwell, C. E., Adams, S., & Barnhart, E. P. (2016). Radiation-resistant bacteria and their implications for nuclear waste disposal. *Reviews in Environmental Science and Biotechnology, 15*(3), 353–364. DOI: 10.1007/s11157-016-9401-2

Banerjee, S., Mittal, A., & Kumar, S. (2017). Microbiologically influenced corrosion: Perspectives and approaches. *Journal of Materials Engineering and Performance, 26*(7), 3113–3125. DOI: 10.1007/s11665-017-2744-1

Beech, I. B., & Sunner, J. (2004). Biocorrosion: Towards understanding interactions between biofilms and metals. *Current Opinion in Biotechnology, 15*(3), 181–186. DOI: 10.1016/j.copbio.2004.05.001 PMID: 15193324

Binnig, G., Quate, C. F., & Gerber, C. (1986). Atomic force microscope. *Physical Review Letters, 56*(9), 930–933. DOI: 10.1103/PhysRevLett.56.930 PMID: 10033323

Briggs, D., & Grant, J. T. (2003). *Surface analysis by auger and X-ray photoelectron spectroscopy*. IM Publications.

Brune, K. D., & Bayer, T. S. (2012). Engineering microbial consortia to enhance biomining and bioremediation. *Frontiers in Microbiology, 3*, 1–6. DOI: 10.3389/fmicb.2012.00203 PMID: 22679443

Cao, S., Wang, J., & Chen, H. (2015). Progress in biopolymer-based anticorrosive coatings. *Progress in Organic Coatings, 87*, 1–8. DOI: 10.1016/j.porgcoat.2015.04.019

Chakraborty, J., Das, S., & Sarkar, P. K. (2020). Enzyme-based green approaches for the mitigation of microbiologically influenced corrosion. *Journal of Materials Engineering and Performance, 29*(5), 2795–2806. DOI: 10.1007/s11665-020-04799-6

Chavez, M. L., Kato, C., & Zhang, D. (2019). Application of scanning electron microscopy in microbiologically influenced corrosion investigations. *Materials and Corrosion, 70*(10), 1725–1733. DOI: 10.1002/maco.201910758

Chen, X., Schippers, A., & Sand, W. (2011). Progress in bioleaching: Fundamentals and mechanisms of microbial metal sulfide oxidation—part A. *Applied Microbiology and Biotechnology*, *97*(17), 7529–7541. DOI: 10.1007/s00253-011-3556-0

Dang, H., & Lovell, C. R. (2016). Microbial surface colonization and biofilm development in marine environments. *Microbiology and Molecular Biology Reviews : MMBR*, *80*(1), 91–138. DOI: 10.1128/MMBR.00037-15 PMID: 26700108

Dinh, H. T., Kuever, J., Mussmann, M., Hassel, A. W., Stratmann, M., & Widdel, F. (2004). Iron corrosion by novel anaerobic microorganisms. *Nature*, *427*(6977), 829–832. DOI: 10.1038/nature02321 PMID: 14985759

Doyle, L. E., Marsili, E., & Bond, D. R. (2021). Metagenomic analysis of electroactive microbial communities in marine sediments. *Frontiers in Microbiology*, *12*, 1–15. DOI: 10.3389/fmicb.2021.705004

Enning, D., & Garrelfs, J. (2014). Corrosion of iron by sulfate-reducing bacteria: New views of an old problem. *Applied and Environmental Microbiology*, *80*(4), 1226–1236. DOI: 10.1128/AEM.02848-13 PMID: 24317078

Enning, D., Venzlaff, H., Garrelfs, J., Dinh, H. T., Meyer, V., Mayrhofer, K., & Widdel, F. (2012). Marine sulfate-reducing bacteria cause serious corrosion of iron under an electroconductive biogenic mineral crust. *Proceedings of the National Academy of Sciences of the United States of America*, *109*(30), 11950–11955. DOI: 10.1073/pnas.1200072109 PMID: 22616633

Flemming, H. C., & Wuertz, S. (2019). Bacteria and archaea on Earth and their abundance in biofilms. *Nature Reviews. Microbiology*, *17*(4), 247–260. DOI: 10.1038/s41579-019-0158-9 PMID: 30760902

Gonzalez, R., Curtis, K., & Bixler, G. (2020). Metagenomics and metabolomics approaches to microbiologically influenced corrosion. *Current Opinion in Biotechnology*, *62*, 1–7. DOI: 10.1016/j.copbio.2019.09.005

Guo, X., Liu, J., & Wang, J. (2018). XPS analysis of the chemical composition of the corrosion product layers of carbon steel in the presence of sulfate-reducing bacteria. *Journal of Materials Engineering and Performance*, *27*(11), 5595–5602. DOI: 10.1007/s11665-018-3636-8

Hamilton, W. A. (2003). Microbially influenced corrosion as a model system for the study of metal microbe interactions: A unifying electron transfer hypothesis. *Biofouling*, *19*(1), 65–76. DOI: 10.1080/0892701021000041078 PMID: 14618690

Handelsman, J. (2004). Metagenomics: Application of genomics to uncultured microorganisms. *Microbiology and Molecular Biology Reviews : MMBR*, *68*(4), 669–685. DOI: 10.1128/MMBR.68.4.669-685.2004 PMID: 15590779

Iravani, S. (2011). Green synthesis of metal nanoparticles using plants. *Green Chemistry*, *13*(10), 2638–2650. DOI: 10.1039/c1gc15386b

Islander, R. L., Devinny, J. S., Mansfeld, F., Postyn, A., & Shih, H. (1991). Microbial ecology of crown corrosion in sewers. *Journal of Environmental Engineering*, *117*(6), 751–770. DOI: 10.1061/(ASCE)0733-9372(1991)117:6(751)

Jia, R., Yang, D., & Xu, D. (2019). Electrochemical impedance spectroscopy (EIS): A powerful tool for the study of microbiologically influenced corrosion (MIC). *Journal of Materials Science and Technology*, *35*(10), 2113–2119. DOI: 10.1016/j.jmst.2019.05.007

Jiang, G., Keller, J., & Bond, P. L. (2016). Determining the long-term effects of H2S concentration, relative humidity, and air temperature on concrete sewer corrosion. *Water Research*, *88*, 136–144. DOI: 10.1016/j.watres.2015.10.002 PMID: 25108169

Kato, S. (2016). Microbial extracellular electron transfer and its environmental application. *Journal of Bioscience and Bioengineering*, *122*(1), 1–8. DOI: 10.1263/jbb.106.1 PMID: 26777236

Kato, S., Hashimoto, K., & Watanabe, K. (2012). Microbial interspecies electron transfer via electric currents through conductive minerals. *Proceedings of the National Academy of Sciences of the United States of America*, *109*(25), 10042–10046. DOI: 10.1073/pnas.1117592109 PMID: 22665802

Kiran, M. G., Pakshirajan, K., & Das, G. (2020). A comprehensive review on biosurfactant assisted bioelectrochemical systems for microbially influenced corrosion control. *Journal of Hazardous Materials*, *402*, 123795. DOI: 10.1016/j.jhazmat.2020.123795

Kubista, M., Andrade, J. M., Bengtsson, M., Forootan, A., Jonák, J., Lind, K., Sindelka, R., Sjöback, R., Sjögreen, B., Strömbom, L., Ståhlberg, A., & Zoric, N. (2006). The real-time polymerase chain reaction. *Molecular Aspects of Medicine*, *27*(2-3), 95–125. DOI: 10.1016/j.mam.2005.12.007 PMID: 16460794

Lee, A. K., Newman, D. K., & Coates, J. D. (2019). Microbial corrosion of iron-based materials: A review. *Corrosion*, *75*(4), 339–359. DOI: 10.5006/3129

Logan, B. E., & Regan, J. M. (2006). Microbial fuel cells—Challenges and applications. *Environmental Science & Technology*, *40*(17), 5172–5180. DOI: 10.1021/es0627592 PMID: 16999086

Lovley, D. R. (2006). Bug juice: Harvesting electricity with microorganisms. *Nature Reviews. Microbiology*, *4*(7), 497–508. DOI: 10.1038/nrmicro1442 PMID: 16778836

Macdonald, D. D. (2006). Reflections on the history of electrochemical impedance spectroscopy. *Electrochimica Acta*, *51*(8-9), 1376–1388. DOI: 10.1016/j.electacta.2005.02.107

Mahadevan, R., Palsson, B. Ø., & Lovley, D. R. (2016). In situ to in silico and back: Elucidating the physiology and ecology of Geobacter spp. using genome-scale modelling. *Nature Reviews. Microbiology*, *9*(1), 39–50. DOI: 10.1038/nrmicro2456 PMID: 21132020

Mand, J., Park, H. S., Jack, T. R., & Voordouw, G. (2014). The role of acetogens in microbially influenced corrosion of steel. *Frontiers in Microbiology*, *5*, 1–14. DOI: 10.3389/fmicb.2014.00268 PMID: 24917861

Marsili, E., Baron, D. B., Shikhare, I. D., Coursolle, D., Gralnick, J. A., & Bond, D. R. (2008). Shewanella secretes flavins that mediate extracellular electron transfer. *Proceedings of the National Academy of Sciences of the United States of America*, *105*(10), 3968–3973. DOI: 10.1073/pnas.0710525105 PMID: 18316736

Moura, M. C., Pontes, F. V., & Lutterbach, M. T. (2018). Biofilm formation on corroded surfaces and its influence on corrosion resistance. *Journal of Materials Science*, *53*(14), 10145–10159. DOI: 10.1007/s10853-018-2287-3

Mullis, K., Faloona, F., Scharf, S., Saiki, R., Horn, G., & Erlich, H. (1986). Specific enzymatic amplification of DNA in vitro: The polymerase chain reaction. *Cold Spring Harbor Symposia on Quantitative Biology*, *51*(0), 263–273. DOI: 10.1101/SQB.1986.051.01.032 PMID: 3472723

Muthusamy, K., Gopalakrishnan, S., & Ravi, T. K. (2008). Biosurfactants: Properties, commercial production and application. *Current Science*, *94*(6), 736–747.

Muyzer, G., & Stams, A. J. (2008). The ecology and biotechnology of sulphate-reducing bacteria. *Nature Reviews. Microbiology*, *6*(6), 441–454. DOI: 10.1038/nrmicro1892 PMID: 18461075

Nielsen, A. H., Hvitved-Jacobsen, T., & Vollertsen, J. (2005). Effects of pH and iron concentrations on sulfide precipitation in wastewater collection systems. *Water Environment Research*, *77*(1), 113–120. DOI: 10.2175/106143005X41645 PMID: 18536490

Nitschke, M., & Costa, S. G. (2007). Biosurfactants in food industry. *Trends in Food Science & Technology*, *18*(5), 252–259. DOI: 10.1016/j.tifs.2007.01.002

Oliveira, D. R., Zaiat, M., & Foresti, E. (2017). Biopolymer-based materials for the control of biofouling in water systems: A review. *Environmental Technology*, *38*(13-14), 1599–1613. DOI: 10.1080/09593330.2017.1334708 PMID: 27735215

Parkes, R. J., Cragg, B. A., & Wellsbury, P. (2011). Recent studies on bacterial populations and processes in subseafloor sediments: A review. *Hydrogeology Journal*, *8*(1), 11–28. DOI: 10.1007/PL00010971

Prosser, J. I. (2015). Dispersing misconceptions and identifying opportunities for the use of 'omics' in soil microbial ecology. *Nature Reviews. Microbiology*, *13*(7), 439–446. DOI: 10.1038/nrmicro3468 PMID: 26052662

Reguera, G., McCarthy, K. D., Mehta, T., Nicoll, J. S., Tuominen, M. T., & Lovley, D. R. (2005). Extracellular electron transfer via microbial nanowires. *Nature*, *435*(7045), 1098–1101. DOI: 10.1038/nature03661 PMID: 15973408

Sreekumari, K. R., & Ohtsu, Y. (2010). Atomic force microscopy: A powerful tool for studying microbiologically influenced corrosion. *Corrosion Science*, *52*(3), 775–781. DOI: 10.1016/j.corsci.2009.11.007

Stetter, K. O., Huber, R., Blochl, E., Kurr, M., Eden, R. D., Fielder, M., Cash, H., & Vance, I. (1993). Hyperthermophilic archaea are thriving in deep North Sea and Alaskan oil reservoirs. *Nature*, *365*(6448), 743–745. DOI: 10.1038/365743a0

Stewart, P. S., & Costerton, J. W. (2001). Antibiotic resistance of bacteria in biofilms. *Lancet*, *358*(9276), 135–138. DOI: 10.1016/S0140-6736(01)05321-1 PMID: 11463434

Tan, C. H., Lee, K. W., & Burmølle, M. (2016). The role of quorum sensing in microbial corrosion. *Biofouling*, *32*(5), 537–547. DOI: 10.1080/08927014.2016.1143394

Uchiyama, T., Ito, K., Mori, K., Tsurumaru, H., & Harayama, S. (2010). Iron-corroding methanogen isolated from a crude-oil storage tank. *Applied and Environmental Microbiology*, *76*(6), 1783–1788. DOI: 10.1128/AEM.00668-09 PMID: 20118376

Venkidusamy, K., & Megharaj, M. (2016). A novel electrophototrophic bacterium, *Rhodopseudomonas palustris* strain RP2: Application in microbial fuel cells. *Bioresource Technology*, *218*, 602–608. DOI: 10.1016/j.biortech.2016.07.007

Videla, H. A., & Herrera, L. K. (2005). Microbiologically influenced corrosion: Looking to the future. *International Microbiology*, *8*(3), 169–180. DOI: 10.2436/im.v8i3.9490 PMID: 16200495

Vigneron, A., Head, I. M., & Tsesmetzis, N. (2016). Damage to offshore production facilities by corrosive microbial biofilms. *Applied Microbiology and Biotechnology*, *100*(6), 2527–2537. DOI: 10.1007/s00253-016-7323-0 PMID: 29423635

Zhang, P., Zhang, Z., & Wang, J. (2014). Nanoscale investigation of microbiologically influenced corrosion by atomic force microscopy. *Electrochimica Acta*, *133*, 303–310. DOI: 10.1016/j.electacta.2014.04.051

Zhang, T., Fang, H. H., & Ko, B. C. (2018). Corrosion of mild steel in the presence of marine biofilms. *Biofouling*, *34*(1), 11–24. DOI: 10.1080/08927014.2017.1409345

Zhang, T., Fang, H. H., & Ko, B. C. (2019). Methanogen population in a marine biofilm corrosive to mild steel. *Journal of Applied Microbiology*, *115*(4), 1071–1081.

ADDITIONAL READING

Beech, I. B., & Sunner, J. (2007). Biocorrosion: Towards understanding interactions between biofilms and metals. *Advances in Microbial Physiology*, *52*, 213–277.

Kip, N., & van Veen, J. A. (2015). The dual role of microbes in corrosion. *The ISME Journal*, *9*(3), 542–551. DOI: 10.1038/ismej.2014.169 PMID: 25259571

Little, B. J., & Lee, J. S. (2014). *Microbiologically Influenced Corrosion*. John Wiley & Sons.

Moura, M. C., & Lutterbach, M. T. (2020). Biofilms and corrosion: An integrative approach using omics technologies. npj Biofilms and Microbiomes, 6(1), 1–10.

Rasheed, P. A., Jabbar, K. A., & Sandhyarani, N. (2019). Biosurfactants as green corrosion inhibitors: Recent advances and future trends. *Journal of Cleaner Production*, *230*, 139–154.

Santoro, C., Arbizzani, C., Erable, B., & Ieropoulos, I. (2017). Microbial fuel cells: From fundamentals to applications. A review. *Journal of Power Sources*, *356*, 225–244. DOI: 10.1016/j.jpowsour.2017.03.109 PMID: 28717261

Sethuraman, M. G., & Natesan, M. (2019). Biopolymer-based coatings for corrosion protection: A review. *Progress in Organic Coatings*, *137*, 105356.

Videla, H. A., & Herrera, L. K. (2009). Understanding microbial inhibition of corrosion: A comprehensive overview. *International Biodeterioration & Biodegradation*, *63*(7), 896–900. DOI: 10.1016/j.ibiod.2009.02.002

Wu, Y., & Zhang, T. (2021). Application of CRISPR/Cas in microbial corrosion research. *Frontiers in Microbiology, 12*, 689306.

Xu, D., Li, Y., & Gu, T. (2016). Mechanistic modeling of biocorrosion caused by biofilms of sulfate-reducing bacteria and acid-producing bacteria. *Bioelectrochemistry (Amsterdam, Netherlands), 110*, 52–58. DOI: 10.1016/j.bioelechem.2016.03.003 PMID: 27071053

KEY TERMS AND DEFINITIONS

Microbiologically Influenced Corrosion (MIC): A corrosion process accelerated by the metabolic activities of microorganisms, such as bacteria and archaea, within biofilms on metal surfaces. It involves electrochemical interactions and the production of corrosive metabolites like hydrogen sulfide.

Biofilm: A structured community of microbial cells enclosed in a self-produced extracellular polymeric substance (EPS) matrix that adheres to surfaces. Biofilms create microenvironments that can accelerate corrosion through metabolic by-products and electron transfer.

Extracellular Electron Transfer (EET): A mechanism by which certain microorganisms transfer electrons to or from solid surfaces, such as metals, via conductive pili, cytochromes, or soluble electron shuttles. This process is central to many MIC mechanisms.

Quorum Sensing (QS): A cell-to-cell communication system used by bacteria to coordinate group behaviors such as biofilm formation, virulence, and metabolite production based on population density.

Biosurfactants: Surface-active compounds produced by microorganisms that reduce surface tension, inhibit microbial adhesion, and disrupt biofilms. Examples include rhamnolipids and surfactin.

Biogenic Nanoparticles: Nanoparticles synthesized using biological entities such as microbes or plant extracts. They exhibit antimicrobial and anticorrosive properties and are considered environmentally benign.

Synthetic Biology: An interdisciplinary field that applies engineering principles to biological systems, enabling the design and construction of novel microbial functions, such as biosensors or anti-corrosive biofilms.

Bioelectrochemical Systems (BES): Systems that harness microbial electron transfer processes to control corrosion or generate energy. Examples include microbial fuel cells (MFCs) and biocathodes.

Metagenomics: The genomic analysis of microbial communities directly from environmental samples, bypassing the need for cultivation. It is used to identify corrosive taxa and functional genes in MIC studies.

Electrochemical Impedance Spectroscopy (EIS): A non-destructive electrochemical technique used to monitor corrosion rates and evaluate the effectiveness of inhibitors by measuring the impedance of metal surfaces in response to an alternating current.

Chapter 5
Microbial Corrosion of Steel in Aquatic and Soil Environments

K. Kesavardhini
Sacred Heart College, Tirupattur, India

S. Dayana
Sacred Heart College, Tirupattur, India

P. Saranraj
https://orcid.org/0000-0002-5052-6532
Sacred Heart College, Tirupattur, India

K. Gayathri
https://orcid.org/0009-0007-3786-2009
Sacred Heart College, Tirupattur, India

Ashish Kumar Nayak
Prasad V. Potluri Siddhartha Institute of Technology, Vijayawada, India

K. Arthi
Sacred Heart College, Tirupattur, India

ABSTRACT

Microbiologically Influenced Corrosion (MIC) presents a substantial threat to the integrity and efficiency of steel in marine and non-marine settings. This occurs due to microorganisms that, under specific conditions, alter their electrochemical environment on the metal's surface. Sulfate-Reducing Bacteria, Iron-Oxidizing Bacteria, and biofilm-forming species in aquatic settings catalyze redox processes that result in pitting, scaling, and localized corrosion of submerged steel structures. In contrast, microbial communities in soil systems react to corrosive metabolite such hydrogen sulphide and organic acids, that are intensified by moisture and organic matter under different redox conditions, thereby markedly affecting corrosion processes. The generation of biofilm was considered an important aspect in these processes because it creates zones of varying oxygenation that concentrate the corrosive chemicals in the interface of the metal.

DOI: 10.4018/979-8-3373-2617-7.ch005

1. INTRODUCTION

Microbial corrosion, also known as microbiologically influenced corrosion, pertains to the complex interaction between microbes and metal surfaces that accelerates the deterioration of steel structures in both aquatic and terrestrial environments. In contrast to chemical or electrochemical corrosion, however, Microbiologically Influenced Corrosion (MIC) is begun or significantly influenced by microbial metabolic activity, including production of corrosive intermediates like hydrogen sulfide, organic acid, and even siderophores. It has effects on diverse steel infrastructure, include pipes, storage tanks, bridge foundations, and marine vessels; therefore posing considerable financial, ecological and safety threats. Globally, MIC was estimated to contribute a high percentage of losses associated with corrosion in the major sectors such as oil and gas production, water transport and energy generation therefore of a crucial status in terms of maintenance and sustainability of infrastructures. The relationships between microbial communities and biofilm formation and environmental factors provide extremely localized and heterogeneous corrosion patterns that are usually hard to observe and predict, hence the significance of comprehending MIC mechanisms (Zhang et al., 2015).

Microbiologically Influenced Corrosion (MIC) of steel in aquatic structures is a significant and escalating international concern, presenting undeniable risks to engineering and transport, and resulting in enormous financial costs because of growing maintenance costs. The annual price of corrosion is estimated at over 4 trillion globally, however China claimed that it spent 2,172.8 billion yuan, which is 3.34 percent of its GDP during the year 2014. Therefore, considering the recent gradual rise of demand on marine steel and the fact that 20 percent of corrosion in marine infrastructures has been attributed to microorganisms, there is an urgent need to investigate Microbiologically Influenced Corrosion ("MIC") of the aquatic steel and implement preventive and control strategies (Li et al., 2015).

Microorganisms participating in MIC activities in marine environments encompass Fungus, Archaea, Algae, and several bacterial species. Sulfate, a prevalent electron acceptor found in the marine atmosphere, supports a substantial bacterial community on steel, participating in both the aquatic sulfur and iron cycles; certain corrosive bacteria modify cathodic and anodic processes on the metal surface (Yuan et al., 2013). Marine bacteria involved in iron biocorrosion can be categorized as Sulfate-Reducing Bacteria (SRB), Sulfur-Oxidizing Bacteria (SOB), Iron-Reducing Bacteria (IRB), Iron-Oxidizing Bacteria (IOB), Acid-Producing Bacteria (APB), and also Slime-Producing Bacteria (SPB). Sulfate-reducing bacteria are considered the principal agents of Microbiologically Influenced Corrosion (MIC) due to their significant electron transfer capacity. Approximately fifty percent of the overall

MIC loss is attributed to SRB-induced MIC in steel materials. The specified species degrade steel materials via several erosion mechanisms (Yu et al., 2013).

Biofilms bacterial and Extracellular Polymeric Substances (EPS) in the marine environment cause steel biocorrosion through electrochemical processes. Biofilms have differing acidity, dissolved oxygen, ion concentration, and organic matter than seawater. The biofilm's microorganisms' metabolic activity affects the rust layer's structure and electrochemical reactivity on the metal substrate due to their direct contact. Many biomolecules can selectively chelate steel cations, making EPS electronegative. Thus, EPS can conduct electrons between biofilm and metal. Cathode depolarization during the redox process increases corrosion with an appropriate electron acceptor. Numerous studies show that microorganisms and biofilms can promote or prevent corrosion. This is microbiologically influenced corrosion suppression. Most MICI research use single-species aerobic biofilms like *Bacillus, Pseudomonas*, and *Shewanella*, as well as facultative anaerobic biofilms like *Escherichia coli*, which may grow on mild steel in labs (Kip & van Veen, 2014). Thus, the relationship between MIC/MICI and microbial ecology is unclear. Successful studies have found two methods of Microbially Induced Corrosion Inhibition (MICI): a biofilm with microorganisms that mitigate corrosion through their metabolic processes or chemical precipitation. MIC and MICI can be altered by several microbial species and activities, such as biofilm electron transport. Microbial molecular analysis and omics technologies in MIC/MICI studies help understand steel surface corrosion in natural environments (Javed et al., 2015).

Aquatic systems, like marine and estuarine freshwater systems, provide certain conditions, such as constant moisture, dissolved ions, and fluctuating physicochemical conditions, that speed up microbial corrosion. The high salinity marine environments especially promote the growth of halotolerant and halophilic microbes; these are normally sulphate-reducing and iron-oxidizing bacteria that initiate and accelerate the localized corrosion in steel (Duan et al., 2008). These estuarine ecosystems possess dynamic salinity and nutrient gradients that produce complex microbial consortia that synergistically interact in corrosion processes. Despite the low salinity of freshwater systems, the abundant nutrients and organic matter facilitate microbial colonization and biofilm formation on the steel surface. Biofilms contribute to corrosive interactions by facilitating the transport of corrosive substances in microenvironments, altering oxygen and also redox gradients, and creating differential aeration cells, all of which accelerate metal dissolution and also localized corrosion (Hou et al., 2017).

In contrast, soil environments are heterogeneous in terms of physicochemical conditions. Variability in Moisture Content, Organic Matter, pH, and Redox Potential exists with regard regarding the makeup and function of microbial communities involved in corrosion. Diverse categories of buried steel structures, including Pipelines and Storage Tanks, may be exposed to aerobic, facultative anaerobic, and strictly anaerobic microorganisms that can produce corrosive metabolites that attack the metal surface. Soil microbial communities can develop thick biofilms and layers of EPS, which protect anaerobic microbes and support localized zones of corrosion and changes to the chemistry of the metal surface. Poor drainage, water-logging, and high organic content promote the development of anaerobic conditions that support the activity of sulfate-reducing bacteria, leading to localized pitting and crevice corrosion. Dynamic interactions involving microbial diversity, soil chemistry, and environmental factors make MIC in soil systems particularly complex and difficult to predict (Chung et al., 2010).

2. ADVANCEMENTS IN THE STUDY OF TYPICAL CORROSIVE BIOFILMS ON MARINE STEEL

In marine conditions, microbes and other metabolic byproducts may directly affect the microenvironment at the border of fully submerged iron. The initial ocean water/metal interaction might occur changed to create a metal/biofilm interaction for the expedite the microbiologically influenced corrosion (MIC) process, an crucial aspect of MIC throughout the difficult coastal setting, like an various biofilm functions are essential for the microbial colonization and the sustenance of the subsequent developments. The polysaccharide found in the biofilm provides protection for Epibiotic species are brought to the surface and extra minerals are sequestered to nourish the internal bacterial population (Loto, 2017). The sequential growth of marine forms of biofilm adheres to a chronological chain of colonies of bacteria events: an early "Conditioning Film" pioneering microbial organisms (frequently bacteria), microcolonies, developed architecture, and biofilm communities. In respect to steel Microbiologically Influenced Corrosion (MIC), both biofilm and mineral deposits (for example, FeS, $FePO_4$, $FeCO_3$, and $FeCl_2$) are composed of integral parts in regards of the deterioration process. In addition to time, the biofilm changes from an aerobic to a anaerobic status on a steel surface that could end in the formation of active balances of metabolism, fostering erosion within the biofilm ecosystems (Fischer et al., 2014).

The EPS in biofilms functions as an adhesive, securing corrosive bacteria to steel surfaces and enhancing metabolic and respiratory efficiency. Unconventional bacterial consortia can induce localized rusting. Advancements in molecular biology, particularly omics methodologies, permit accurate in situ examination of bacterial communities, enabling the observation of structural and compositional alterations in microbial populations over time and location. Previous in vivo studies have identified antibiotic-resistant bacteria and Bacteroidetes as the predominant microbial phylum in degradation specimens in maritime structures, regardless of the employed methodology. Steidl et al. (2016) explain why phylum Proteobacteria is the most dominant in marine corrosive rust samples: (a) There is evidence that Proteobacteria has now emerged as the most widespread bacteria species exhibiting the greatest prevalence in aquatic environments including saltwater and debris, specifically the alpha and gamma classes subdivisions make up 75% of culturable corrosion bacteria. Most organisms in this phylum serve as pioneer colonizers and efficient biofilm formers (Cui et al., 2015). *Roseobacter* belongs to *Alphaproteobacteria* and is recognized as a pioneer colonizer in marine environments. It carries genes that encode Barrier proteins, pores, and fibrous protein essential for biofilm development. The central microorganisms involved in sulfur and iron redox-cycles belong to the Proteobacteria, including significant Sulfate- Reducing Bacteria also known as SRB in *Deltaproteobacteria*, Iron- Oxidizing Bacteria (IOB) in *Betaproteobacteria* and *Zetaproteobacteria*, as well as Sulfur- Oxidization Bacterial (SOB) in *Beggiatoaceae* and *Leucotrichaceae*. Bermont-Bouis et al. (2007) indicated that the composition of bacterial communities that form biofilms on steel coupons after brief immersion was subjected to the most significant changes, which afterwards was stabilized after many intermittent fluctuations. In the end, SRB holds the paramount position. An investigation of diversity of bacteria in the saltwater induction method in service on an offshore oil platform presented several microorganisms, particularly sulfate-reducing bacteria SRB in sterile saltwater. Additionally, it has been demonstrated that SRB and SOB can synergistically enhance the low water content corrosion related to carbon steels (Lanneluc et al., 2015).

EPS behaves like cement in concrete that is firmly attached to substrate surfaces. In addition, EPS could specifically connect to ions of metal due to the presence of polysaccharides and Proteins containing negatively charged functional groups such as Carboxyl, phosphate, sulfate, glycerate, and pyruvate. Dong et al. (2015) posited in which the Extracellular Polymeric Substances (EPS) generated by *Proteus natriegens* in carbon steel surfaces may interact with the ions of metal to facilitate corrosion caused by microorganisms (MIC) (Dong et al., 2015). The combination of high-valent metal ions like Ca^{2+}, Cu^{2+}, Mg^{2+}, and Fe^{3+} within the substance layer facilitates the EPS's function as carriers of electrons that connect biofilm and metallic substances, enabling a distinctive redox process pathway. In the presence of appropriate electron acceptor, a cathode depolarization will be triggered by the redox process, thus accelerating MIC. Besides, biofilms have different stages of microbial growth and live in special ecological niches, thus having special characteristics that may enhance or decrease the corrosion rate of steel made from carbon. Dong et al. (2015) found that with minimal quantities, Polymeric Extracellular Substances (EPS) derived by Sulfate-Reducing Organisms (SRB) adsorbed into carbon steel surfaces might inhibit Microbiologically Influenced Corrosion (MIC), however at elevated concentration, EPS may exacerbate the progression of MIC. In the presence of IOB, the pH level of the rust layer, comprising the biofilm EPS and corrosion product progressively diminished, resulting in increased porosity of the rust layer. This demonstrates that IOB is more corrosive towards carbon steel (Vigneron et al., 2016). Conversely, many bacteria, such as *Shewanella algae*, *Shewanella oneidensis, Pseudomonas flava, Bacillus subtilis*, and *Bacillus brevis* are documented to possess the ability to suppress steel corrosion. *Pseudoalteromonas lipolytica* may safeguard steel from corrosion by creating Self-treatment hybrids barrier made of inorganic and organic components capabilities, consisting of multilayers of calcite and also extracellular polymeric compounds on the steel surface. *Vibrio neocaledonicus*, extracted with sea debris can lower the carbon steel's rate of degradation by as much as 60 times by oxygen scavenging facilitated by extracellular polymeric compounds in biofilms (Moradi et al., 2015). The influence of biofilm on corrosion, whether accelerating or inhibiting it, is significantly determined by the microbial populations adhering to the steel surface and also their extracellular polymeric substances. Enhanced molecular biology research and genomic evaluation are essential to clarify the regulatory systems of the metabolic processes on biofilm interfaces on steel.

3. IMPACT OF STEEL ALLOY TYPE AND COMBINATION ON MARINE MICROORGANISM ADHESION AND CORROSION

Microorganisms selectively attach to metallic surfaces for sustenance, recognizing these materials as energy sources. Concurrently, alloying Surface texture and features have affected biofilm production and its Polymeric Extracellular Substances (EPS). In conventional compounds, including stainless steel, steel made from carbon, and various metal-based alloys, scientists have found the components including Iron (Fe), Chromium (Cr), Molybdenum (Mo), Nickel (Ni), as well as Copper (Cu) affect the rate of Microbiologically Influenced Corrosion (MIC) and the initiation of Microbiologically Influenced Corrosion (MICI) of steel in marine environments in diverse ways. Iron exerts a unique influence on microbial-induced corrosion in metabolic processes and elemental cycling. Javed et al. (2015) found that Sulfate-Reducing Bacteria (SRB) in biofilms within iron-rich environments produced higher amounts of Extracellular Polymeric Substances (EPS) than those in iron-deficient environments during the later stages of attachment, suggesting that increased ferrous ion concentrations significantly amplify corrosion caused by SRB. Researchers find out that the Extracellular Polymeric Substances (EPS) of Sulfate-Reducing Bacteria (SRB), which revealed that Lipopolysaccharides (LPS) within the gram-negative microbial exterior surfaces, such as SRB in *Desulfovibrio sp*., have the ability to interact exclusively with ferric ions (Fe^{2+}). An analysis of the dsrAB gene's level of expression was conducted in our lab in a standard Sulfate-

Reducing Bacteria (SRB) strain previously obtained from seawater. The findings demonstrated that the incorporation of iron or an elevation in Fe^{2+} concentration significantly increased the expression of the dsrAB gene (Li et al., 2018).

Although both are iron-based materials, the Microbiologically Induced Corrosion (MIC) on the outside surfaces of steel made of carbon and stainless steel are quite distinct. Elements like Chromium (Cr) along with Iron (Mo) have a major effect on the biofilm that grows on stainless steel. They may form a layer of passivation that exists between the biofilm along with the substrate, which lowers the amount of corrosion of stainless steel significantly in comparison with carbon steel in seawater. All passive iron-based metal complexes, especially stainless steel, undergo ennoblement in marine saltwater, which could result in a 200–400 mV change. According to research, the amount of chromite in the coating of passivation has a significant impact on how resistant stainless steel is to corrosive in seawater. The modification of chromium concentration within the MIC process is still debated; nevertheless, it is clear that chromium levels are affected by corrosive biological films (Moradi et al., 2014). Stainless steel, particularly duplex stainless steel with improved resistance to chloride and sulfide corrosion, may undergo Microbiologically Influenced Corrosion (MIC) due to *Vibrio natriegens* and also *Desulfovibrio desulfuricans*, facilitated by loose porous biofilms and also metabolic sulfides in marine settings. Dec et al. (2017) argued that the biofilm of the *Desulfovibrio desulfuricans* formed strong chemical interactions with EPS-Fe, facilitating the anodic dissolution onto the surface of DSS. Copper is frequently employed due of its toxic nature to microbes, it is used in aquatic corrosion agents. Research indicates that ions of Extracellular Polymeric Substances (EPS) within the biofilm can engage with nickel, tin, and copper in low alloy steel immersed in seawater, thus reducing the corrosion rate caused by bacteria (Chen et al., 2014). As a result, in seawater, the low-alloyed steel incorporating copper has a higher resistivity to corrosion than the original copper alloy. Similar to copper in biofilms, zinc can significantly deteriorate biofilms in aquatic ecosystems. The results of an earlier genome analysis of bio films grown on Zn-coated metallic substrates showed considerable microbial diversity and a variety of metal ion transporter and transposase genes, enabling microbial adaptation to toxic Zn surfaces (Ding et al., 2019).

4. STEEL'S DYNAMICAL CORROSION IN AN AQUATIC ENVIRONMENT

The immersion of the carbon steel within natural seawater will result in changes to its chemical or physical characteristics and the formation of a dynamic biofilm, both of which will impact the steel surface's oxidation process. Under saline conditions, protons gradually increase the iron-containing oxidation. Thus, organic corrosion is judged technically negligible in oxygen-depleted maritime conditions, implying that biological mechanisms play an significant part in the degradation of steel. Melchers and Jeffrey's (2012) phenomenological framework for carbon steel in saltwater indicates that dynamic corrosion is primarily driven by the depletion of dissolved oxygen (DO) during the initial phase, while the subsequent anoxic phase represents the principal stage of Microbiologically Influenced Corrosion (MIC). This concept has been validated through laboratory investigations, indicating that carbon steel will undergo considerable corrosion due to sulfate-reducing bacteria following the depletion of dissolved oxygen (Wu et al., 2016). Subsequent extensive studies have shown that corrosion rates during anaerobic periods have escalated. In the absence of dissolved oxygen at the material interfaces microenvironment, the degree of corrosion may be influenced by the density of the iron oxide layers and the pace at which nutrients are transferred (Enning et al., 2012).

The latest work has demonstrated that the electroactive microorganisms, including as Sulfate-Reducing Bacteria (SRB), may significantly destroy steel substrates by using direct electron transfer to acquire energy (Venzlaff et al., 2013). In contrast to the corrosion induced through Metabolic Chemical Agents (CMIC), electrically microbially influenced corrosion (EMIC) is indicated by equation (1).

$$4\ Fe^0 + SO_4^{2-} + 3\ HCO_3^- + 5\ H^+ \rightarrow FeS + 3\ FeCO_3 + 4\ H_2O \quad (1)$$

The excess electrons will be released as H_2 produced through a hydrogenase-catalyzed process. The EMIC mechanism through direct electron consumption should be referred to as Extracellular Electron Transfer (EET), which originated from the studies of Microbial Fuel Cells (MFC). These microorganisms normally transmit ions *via* the outermost oxidase protein layer, including Cytochrome C, Nanowires, or solubility mediators of redox (Wang et al., 2019). During carbon feeding shortage, an established electrically active bacterium on metal might turn out considerably more caustic due to the extracellular electron transfer pathway. The process of EET explains EMIC, as Sherar et al. (2011) have proved the existence of nanowires through scanning electron microscopy when investigating SRB-induced carbon steel corrosion in an environment without specific organic carbon sources. This indicates that SRB can transfer electrons to the cells over long distances, enhancing the ability to survive under conditions of carbon source scarcity and increasing the corrosion rate accordingly. Several other studies have already been able to establish that the process of EMIC may also be observed for the archaeological remains of corrosion the nitrate-removing microbe *Methanococcus maripaludis* in particular *Prolixibacter sp.*, as well as the acetogenic bacterium *Sporomusa sp*. The molecular mechanisms through which microorganisms derive is unknown how ions move from iron-containing substrates, however new studies have revealed for the first time that the electron transfer intermediary conveyed by the phzH gene within *Pseudomonas aeruginosa* is implicated in Extracellular Electron Transfer (EET) during microbial iron corrosion that the genetic level. Gu et al. (2019) presented ideas on Biocatalytic Cathodic Sulfate Reduction (BCSR) to clarify the mechanism by which Microbiologically Induced Corrosion (MIC) through Sulfate-Reducing Bacteria (SRB) affects steel, analyzed through the perspective of bioenergetics. It suggests that Iron (Fe0) may supply because this reaction is economically advantageous for tiny molecules of carbon that is organic (acetate), electrons for sulfate decrease, which causes the steel to corrode.

5. MICROBES ASSOCIATED WITH MARINE MIC

Corrosive biofilms consist of diverse species, including bacteria, fungi, and archaea, which can directly or indirectly contribute to the corrosion process. The primary bacterial types associated with steel corrosion vary according to their metabolic activities, including, but not restricted to Sulfate-Reducing Bacteria, Sulfur-Oxidizing Bacteria, Metal-Oxidizing Bacteria, and Metal-Reducing Microorganisms along with different microorganisms that can release acidic substances and producing extracellular substances (Zuo, 2007). Understanding the different roles of various microorganisms in corrosion and their respective mechanisms of induction in corrosion on steel should strongly underpin effective methods of corrosion prevention.

5.1. Microbes That Dissolve Marine Sulphate

Microbes with anaerobic metabolism that use organic matter as a source of electrons during respiration are known as marine sulfate-reducing bacteria. About half of the sulfur reductions in coastal sediments are caused by SRB, which is crucial to the biogeochemical element cycle. In addition, SRB dominates in seawater sediments and is considered a critical contributor to metal corrosion in marine settings. The acidic metabolic byproduct produced by sulfate-reducing bacteria shows severe corrosion on the metal surface. As a result of the substantial study of saltwater sediments and rocks, many species that belonged to the marine sulfate-reducing bacteria have been identified. Currently, about 60 species of Sulfate-Reducing Bacteria are documented in the marine environment Huang et al. (2017). Recent development in genomics and metabolic pathways explained the eco-physiology and geographical distribution of the Sulfate-Reducing Bacteria in marine environments. Most of the studies have proved that sulfate-reducing bacteria act as the primary microorganisms responsible for long-term marine immersion processes. For instance, substantial SRB species, like *Desulfovibrio* and *Desulfobacter*, exist in the carbon-based steel's innermost barrier, which has a significantly greater abundant value compared with other types of bacteria Zhang et al. (2019). Sulfate-reducing bacteria produce hydrogen sulfide as the final metabolic end product, which further reacts with the metal surface, forming a biofilm with metal sulfide production. Protons are let out during the process, lowering the environment's pH level area, developing an acidified microenvironment that accelerates corrosion at a higher rate by Zhang et al. (2022).

5.2. Marine Sulfur-Oxidizing Bacteria

Marine SRB behave differently from marine SOB because of their ability to oxidize a range of reduced sulfur compounds, including hydrogen sulfide and thiosulfate, and elemental sulfur into sulfuric acid, at a pH that is more acidic. Metal corrosion can be caused by a variety of SOB, including both anaerobic and aerobic bacteria. Sulfide-oxidizing bacteria, especially the genera *Sulfuricurvum* and *Thiomicrospira*, have been the subject of considerable recent study (Kodama & Watanabe, 2004). Other important SOB associated with marine corrosion include *Thioalkalivibrio*, *Sulfurimonas*, and *Thiomicrorhabdus*. Okabe et al. (2007) identified the fact that SOB can oxidize H_2S produced by SRB into H_2SO_4. Their methodology achieved a pH decline in the concrete surface from 12.0 to 1.6 within 102 days. Hence, by increasing the acidity around the metal, SOB can also lead to the increased corrosion of metals.

The corrosion induced by marine sulfate-reducing bacteria can be explained by two processes: corrosion by electricity and a direct chemical response. Sulfuric acid formed from Metals may be physically corroded by the oxidization of sulfur-based substances. For example, SOB oxidizes to produce sulfate and hydrogen ion particles, which then react with metals to generate metal sulfates and release further ions containing hydrogen. When SOB is present, localized electrical cells form on a metal's surface, causing chemical corrosion. In such a case, SOB oxidizes sulfur compounds and thus gives rise to electron flow across the surface of the metal and, as a result it activates both cathode and anode regions. Consequently, metal ions are progressively depleted, resulting in the development of pits or as cavities on the metallic surfaces (Liu et al., 2022).

5.3. Microbes That Oxidate Aquarium Metals

Marine organisms, particularly Iron-oxidizing microbes use metals as electron donors for energy metabolism, which causes metal-based materials in aquatic settings to corrode. A multitude of metal-oxidizing bacteria has been identified as factors in the corrosion of the metallic substances found in maritime habitats. By oxidizing the ferrous ions, *Acidimicrobium* and *Mariprofundus* aid in the deterioration of iron-based products. Iron-oxidizing bacteria are the two primary types of metallic-oxidizing microbes and manganese-oxidizing bacteria. Iron-oxidizing bacteria may stimulate the oxidation activities of iron by oxidizing the ferrous ions and creating acidic and electronic waste products (Little et al., 1991). The corrosion of metals can be enhanced by the conversion of solubility divalent manganese (II) into insoluble manganese dioxide can be facilitated by iron-oxidizing bacteria. Numerous metallic substances may corrode more easily as a result of these oxidizing metallic organisms, especially Iron, Steel, Copper, and Aluminum. In marine habitats, metal-oxidizing bacteria mostly facilitate metal oxidation via their metabolites. The Oxidative Sulfur–Iron Bacteria (OISSB) facilitates metal corrosion through the oxidation of ferrous ions and sulfide minerals, leading to the generation of acidic sulfur compounds (Xu et al., 2023).

5.4. Microbes That Dissolve Marine Metals

The Marine Metal-Reducing Bacterium (MRB) is a broad group of autonomous or purely anaerobic organisms. In actuality, iron-reducing and manganese- reducing bacteria is members of metal-reducing bacteria. These bacteria use metal ions as electron donors for the breakdown of energy, which leads to metal decreases in contrast to metal-oxidizing bacteria and subsequent corrosion of metallic objects. Metal-reducing bacteria, including *Geobacter* and *Shewanella*, have been extensively researched for their varied electron transfer capacities. *Geobacter* was discovered to immediately take in electrons *via* surfaces of metal using an external electron exchange mechanism made possible by cytochrome C protein in the exterior membrane and nanowires (Rawlings, 2005).

The microorganisms that reduce metals take part in the corrosion process by several mechanisms that can be classified into three categories: Firstly, They change the metallic substrate's physicochemical characteristics by generating an atmosphere that oxidizes hydrogen using metabolic processes and producing tiny perforations. According to Snider et al. (2012), they turned insoluble metal oxides such as Fe (III) oxides and Cr (IV) oxides into solubility metal ions and hence promoted material deterioration. Lastly, the A homogeneous biofilm of reducing metals microbes regulated electrically reactions and provided an environment promoting localized corrosion by Loto (2017).

5.5. Microbes That Decrease Marine Nitrate

The organisms that employ nitrate as an electron acceptor are known as marine nitrogen-reducing microbes. Ocean water and marine sediments are typically the primary sources of these microorganisms. These are known to cause biocorrosion of aquatic structures and ship hulls, posing a significant threat to safety and reliability in marine engineering. *Pseudomonas aeruginosa* is a type of non-respiring bacteria, Lovley et al. (2004) comprehensively investigated its electroactive biofilm corrosion. Chugh et al. (2022) found that the biofilm of *Bacillus licheniformis* on carbon steel showed higher aggressiveness than that of conventional *desulfurizing bacteria*, resulting in serious corrosion.

The Biological and Electrochemical processes involved in the kinetics and mechanisms of microbiologically induced corrosion through nitrate-reducing bacteria are complex. Biologically, these include the adhesion and biofilm formation by NRB on the surface of metals, the reduction of nitrate to nitrite or other nitrogenous compounds by NRB, and corrosive metabolite generation (Xu et al., 2013). These include the galvanic cell formation between the different metal regions or interaction metal and biological film, how NRB intermediates alter anodic and cathodic procedures, and how metal ions disintegrate in acidic or alkaline conditions (Gu et al., 2021). Such complex interactions may cause serious corrosion of aquatic structures and machinery.

5.6. Marine Acid-Producing Bacteria

Acidogenic bacteria metabolically synthesize organic and inorganic acids that are directly attacking the metallic surface exposed in a marine environment (Little et al., 2020). The oxidized species of nitrogen, nitrite (NO_2), serves as the electron donor for aerobic nitrogen-oxidizing bacteria to form the corrosive nitric acid. This group belongs to the nitrifying bacteria which include *Nitrobacter, Nitrococcus*, and *Nitrospira. Thiobacillus* species can oxidize reduced sulfides, such as sulfur, sulfites, and thiosulfates, producing metabolites H_2SO_3 or H_2SO_4 that corrode metals. Acid-producing bacteria, like those from the genus *Desulfurococcus*, have been implicated in MIC of marine steel infrastructure (Chaudhary et al., 2019). There microorganisms that produce propionic acid as the sole product of their electron transport cycle. The metabolic end products of marine acid-producing bacteria participate in redox reactions that advance corrosion. Unlike in electromicrobial corrosion, metabolic products do not participate in the corrosion process *via* the catalytic action of biocatalysts; rather, they are directly reduced on the surface of the metal. At low pH, the proton reduction is coupled with the oxidation of metallic elements, which enhances the dissolution of metals and, subsequently advances the corrosion process (Khan et al., 2022).

5.7. Marine Fungi

Marine fungi interact with both biotic and abiotic components. Recent studies have shown that they play essential roles in marine food systems as well as in corrosion processes (Starosvetsky et al., 2001). Marine fungus breakdown and consume organic compounds, improving organic matter cycling and ecosystem stability. Some cause metal corrosion, reducing marine structure life and safety. Fungi metabolism releases acidic chemicals and metallic ions, causing this rusting. *Aspergillus*, *Candida*, and *Paecilomyces* produce acidic metabolic byproducts that corrode iron and copper. Marine fungus produce exo-enzymes such proteases and cellulases, which accelerate metal corrosion by producing acid from organic detritus and proteins degrading on marine steel structures (Edyvean & Videla, 1991).

Primary agents of corrosion by marine fungi are the acidic metabolites that are produced in fungal metabolic processes and release metal ions. *Aspergillus* metabolic waste products contain acids that corrode copper and iron, lowering their marine construction durability (Silber et al., 2016). A few marine fungus produce exogenous enzymes that promote metal corrosion. Proteases and cellulases decompose organic materials and proteins on marine structures, forming more acidic compounds and accelerating corrosion (Liu et al., 2023).

5.8. Marine Archaea

Recent research into microbial corrosion has focused on archaeal groups, especially those in the marine environments. This is primarily because its involvement Corrosion of pipelines, ships, and oil rigs causes economic losses and environmental issues (He et al., 2022). Marine ecological corrosion was dominated by methanogenic and sulfur-oxidizing archaea. Methanogens reduce carbon dioxide or organic molecules in sedimentary or aquatic environments to create methane. The creation of cathodic sites may increase metallic structural corrosion. Sulfur oxidizers employ sulfide and elemental sulfur as energy metabolism electron suppliers, producing sulfuric acid, which can corrode metals (Melchers, 2005).

By forming biofilm or rust nodules, archaea can change the pH, potential, oxygen dispersion, and ion levels of material surfaces. These modifications affect material electrochemical corrosion. Additionally, archaea metabolic products can locally cause metal corrosion or damage through chemical reactions (Usher et al., 2014). Methanogenic Archaea (MA) use metals or hydrogen to convert CO_2 to methane. Sulfate- Reducing Archaea (SRA) use sulfate as an electron acceptor to convert organic compounds or hydrogen into hydrogen sulfide. SOA convert sulfur or sulfide into sulfate using oxygen or nitrate as electron acceptors. Metabolic processes alter surface electron density and pH, facilitating metal breakdown or pitting corrosion. The metal may react with Methane, Hydrogen Sulfides, and Sulfate, causing stress corrosion or passive layer breakdown and cracking (Hornibrook et al., 2000). Table 1 about Steel structures that are used in aquatic and soil setting are very vulnerable to attack by microbial colonies that greatly promote the rate of corrosion. This is commonly referred to as microbially influenced corrosion (MIC) that is a result of interactions between microbial communities, environmental factors, and metal surfaces. Biofilms, release of corrosive metabolites and electrochemical change of electrons and environments at the metal-environment interface, can occur as microorganisms, leading to localized or generalized corrosion. It is essential to understand the key microbial groups concerned and their particular mechanisms in order to create viable mitigation measures in the marine and terrestrial systems.

Table 1. Major Microbial Groups and Mechanisms Involved in Steel Corrosion in Aquatic and Soil Systems

Category	Microorganisms Involved	Mechanism of Corrosion	Environmental Conditions	Corrosion Products / Effects
Sulfate-Reducing Bacteria (SRB)	*Desulfovibrio, Desulfotomaculum*	Anaerobic reduction of sulfate to sulfide; formation of FeS leading to pitting corrosion.	Anoxic aquatic sediments, water pipelines, and soil with high sulfate content.	FeS, H_2S, black crust formation, hydrogen embrittlement.
Iron-Oxidizing Bacteria (IOB)	*Gallionella, Leptothrix*	Oxidation of Fe^{2+} to Fe^{3+}, forming ferric hydroxides that deposit as tubercles.	Aerobic or microaerophilic environments such as rivers and soil-water interfaces.	$Fe(OH)_3$, rust tubercles, localized under-deposit corrosion.
Iron-Reducing Bacteria (IRB)	*Geobacter, Shewanella*	Reduction of Fe^{3+} to Fe^{2+}, destabilizing passive films on steel surfaces.	Anaerobic environments with organic matter and reducible iron oxides.	Accelerated corrosion beneath biofilms.
Sulfur-Oxidizing Bacteria (SOB)	*Thiobacillus, Acidithiobacillus*	Oxidation of sulfide or elemental sulfur to sulfuric acid causing acid attack on steel.	Aerobic conditions in soils, sewage systems, and marine environments.	Sulfuric acid production, extensive surface etching.

continued on following page

Table 1. Continued

Category	Microorganisms Involved	Mechanism of Corrosion	Environmental Conditions	Corrosion Products / Effects
Manganese-Oxidizing Bacteria (MOB)	*Leptothrix discophora, Pseudomonas putida*	Oxidation of Mn^{2+} leading to deposition of Mn oxides, creating differential aeration cells.	Freshwater and soil systems rich in manganese.	MnO_2 deposits, uneven corrosion attack.
Biofilm-Forming Heterotrophs	*Pseudomonas, Bacillus, Clostridium*	Biofilm formation alters electrochemical potential and enhances localized corrosion.	Common in both marine and terrestrial environments.	Localized pitting, differential aeration corrosion.

6. MECHANISMS OF MIC IN THE MARINE ENVIRONMENT

The specific adhesion of microbes to metallic surfaces illustrates their pursuit of existence. The microbes utilize metal elements as energy sources and adhere to them for existence. Researchers have discovered that elements of alloying and surface microstructure influence the rate of microbiologically induced corrosion in steel in various manners (Qian et al., 2022). Stainless steel and steel made from carbon are both composed of iron-containing alloys. However, their surface microbially induced corrosion differs significantly. The passivation layer onto stainless steel prevents microbial corrosion, compared to carbon steel in seawater, is significantly lower. The corrosion mechanisms involved in stainless steel, alloys of steel, and also iron-containing steel differ, but bacterial degradation processes are comparable (Liamleam & Annachhatre, 2007).

Microbes have been involved in corrosion for over a century, but the risks of microbiologically induced corrosion have been disregarded. Historically, microbe corrosion was studied by examining their corrosion behaviors, but their molecular mechanisms have not been studied. Traditional theories like the concentration of oxygen gradient theory of cells and the corrosion byproduct concept held that bacteria do not directly cause corrosion. These hypotheses are restricted and lack proof to explain rusting. In 2009, Gu et al. (2019) proposed the Bio Catalytic Cathodic Sulfates Reaction (BCSR) hypothesis based on bioenergetics and bioelectrochemistry, advancing MIC research. This theory states that bacteria directly absorb electrons from metal surfaces to cause microbial corrosion. Research indicated that sulfate-reducing bacteria use iron as an electron donor to meet energy needs when carbon sources or external electrons are absent to erode metals (Pimenov & Neretin, 2006).

Electron mobility is the movement of protons through cytochrome protein on the membrane's outer layer or conductive biological nanowires. The sulfate-lowering bacterium *Desulfovibrio ferrophilus* IS5 directly transfers electrons from solid sulfides through a greatly effective cytochrome-based pathway similar to OMCs, enabling respiration and also survival during starvation. Tang et al. (2021) established that the orientation of electron transfer pathways significantly influence microbial corrosion control. They showed that genetic mutations connected with direct electron flow within *Geobacter sulfurreducens* and *Geobacter metallireducens* can directly transfer electrons from iron towards organisms, degrading stainless steel. The e-pili also helped *Geobacter* and metal exchange electrons, boosting microbial corrosion. When microorganisms create soluble electron transfer mediators including Phenazines, the Flavins, Quinones, Humic Acids like Anthraquinone-2, 6-Disulfonate, and neutral red, they absorb electrons from metal surfaces indirectly. These mediators transport electrons to the cell membrane and discharge them *via* c-type cytochrome. Riboflavin and FAD increase *Desulfovibrio vulgaris*-induced corrosion of 304 stainless steel and carbon steel (Ma et al., 2020). Cross-fusion molecular biology has been used to study how phenazine chemicals regulate marine *P. aeruginosa* bacterium from a genetic perspective on the corrosion (Javed et al., 2017). Gene knockouts showed that phenazine molecules from phzH and other genes transport extracellular electrons (Zhou et al., 2013). Figure 1 Schematic representation of the major microbial groups and their mechanisms contributing to steel corrosion in marine environments, illustrating biofilm formation, metabolic activities, and localized corrosion processes.

Figure 1. Mechanism of MIC in Marine Environment

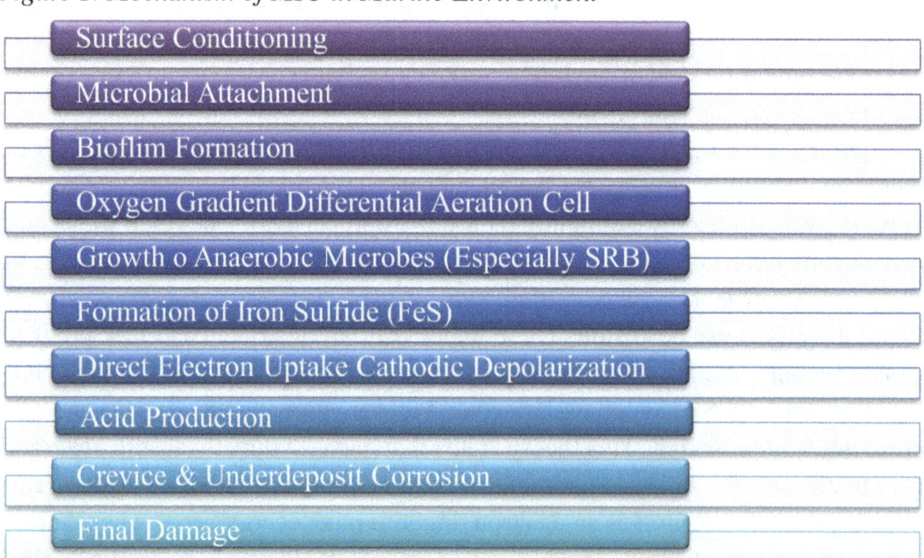

7. MARINE MIC INHIBITION

7.1. Surface Treatment

7.1.1 Physical Treatment

Adhesion and proliferation of marine microorganisms on the material surface predominantly induce MIC. The development of biofilms is an essential step in triggering these events. In general, regular cleaning is one of the common strategies developed to mitigate MIC in the marine environment. This kind of strategy requires rigorous procedures, and cleaning efficacy in micro-areas is not prominent. Hence, it is necessary to emphasize the challenge in eliminating established biofilms (Li et al., 2018).

7.1.2 Cathodic Protection

CP has proved extensively employed in nautical engineering and naval construction, significantly prolonging the service life of metallic components while minimizing the occurrence of MIC. Certain studies have concentrated on the impact of CP in suppressing MIC (Tang et al., 2021). Cathodic protection is an electromagnetic process technique employed to protect the corrosion of metal structures by imparting an externally applied current. It ensures that the metal surface behaves like a Cathode, which delays electron migration during corrosion and thus decreases the rate of corrosion. The cathode produced by CP has characteristics that are not favorable for microbial life and biofilm building and thus ultimately inhibits the development of MIC. In specific cases, metal corrosion is favored by cathodic protection. For instance, Cathodic defense versus Sulfate-Reducing Bacteria (SRB) or species of *Vibrio* necessitates an augmentation of the conductive electric upon application, and elevated cathodic currents heighten the risk of protection from corrosion failure (Zhang et al., 2015).

7.2. Coating

Various methods have been adopted for controlling corrosion, coating technology is one of the most effective methods for preventing the entry of corrosive agents physically. Various anti-corrosion coatings have been developed, including Organic, Inorganic, and Hybrid Systems.

7.2.1 Organic Coatings

Organic coatings are popular due to their Adhesion, Adaptability, and Corrosion Resistance. Marine applications include Hulls, Pipes, and Offshore constructions that use Spray, Brush, or as Roller Coatings. The most common organic marine anticorrosion ingredient is epoxy (Huang et al., 2018). The common polymer epoxy coating protects metals from microbial corrosion and is robust and durable. Other polymer coatings include Polyurethane, Polyethylene, and Polypropylene. Polyurethane coatings resist marine Biofouling and Biological Deterioration better. The coatings smooth the surface and prevent microbial adherence. Additionally, polyethylene and polypropylene coatings resist mechanical wear and abrasion (Zhou et al., 2022).

7.2.2 Inorganic Coatings

Oxides of Metal, Nitrides, and Carbides make inorganic coatings, also called metallic coatings, corrosion-resistant and stable under harsh conditions. These are usually made of harmful metals like Copper, Zinc, Zinc, Nickel, and Aluminum. Superior antifouling qualities make copper-based coatings popular. Copper ions from such coatings limit bacterial development (Simões et al., 2010). Zinc-based coatings are promising antifouling, whereas nickel-based coatings kill sulfate-reducing microorganisms. Due to its corrosion resistance, aluminum and its alloys are useful in numerous marine applications (Googan, 2022).

7.2.3 Mixed Coatings

Mixed coatings of organic and inorganic materials may improve microbial corrosion prevention. Mixed coatings have superior corrosion resistance and mechanical qualities than single-material coatings. Due to its self-healing and thermal stability, it has been a popular issue in research (Zuo, 2007). In addition to investigating Coating Materials, Plasma Treatment, Micro- Arc Oxidation, and Sol-Gel Coatings have been developed to improve coatings efficacy against the bacterial corrosion by boosting adhesion and durability. Biocides, Enzymes, and Peptides can also be added to such coatings to limit microbial growth and biofilm development (Figueira, 2020).

7.3 Material Design

A major obstacle to microbial corrosion research has always been trying to establish a relationship between microorganisms and metals. A bacterial community uses selective adhesion to metal surfaces as a survival strategy, although these materials are often thought of as energy sources (Armelin et al., 2009).

7.3.1 Incorporation of Antibacterial Elements and Tuning of Texture

Antibacterial alloying elements improve steel corrosion resistance reliably and effectively. Metals like Copper, Silver, and Chromium can hinder bacterial adhesion. Metals have three antibacterial mechanisms: Bacterial membranes and cell walls interact with positively charged metal ions. This kills germs by destroying their habitat. Metal ions inactivate enzymes and denaturate membrane proteins in bacteria by binding to protein groups. Bacteria division and proliferation were stopped. Finally, metal ions caused air and aqueous oxygen to generate reactive oxygen species. Bacteria died from oxidative damage to Proteins, Nucleic Acids, and Lipids. Compared to 317L stainless steel, 317L-Cu stainless steel reduced sessile cell counts by 98.3% against *Staphylococcus aureus*. 304L-Cu stainless steel pitted less and lost less weight than 304 in *Escherichia coli* medium. Its corrosion current density was four times lower than that 304 stainless steel. The antibacterial activity of copper caused its fair corrosion resistance. HNS-Cu and 2205-Cu DSS also performed similarly (Pistone et al., 2021). Recently, microbial corrosion researchers advocated modifying metallic composition contrast or grain size at the microstructural level to increase microbiological corrosion resistance in metallic objects. Antibacterial activity against sulfate-reducing bacteria and *Pseudomonas aeruginosa* and pitting corrosion resistance were stronger in acicular ferro pipeline steel than in X80 steel (Popescu et al., 2022).

7.4. Biological Protection

Due to their complex metabolism, microbes can serve two roles in the environment. Microbial corrosive inhibition outperforms existing anticorrosion treatments and is environmentally beneficial. Some corrosive microorganisms change biofilm conditions to safeguard metals. The maritime environment makes microbial corrosion inhibition more complicated than conventional approaches. MICI involves corrosion inhibitor secretion, biofilm formation, corrosion product shielding, microenvironment change, and anodic and cathodic process abnormalities (Galedari et al., 2019).

7.4.1 Antimicrobial Active Substance

Certain microorganisms release compounds that prevent metal corrosion. These include biosurfactants, the Proteases, Alkaline Phosphatases, Carbonic Anhydrases, and Amino Acids like Polyaspartic and Polyglutamic Acids. Proteases change pH and settle ions, protecting metals. *Bacillus sp.* and *Pseudomonas fluorescens* surfactants inhibited corrosion on stainless steel 304 and carbon steel ST37 (Kaur et al., 2015). In addition, acidic extracellular discharges on microorganisms include numerous cations that can adsorbs and provide a natural barrier layer on metals, reducing corrosion (Kugel et al., 2011).

7.4.2 EPS Protection

Biofilms form when bacteria and their exopolymers stick to surfaces. Biomes vary in Strength, Viscosity, and Characteristics. Qu et al. (2015) found that *Bacillus subtilis* biofilms grew on cold-rolled steel surfaces in a nutrient-enriched simulated seawater media. *Bacillus subtilis* reduced OCP compared to sterile artificial saltwater media. Biofilm development inhibited rusting. The huge number of bacterial species in biofilms and metal interaction is complicated and should be researched. Biofilm contains polysaccharide EPS with C=O and C-O linkages for carboxyl groups (Ma et al., 2020). These might form a compact protective coating with iron and other metallic ions. Recent research suggests that seawater biofilm-forming microorganisms decrease corrosion (Murcar-Evans et al., 2017). The biological coating on low-alloy steel surfaces was connected to *Bacillus subtilis*, the dominant species. A pure bacterial subtilis culture was grown and developed to study its effect on low-alloy steel corrosion. *Bacillus subtilis* forms a homogenous layer of polysaccharides and starch fibers that survive harsh chemical and physical conditions, giving the biofilm structural stability. Scan vibrating electrode results show consistent current distribution on sample surface, indicating *Bacillus subtilis* creates a uniform layer (Luo et al., 2021).

8. MICROBIAL CORROSION IN AQUATIC ENVIRONMENTS

8.1.1 MIC in Marine Systems

Marine habitats are prone to microbial corrosion because high salinity, abundant dissolved ions, and permanent dampness enhance microbial colonization of steel surfaces. The dominance of SRB in anoxic niches and synthesis of hydrogen sulfide react with iron to generate iron sulfides, which induce localized pitting and tubercles. IOB and other microbes develop heterogeneous corrosion zones and rust layers under aerobic circumstances (Zhang et al., 2019). Biofilm on submerged steel surfaces protects anaerobic bacteria from oxygen, and microbial communities EPS collects salts, ions, and metabolites to accelerate localized corrosion (Khan et al., 2022). Seasonal fluctuations in temperature, wave action, and salinity alter biofilm structure due to microbial community composition and activity. Hydrocarbon and heavy metal pollution in the marine ecosystem can provide substrate or stressors and alter microbial metabolic activity, increasing corrosion rates. This contact makes marine steel structures very susceptible to rapid corrosion, requiring anti-fouling paint and cathodic protection systems (Seyeux et al., 2015).

8.1.2 MIC in Estuarine Systems

The mixing of freshwater with seawater causes estuarine regions, which form special conditions leading to Varying Salinity, Oxygen Gradient, and Nutrient Availability, which have an effect on the microbial corrosion processes. Estuary microbial communities are commonly richer and contain halotolerant SRB, Iron-Reducing, Nitrate-Reducing Microorganisms, and these jointly enhance the breakdown of steel by their synergetic metabolic interplay. The formation of biofilms is very dynamic, and the existence of EPS provides microhabitats to anaerobic microorganisms even in waters with a high level of oxygen (Liu et al., 2018). The changes in salinity will selectively stress some populations of microbes and enrich other ones giving rise to changes in dominant corrosion causing taxa. Localized electrochemical reactions in biofilms are also triggered by the presence of differential aeration and so, pitting and crevice corrosion is also encouraged. The terrestrial runoffs of nutrients including nitrogen, organic carbon also enhance the microbial activity and generate the corrosive metabolites. It is these environmental changes that have induced intricate corrosion patterns in estuarine steel structures, including bridges, docks, and pipeline, with a heavy stress on adaptive strategies as far as corrosion monitoring and mitigation is concerned (Sheng et al., 2013).

8.1.3 MIC in Freshwater Systems

Low salinity Freshwater systems (such as rivers, lakes, and artificial reservoirs) normally possess insufficient salinity relative to the marine or estuarine habitat but possess adequate amounts of moisture and dissolved minerals and organic matter to support an active microbial community. Typically, aerobic and facultative anaerobic microorganisms (e.g. iron-oxidizing bacteria, heterotrophic bacteria) control the corrosion process in freshwater systems. The formation of biofilm on the steel surface into localized microenvironments of varying oxygen concentration facilitates electrochemical corrosion and pitting (Enning et al., 2012). Bioavailability of organic Carbon, Nitrogen and Trace Elements promotes metabolism of microbes to generate organic acids and other corrosive metabolites, which enhances the corrosion of steel. Seasonal variations, the velocity of water flow, human activities (pollution), and runoff of nutrients can alter the microbial diversity and activity thereby altering the rate and distribution of corrosion. In general, MIC causes a considerable impact on the freshwater pipelines, storage tanks, and bridge foundations; hence, preventive interventions, including the use of Biofilm- Targeting Coating, Regular Cleaning, and Water Chemistry Control, are necessary to prolong the service life of steel infrastructure (Dec et al., 2017).

8.2 Changes in Salinity, Dissolved Oxygen and Nutrient Gradients in Microbial Corrosion in Aquatic Environment.

8.2.1 Role of Salinity

Salinity is regarded as one of the fundamental forces of microbial corrosion because it determines both the structure and the composition of Microbial Communities, Metabolic Activity, and Electrochemical Interactions at the steel interface. The elevated salt levels in marine environments cause osmotic pressures that selectively enrich Halotolerant and Halophilic microorganisms, such as Sulfate-Reducing Bacteria (SRB) and some Iron Oxidizing Bacteria (IOB) which are considered key participants in MIC. These microbes grow in salty environments, generating corrosive metabolites including hydrogen sulfide, organic acids and siderophores that speed up localized metal dissolution and pitting. Conversely, in freshwater environments with low salinities, facultative anaerobes and heterotrophic bacteria are selected and the mechanisms of corrosion and biofilm morphology can also vary. The salinity of estuaries tends to be unstable, so dynamic microbial succession (where only the most resilient species can survive) tends to dominate the environment, produces complex and heterogeneous corrosion flux patterns. Salinity is also an ionic strength controller of water and thus it affects the electrochemical potential, conductivity and solubility of corrosion product that affect the morphology and adhesion characteristics of rust as well as overall degradation kinetics in steel. Explanation of these associations is essential to successful and context-dependent strategies in the reduction of MIC (Li et al., 2017).

8.2.2 Role of Dissolved Oxygen

The dissolved oxygen is an essential factor that determines Metabolism, Spatial Distribution, and Corrosion possibility on microbial consortia of aquatic environments. Thus, aerobic areas are selective in the growth of iron-oxidizing bacteria, which accelerate the oxidation of Fe^{2+} to Fe^{3+} to form dense rusty cap, tuberculi and discontinuous corrosion sites. In line with this, biofilm-generated oxygen-depleted microenvironment offers a niche to anaerobic microbes such as SRB to generate hydrogen sulfide and other corrosive metabolites that attack directly on steel surfaces. The Xu and Gu (2014) proposed that the localized electrochemical cells established by the differential aeration in the biofilm, therefore, Promotes Pitting,

Crevice Corrosion and an increased metals dissolution rate. The hydrodynamic conditions such as Velocity of Flow, Turbulence and Diffusion Rates control the transport of oxygen to the metal interface and regulate the thickness and activity of the biofilm. Also, seasonal temperature variations and stratification of water bodies create vertical and temporal DO gradients that lead to the change in the metabolic activity and corrosion severity of microorganisms. That is why, accurate control and monitoring of dissolved oxygen is among the most crucial aspects as far as the severity of MIC prediction and the effective mitigation measures are concerned by McBeth and Emerson (2016).

8.2.3 Role of Nutrient Gradients

Nutrient gradients determine Microbial Activity and, subsequently, the dynamics of biofilm development and corrosion. Nutrients in the form of Carbon Sources, Nitrogen, Phosphorus, and Trace Elements stimulate microbial metabolism for the production of extracellular polymeric substances, enhancing the stability of biofilms and concentrating corrosive metabolites at the steel interface. Under nutrient-rich conditions, the maturation of biofilm and microbial respiration occur quicker with a higher production of metabolites, therefore increasing localized pitting and crevice corrosion (Javed et al., 2017). Nutrient-limiting conditions may restrain microbial growth, hence reducing the rate of progress within MIC. The heterogeneous distribution of nutrients due to anthropogenic inputs like Agricultural Runoff, Industrial Effluents, and Organic Matter Deposition may, therefore, provide variable corrosion patterns in space where the microbial consortia act synergistically with primary and secondary nutrient sources. The secondary metabolic products that are involved in the direct dissolution of metals include organic acids and sulfides; mass transport and oxygen gradients, which control the enhanced localized attack, are modulated by the architecture of the biofilm. The processes based on nutrients will be significant in formulating the proactive measures in controlling corrosion, how to cope with the environment, the use of biocides and also how to control the biofilm-based interventions (Vigneron et al., 2016).

9. MICROBIAL CORROSION OF SOIL ENVIRONMENT

9.1 Microbial Community of Soils and Their Corrosion Potential

Such soil environments are very diverse with the microbial communities and these are likely to affect the corrosion of buried steel infrastructures in a great way. These consist of aerobic, facultative anaerobic, and strictly anaerobic microorganisms in the participation of symbiotic metabolic processes collectively, which increase metal degradation: SRB, IOB, and heterotrophic bacteria. The biofilm formation in soil on the steel surface creates micro environments which protects the anaerobic microbes and concentrates corrosive metabolites like Hydrogen Sulfide, Organic Acids and Extracellular Enzymes. The microbial consortia composition relates to Soil Type, Texture, pH, and moisture content; in general, clay rich soils exhibit slower diffusions but may provide higher metabolite accumulation at the metal surface (Li et al., 2021). Microbial metabolism causes localized redox changes, therefore creating differential aeration cells that intensify pitting corrosion. Moreover, the breakdown of protective oxide layers is accelerated through synergistic interactions between microbial populations, which further aggravates the corrosion process. Seasonal and spatial variations within microbial activities introduce heterogeneity into the corrosion patterns, thus rendering buried pipelines, tanks, and foundations particularly susceptible. Predictions of the severity of MIC and implementation of effective prevention strategies depend upon knowledge of the taxonomic and functional diversity of soil microbes (Sun et al., 2016).

9.2 Redox Potential Effect

Redox potential is a main driver underlying microbial corrosion in soil environments because it determines the electron acceptors available and influences metabolic pathways of corrosion-causing microbes. Thus, with higher redox potentials due to well-aerated soils, for example, there is a possibility for more active aerobic microbial metabolism and, therefore, growth and activity by iron-oxidizing bacteria that are involved in the deposition of rust and tuberculation on the surface of steel. In waterlogged and compacted soils, anaerobic microzones have a low redox potential that creates ecological niches for sulfate-reducing bacteria and methanogens whose corrosive metabolites include hydrogen sulfide and methane (Zhou et al., 2022). These react with the steel surface to develop localized dissolution and pitting corrosion. Changes in redox potential associated with seasonal changes in the Water Table, Rainfall, or soil Management are linked to dynamic changes in the structure of microbial communities and their metabolic activities. The differential redox zones in soil biofilms develop localized electrochemical cells which are enhancing heterogeneity in corrosion. Thus, monitoring and management of the soil redox potential through optimization of drainage and soil aeration are of great importance in mitigation of MIC in subsurface steel structures (Shi et al., 2018).

9.3 Effects of Water-Logging and Organic Matter

Water-logging and organic matter content are prime importance in affecting microbial corrosion in soils. In saturated soils, the rate of oxygen diffusion is significantly reduced, creating anaerobic conditions that help in the proliferation of SRB and other anaerobes and further enhance the degradation of steel. Water-logged zones in soil maintain high moisture content and hence enhance microbial colonization and stabilize the biofilm on the metal surface. Organic matter supports heterotrophic and fermentative microorganisms by providing a source of carbon and energy for growth and the production of metabolic end products (Wang et al., 2022). Complex water retention characteristics, along with Soil Texture and Organic Matter, control the buildup of corrosive metabolic products at the metal interface, thereby increasing localized pitting and crevice corrosion. High-nutrient soils may increase microbial activity, while low-nutrient soils could limit microbial growth and reduce MIC progress. Anthropogenic influences-farm runoff, effluent disposal, and irrigation-alter the organic content and moistness of soil and thus the structure of microbial communities and corrosion processes. Further, a complete understanding of these factors would be required to evolve useful strategies for controlling corrosion through, inter alia, soil drainage, optimization of organic matter content, and application of protective coatings or inhibitors judiciously for steel infrastructure that is buried (Purwasena et al., 2019).

10. DETECTION AND CHARACTERIZATION TECHNIQUES

10.1 Electrochemical and Physical Methods

The electrochemical and physical techniques form the basis for the detection and quantification of microbial corrosion of steel in aquatic and soil environments. Potentiodynamic polarization enables the measurement of corrosion current density and potential that allows the determination of the kinetics of metal dissolution associated with microbial activity. EIS allows for the determination of the biofilm resistance, charge transfer resistance, and double-layer capacitance in microbial-induced corrosion and can help differentiate this from abiotic corrosion processes. Weight-loss analysis, a classical method, consists of steel coupons exposed to the environment for periods of time and measures mass loss to quantify the rates of corrosion (Dagbert et al., 2006). Used together, these methods can obtain qualitative and quantitative information on the severity of corrosion, localized attack, and inhibition efficiency. Furthermore, electrochemical methods are very useful in monitoring MIC in real time, under controlled conditions, due to the good reproducibility and sensitivity of the techniques. Moreover, several in situ electrochemical techniques are applied to monitor biofilm development and corrosion progression for a mechanistic understanding of microbial interaction with the metal surface (Purwasena et al., 2019). In any case, however, none of these methods has the capability to identify the microbial taxa responsible for the corrosion process, and thus requires microscopic and molecular approaches. Sophisticated electrochemical arrays in various formats can simulate environmental gradients, for example, of oxygen, salinity, and redox potential, which may provide an improved realism of MIC in the field. Thus, techniques mentioned form the basis for corrosion risk assessment, early detection, and evaluation of mitigation strategies in steel infrastructure (Ghafari et al., 2013).

10.2 Microscopic and Spectroscopic Approaches

Microscopic and spectroscopic techniques enable detailed visualization and chemical characterization of corrosion surfaces and associated microbial biofilms. SEM shows the biofilm architecture, corrosion pits, and morphology of the metal surface in high resolution, while AFM provides three-dimensional topographical information for assessment of the Alteration in surface roughness and microstructure. The surface's chemical composition, encompassing the oxidation state on iron and various other metals, as well as the existence of sulphides, phosphates, or as other microbial compounds, can be ascertained by X-ray photoelectron spectroscopy (XPS). Energy-dispersive X-ray spectroscopy, commonly integrated with scanning electron microscopy, is an elemental mapping of the corroded steel surface to indicate the presence of corrosive elements and microbial deposits on the surface (Guo et al., 2017). These methods give a feeling of microstructural and chemical alterations of MIC. These methods have been in a position to describe how biofilms facilitate localized corrosion, ion concentration, and alterations of the metal surface with time. As part of an association with the measurement of electrochemical, microscopic and spectroscopic tests give an idea of a mechanistic interaction between microbial communities and steel to the formation of efficient corrosion prevention methods. These methods are the ones that enable detection of most primary-stage characteristics of corrosion otherwise invisible to the naked eye and therefore represent them onto proactive monitoring-upkeep terms as stated by Deng et al. (2020). The importance of high-resolution imaging and spectroscopy has emerged as a tool of analysis in different laboratory and field tests that are employed to assess the efficacy of corrosion inhibitors and protective coating on keeping down microbial activities. The spatial distribution of microbes and corrosion products is significant knowledge in the explanation of the dynamics of MIC and optimization of mitigation measures (Okorie & Nwokorie, 2021).

10.3 Molecular and Genomic Tools

The molecular and genomics technology has totally transformed the detection and characterization mode of microbial communities that cause corrosion which are very accurate in terms of taxonomy and functional understanding. Complex biofilms can be analysed for the presence of sulfate-reducing bacteria, iron-oxidizing bacteria, and other taxa significant in microbiologically influenced corrosion (MIC) by 16S rRNA gene sequencing. Quantitative PCR facilitates the identification of microbial abundance and detection of a particular gene related to corrosion metabolisms e.g. dsrAB in SRB (Luef et al., 2013). Metagenomics and Metatranscriptomics enable functioning profiling in detail, characterizing metabolic pathways and enzyme systems, which evidence the hypothetical electron transfer mechanism to drive MIC. It also gives details of uncultivable or low-abundance taxa of microorganisms that otherwise would not be identified using standard culture-based methods. The causality between microorganisms and corrosion patterns can be proven in the association of the microbial community composition with the electrochemical and physical corrosion information (Barrera- Diaz et al., 2012). Furthermore, biofilm dynamics with regard to temporal change can be studied using molecular methods and consequently give valuable data on the succession of microbes and their adaptation to environmental perturbations such as changes in nutrient levels, redox potential, or oxygen levels. The use of sophisticated bioinformatic analyses enables us to forecast the interactions between microbes and enable functional possibilities of specific mitigation efforts. The integration of molecular, electrochemical, and microscopic techniques with multi-dimensionality facilitates MIC, hence allowing for the accurate detection, monitoring, and regulation of the microbial corrosion within aquatic and terrestrial ecosystems (Cattant et al., 2008).

REFERENCES

Armelin, E., Alemán, C., & Iribarren, J. I. (2009). Anticorrosion performances of epoxy coatings modified with polyaniline: A comparison between the emeraldine base and salt forms. *Progress in Organic Coatings*, *65*(1), 88–93. DOI: 10.1016/j.porgcoat.2008.10.001

Barrera-Díaz, C. E., Lugo-Lugo, V., & Bilyeu, B. (2012). A review of chemical, electrochemical and biological methods for aqueous Cr (VI) reduction. *Journal of Hazardous Materials*, *223*, 1–12. DOI: 10.1016/j.jhazmat.2012.04.054 PMID: 22608208

Bermont-Bouis, D., Janvier, M., Grimont, P. A. D., Dupont, I., & Vallaeys, T. (2007). Both sulfate-reducing bacteria and Enterobacteriaceae take part in marine biocorrosion of carbon steel. *The Journal of Applied Bacteriology*, *102*(1), 161–168. DOI: 10.1111/j.1365-2672.2006.03053.x PMID: 17184331

Cattant, F., Crusset, D., & Féron, D. (2008). Corrosion issues in nuclear industry today. *Materials Today*, *11*(10), 32–37. DOI: 10.1016/S1369-7021(08)70205-0

Chaudhary, S., Tanvi, R. D., & Goyal, S. (2019). Different applications of sulphur oxidizing bacteria: A review. *International Journal of Current Microbiology and Applied Sciences*, *8*(11), 770–778. DOI: 10.20546/ijcmas.2019.811.091

Chen, S., Wang, P., & Zhang, D. (2014). Corrosion behavior of copper under biofilm of sulfate-reducing bacteria. *Corrosion Science*, *87*, 407–415. DOI: 10.1016/j.corsci.2014.07.001

Chugh, B., Sheetal, , Singh, M., Thakur, S., Pani, B., Singh, A. K., & Saji, V. S. (2022). Extracellular electron transfer by Pseudomonas aeruginosa in biocorrosion: A review. *ACS Biomaterials Science & Engineering*, *8*(3), 1049–1059. DOI: 10.1021/acsbiomaterials.1c01645 PMID: 35199512

Chung, H. C., Lee, O. O., Huang, Y. L., Mok, S. Y., Kolter, R., & Qian, P. Y. (2010). Bacterial community succession and chemical profiles of subtidal biofilms in relation to larval settlement of the polychaete *Hydroides elegans. The ISME Journal*, *4*(6), 817–828. DOI: 10.1038/ismej.2009.157 PMID: 20090788

Cui, X.-J., Lin, X.-Z., Liu, C.-H., Yang, R.-S., Zheng, X.-W., & Gong, M. (2015). Fabrication and corrosion resistance of a hydrophobic micro-arc oxidation coating on AZ31 Mg alloy. *Corrosion Science*, *90*, 402–412. DOI: 10.1016/j.corsci.2014.10.041

Dagbert, C., Meylheuc, T., & Bellon-Fontaine, M.-N. (2006). Corrosion behaviour of AISI 304 stainless steel in presence of a biosurfactant produced by *Pseudomonas fluorescens*. *Electrochimica Acta*, *51*(24), 5221–5227. DOI: 10.1016/j.electacta.2006.03.063

Dec, W., Mosiałek, M., Socha, R. P., Jaworska-Kik, M., Simka, W., & Michalska, J. (2017). Characterization of *Desulfovibrio desulfuricans* biofilm on high-alloyed stainless steel: XPS and electrochemical studies. *Materials Chemistry and Physics*, *195*, 28–39. DOI: 10.1016/j.matchemphys.2017.04.011

Deng, X., Dohmae, N., Kaksonen, A. H., & Okamoto, A. (2020). Biogenic iron sulfide nanoparticles to enable extracellular electron uptake in sulfate-reducing bacteria. *Angewandte Chemie*, *132*(15), 6051–6055. DOI: 10.1002/ange.201915196 PMID: 31875491

Ding, W., Zhang, W., Alikunhi, N. M., Batang, Z., Pei, B., Wang, R., Chen, L., Al-Suwailem, A., & Qian, P. Y. (2019). Metagenomic analysis of zinc surface–associated marine biofilms. *Microbial Ecology*, *77*(2), 406–416. DOI: 10.1007/s00248-018-01313-3 PMID: 30612183

Dong, Y., He, Z., & Guo, N. (2015). The formation process of biofilm of marine microorganisms and the influence on the corrosion of ship structural steel. *Journal of Marine Science*, *33*(1), 39–44. DOI: 10.3969/j.issn.1001-909X.2015.01.006

Duan, J., Wu, S., Zhang, X., Huang, G., Du, M., & Hou, B. (2008). Corrosion of carbon steel influenced by anaerobic biofilm in natural seawater. *Electrochimica Acta*, *54*(1), 22–28. DOI: 10.1016/j.electacta.2008.04.085

Edyvean, R. G., & Videla, H. A. (1991). Biological corrosion. *Interdisciplinary Science Reviews*, *16*(3), 267–282. DOI: 10.1179/isr.1991.16.3.267

Enning, D., Venzlaff, H., Garrelfs, J., Dinh, H. T., Meyer, V., Mayrhofer, K., Hassel, A. W., Stratmann, M., & Widdel, F. (2012). Marine sulfate-reducing bacteria cause serious corrosion of iron under electroconductive biogenic mineral crust. *Environmental Microbiology*, *14*(7), 1772–1787. DOI: 10.1111/j.1462-2920.2012.02778.x PMID: 22616633

Figueira, R. B. (2020). Hybrid sol–gel coatings for corrosion mitigation: A critical review. *Polymers*, *12*(3), 689. DOI: 10.3390/polym12030689 PMID: 32204462

Fischer, M., Friedrichs, G., & Lachnit, T. (2014). Fluorescence-based quasicontinuous and in situ monitoring of biofilm formation dynamics in natural marine environments. *Applied and Environmental Microbiology*, *80*(12), 3721–3728. DOI: 10.1128/AEM.00298-14 PMID: 24727266

Galedari, S. A., Mahdavi, A., Azarmi, F., Huang, Y., & McDonald, A. (2019). A comprehensive review of corrosion resistance of thermally-sprayed and thermally-diffused protective coatings on steel structures. *Journal of Thermal Spray Technology*, *28*(4), 645–677. DOI: 10.1007/s11666-019-00855-3

Ghafari, M. D., Bahrami, A., Rasooli, I., Arabian, D., & Ghafari, F. (2013). Bacterial exopolymeric inhibition of carbon steel corrosion. *International Biodeterioration & Biodegradation*, *80*, 29–33. DOI: 10.1016/j.ibiod.2013.02.007

Googan, C. (2022). *Marine Corrosion and Cathodic Protection* (1st ed.). CRC Press., DOI: 10.1201/9781003216070

Gu, T., Jia, R., Unsal, T., & Xu, D. (2019). Toward a better understanding of microbiologically influenced corrosion caused by sulfate reducing bacteria. *Journal of Materials Science and Technology*, *35*(4), 631–636. DOI: 10.1016/j.jmst.2018.10.026

Gu, T., Wang, D., Lekbach, Y., & Xu, D. (2021). Extracellular electron transfer in microbial biocorrosion. *Current Opinion in Electrochemistry*, *29*, 100763. DOI: 10.1016/j.coelec.2021.100763

Guo, Z., Liu, T., Cheng, Y. F., Guo, N., & Yin, Y. (2017). Adhesion of *Bacillus subtilis* and *Pseudoalteromonas lipolytica* to steel in a seawater environment and their effects on corrosion. *Colloids and Surfaces. B, Biointerfaces*, *157*, 157–165. DOI: 10.1016/j.colsurfb.2017.05.045 PMID: 28586728

He, J., Tan, Y., Liu, H., Jin, Z., Zhang, Y., He, F., Yan, Z., Liu, H., Meng, G., & Liu, H. (2022). Extracellular polymeric substances secreted by marine fungus *Aspergillus terreus*: Full characterization and detailed effects on aluminum alloy corrosion. *Corrosion Science*, *209*, 110703. DOI: 10.1016/j.corsci.2022.110703

Hornibrook, E. R., Longstaffe, F. J., & Fyfe, W. S. (2000). Evolution of stable carbon isotope compositions for methane and carbon dioxide in freshwater wetlands and other anaerobic environments. *Geochimica et Cosmochimica Acta*, *64*(6), 1013–1027. DOI: 10.1016/S0016-7037(99)00321-X

Hou, B., Li, X., Ma, X., Du, C., Zhang, D., Zheng, M., & Ma, F. (2017). The cost of corrosion in China. *NPJ Materials Degradation*, *1*(1), 1–10. DOI: 10.1038/s41529-017-0005-2

Huang, Y., Liu, S., & Jing, C. (2017). Microbiologically influenced corrosion and mechanisms. *Wei Sheng Wu Xue Tong Bao*, *44*, 1699–1713.

Huang, Y., Zhou, E., Jiang, C., Jia, R., Liu, S., Xu, D., Gu, T., & Wang, F. (2018). Endogenous phenazine-1-carboxamide encoding gene *PhzH* regulated the extracellular electron transfer in biocorrosion of stainless steel by marine *Pseudomonas aeruginosa*. *Electrochemistry Communications*, *94*, 9–13. DOI: 10.1016/j.elecom.2018.07.019

Javed, M. A., Neil, W. C., McAdam, G., & Wade, S. A. (2017). Effect of sulphate reducing bacteria on the microbiologically influenced corrosion of ten different metals using constant test conditions. *International Biodeterioration & Biodegradation*, *125*, 73–85. DOI: 10.1016/j.ibiod.2017.08.011

Javed, M. A., Stoddart, P. R., & Wade, S. A. (2015). Corrosion of carbon steel by sulphate reducing bacteria: Initial attachment and the role of ferrous ions. *Corrosion Science*, *93*, 48–57. DOI: 10.1016/j.corsci.2015.01.006

Kaur, S., Gallei, M., & Ionescu, E. (2015). Polymer–ceramic nanohybrid materials. In *Organic–Inorganic Hybrid Nanomaterials* (pp. 143–185). doi:DOI: 10.1016/B978-0-444-63317-8.00006-3

Khan, M. S., Liang, T., Liu, Y., Shi, Y., Zhang, H., Li, H., Guo, S., Pan, H., Yang, K., & Zhao, Y. (2022). Microbiologically influenced corrosion mechanism of ferrous alloys in marine environment. *Metals*, *12*(9), 1458. DOI: 10.3390/met12091458

Kip, N., & van Veen, J. A. (2014). The dual role of microbes in corrosion. *The ISME Journal*, *9*(3), 542–551. DOI: 10.1038/ismej.2014.169 PMID: 25259571

Kodama, Y., & Watanabe, K. (2004). *Sulfuricurvum kujiense* gen. nov., sp. nov., a facultatively anaerobic, chemolithoautotrophic, sulfur-oxidizing bacterium isolated from an underground crude-oil storage cavity. *International Journal of Systematic and Evolutionary Microbiology*, *54*(6), 2297–2300. DOI: 10.1099/ijs.0.63243-0 PMID: 15545474

Kugel, A., Stafslien, S., & Chisholm, B. J. (2011). Antimicrobial coatings produced by "tethering" biocides to the coating matrix: A comprehensive review. *Progress in Organic Coatings*, *72*(3), 222–252. DOI: 10.1016/j.porgcoat.2011.07.004

Lanneluc, I., Langumier, M., Sabot, R., Jeannin, M., Refait, P., & Sablé, S. (2015). On the bacterial communities associated with the corrosion product layer during the early stages of marine corrosion of carbon steel. *International Biodeterioration & Biodegradation*, *99*, 55–65. DOI: 10.1016/j.ibiod.2015.01.003

Li, M., Nan, L., Liang, C., Sun, Z., Yang, L., & Yang, K. (2021). Antibacterial behavior and related mechanisms of martensitic Cu-bearing stainless steel evaluated by a mixed infection model of *Escherichia coli* and *Staphylococcus aureus* in vitro. *Journal of Materials Science and Technology*, *62*, 139–147. DOI: 10.1016/j.jmst.2020.05.030

Li, X., Duan, J., Xiao, H., Li, Y., Liu, H., Guan, F., & Zhai, X. (2017). Analysis of bacterial community composition of corroded steel immersed in Sanya and Xiamen seawaters in China via method of Illumina MiSeq sequencing. *Frontiers in Microbiology*, *8*, 1737. Advance online publication. DOI: 10.3389/fmicb.2017.01737 PMID: 28955315

Li, X., Zhang, D., Liu, Z., Li, Z., Du, C., & Dong, C. (2015). Materials science: Share corrosion data. *Nature*, *527*(7579), 441–442. DOI: 10.1038/527441a PMID: 26607528

Li, Y., Xu, D., Chen, C., Li, X., Jia, R., Zhang, D. W., Wolfgang, S., Wang, F. H., & Gu, T. (2018). Anaerobic microbiologically influenced corrosion mechanisms interpreted using bioenergetics and bioelectrochemistry: A review. *Journal of Materials Science and Technology*, *34*(10), 1713–1718. DOI: 10.1016/j.jmst.2018.02.023

Li, X., Duan, J., Xiao, H., Li, Y., Liu, H., Guan, F. & Zhai, X. (2017).

Liamleam, W., & Annachhatre, A. P. (2007). Electron donors for biological sulfate reduction. *Biotechnology Advances*, *25*(5), 452–463. DOI: 10.1016/j.biotechadv.2007.05.002 PMID: 17572039

Little, B., Wagner, P., & Mansfeld, F. (1991). Microbiologically influenced corrosion of metals and alloys. *International Materials Reviews*, *36*(1), 253–272. DOI: 10.1179/imr.1991.36.1.253

Little, B. J., Blackwood, D. J., Hinks, J., Lauro, F. M., Marsili, E., Okamoto, A., Rice, S. A., Wade, S. A., & Flemming, H.-C. (2020). Microbially influenced corrosion—Any progress? *Corrosion Science*, *170*, 108641. DOI: 10.1016/j.corsci.2020.108641

Liu, H., Gu, T., Asif, M., Zhang, G., & Liu, H. (2017). The corrosion behavior and mechanism of carbon steel induced by extracellular polymeric substances of iron-oxidizing bacteria. *Corrosion Science*, *114*, 102–111. DOI: 10.1016/j.corsci.2016.10.025

Liu, H., Zhang, W., Zhao, Z., Lai, H., & Liu, H. (2023). Broken passive film and subsequent pitting corrosion behavior of 2205 duplex stainless steel induced by marine fungus *Aspergillus terreus* in artificial seawater. *Corrosion Science*, *218*, 111147. DOI: 10.1016/j.corsci.2023.111147

Liu, L., Wu, X., Wang, Q., Yan, Z., Wen, X., Tang, J., & Li, X. (2022). An overview of microbiologically influenced corrosion: Mechanisms and its control by microbes. *Corrosion Reviews*, *40*(2), 103–117. DOI: 10.1515/corrrev-2021-0039

Liu, T., Guo, Z., Zeng, Z., Guo, N., Lei, Y., Liu, T., & Wang, X. (2018). Marine bacteria provide lasting anti-corrosion activity for steel via biofilm-induced mineralization. *ACS Applied Materials & Interfaces, 10,* 14991. DOI: 10.1021/acsami.8b03317

Loto, C. A. (2017). Microbiological corrosion: Mechanism, control and impact—A review. *International Journal of Advanced Manufacturing Technology*, *92*(9–12), 4241–4252. DOI: 10.1007/s00170-017-0494-8

Lovley, D. R., Holmes, D. E., & Nevin, K. P. (2004). Dissimilatory Fe(III) and Mn(IV) reduction. *Advances in Microbial Physiology*, *49*, 219–286. DOI: 10.1016/S0065-2911(04)49005-5 PMID: 15518832

Luef, B., Fakra, S. C., Csencsits, R., Wrighton, K. C., Williams, K. H., Wilkins, M. J., Downing, K. H., Long, P. E., Comolli, L. R., & Banfield, J. F. (2013). Iron-reducing bacteria accumulate ferric oxyhydroxide nanoparticle aggregates that may support planktonic growth. *The ISME Journal*, *7*(2), 338–350. DOI: 10.1038/ismej.2012.103 PMID: 23038172

Luo, J., Mamat, B., Yue, Z., Zhang, N., Xu, X., Li, Y., Su, Z., Ma, C., Zang, F., & Wang, Y. (2021). Multi-metal ions doped hydroxyapatite coatings via electrochemical methods for antibacterial and osteogenesis. *Colloid and Interface Science Communications*, *43*, 100435. DOI: 10.1016/j.colcom.2021.100435

Ma, Y., Zhang, Y., Zhang, R., Guan, F., Hou, B., & Duan, J. (2020). Microbiologically influenced corrosion of marine steels within the interaction between steel and biofilms: A brief view. *Applied Microbiology and Biotechnology*, *104*(2), 515–525. DOI: 10.1007/s00253-019-10184-8 PMID: 31807887

McBeth, J. M., & Emerson, D. (2016). In situ microbial community succession on mild steel in estuarine and marine environments: Exploring the role of iron-oxidizing bacteria. *Frontiers in Microbiology*, *7*, 767. DOI: 10.3389/fmicb.2016.00767 PMID: 27252686

Melchers, R. E. (2005). The effect of corrosion on the structural reliability of steel offshore structures. *Corrosion Science*, *47*(10), 2391–2410. DOI: 10.1016/j.corsci.2005.04.004

Melchers, R. E., & Jeffrey, R. (2012). Corrosion of long vertical steel strips in the marine tidal zone and implications for ALWC. *Corrosion Science*, *65*, 26–36. DOI: 10.1016/j.corsci.2012.07.025

Moradi, M., Song, Z., & Tao, X. (2015). Introducing a novel bacterium, *Vibrio neocaledonicus* sp., with the highest corrosion inhibition efficiency. *Electrochemistry Communications*, *51*, 64–68. DOI: 10.1016/j.elecom.2014.12.007

Moradi, M., Song, Z., Yang, L., Jiang, J., & He, J. (2014). Effect of marine *Pseudoalteromonas* sp. on the microstructure and corrosion behaviour of 2205 duplex stainless steel. *Corrosion Science*, *84*, 103–112. DOI: 10.1016/j.corsci.2014.03.018

Murcar-Evans, B. I., Cabral, A. D., Toutah, K., de Araujo, E. D., Lai, A., Macdonald, P. M., Berger-Becvar, A., Kraskouskaya, D., & Gunning, P. T. (2017). ProxyPhos sensors for the detection of negatively charged membranes. *Analyst*, *142*(23), 4511–4521. DOI: 10.1039/C7AN00568G PMID: 29098228

Okabe, S., Odagiri, M., Ito, T., & Satoh, H. (2007). Succession of sulfur-oxidizing bacteria in the microbial community on corroding concrete in sewer systems. *Applied and Environmental Microbiology*, *73*(3), 971–980. DOI: 10.1128/AEM.02054-06 PMID: 17142362

Okorie, I. E., & Nwokorie, R. C. (2021). A review of fungal influenced corrosion of metals. *Zaštita Materijala*, *62*(4), 333–339. DOI: 10.5937/zasmat2104333O

Pimenov, N. V., & Neretin, L. N. (2006). Composition and activities of microbial communities involved in carbon, sulfur, nitrogen and manganese cycling in the oxic/anoxic interface of the Black Sea. In *Past and Present Water Column Anoxia* (pp. 501–521). Springer., DOI: 10.1007/1-4020-4297-3_19

Pistone, A., Scolaro, C., & Visco, A. (2021). Mechanical properties of protective coatings against marine fouling: A review. *Polymers*, *13*(2), 173. DOI: 10.3390/polym13020173 PMID: 33418953

Popescu, C., Alain, S., Courant, M., Vardelle, A., Denoirjean, A., & Cavarroc, M. (2022). Thermal spray copper-based coatings against contamination of thermoplastic surfaces: A systematic review. *Engineering Science and Technology, an International Journal*, *101194*, 101194. Advance online publication. DOI: 10.1016/j.jestch.2022.101194

Purwasena, I. A., Astuti, D. I., Fauziyyah, N. A., Putri, D. A. S., & Sugai, Y. (2019). Inhibition of microbial influenced corrosion on carbon steel ST37 using biosurfactant produced by *Bacillus* sp. *Materials Research Express*, *6*(11), 115405. DOI: 10.1088/2053-1591/ab4948

Qian, P.-Y., Cheng, A., Wang, R., & Zhang, R. (2022). Marine biofilms: Diversity, interactions and biofouling. *Nature Reviews. Microbiology*, *20*(11), 671–684. DOI: 10.1038/s41579-022-00744-7 PMID: 35614346

Rawlings, D. E. (2005). Characteristics and adaptability of iron- and sulfur-oxidizing microorganisms used for the recovery of metals from minerals and their concentrates. *Microbial Cell Factories*, *4*(1), 13. DOI: 10.1186/1475-2859-4-13 PMID: 15877814

Seyeux, A., Zanna, S., Allion, A., & Marcus, P. (2015). The fate of the protective oxide film on stainless steel upon early stage growth of a biofilm. *Corrosion Science*, *91*, 352–356. DOI: 10.1016/j.corsci.2014.10.051

Sheng, G. P., Xu, J., Luo, H. W., Li, W. W., Li, W. H., Yu, H. Q., & Hu, F. C. (2013). Thermodynamic analysis on the binding of heavy metals onto extracellular polymeric substances (EPS) of activated sludge. *Water Research*, *47*(2), 607–614. DOI: 10.1016/j.watres.2012.10.037 PMID: 23159005

Sherar, B. W. A., Power, I. M., Keech, P. G., Mitlin, S., Southam, G., & Shoesmith, D. W. (2011). Characterizing the effect of carbon steel exposure in sulfide-containing solutions to microbially induced corrosion. *Corrosion Science*, *53*(3), 955–960. DOI: 10.1016/j.corsci.2010.11.027

Shi, X., Yan, W., Xu, D., Yan, M., Yang, C., Shan, Y., & Yang, K. (2018). Microbial corrosion resistance of a novel Cu-bearing pipeline steel. *Journal of Materials Science and Technology*, *34*(12), 2480–2491. DOI: 10.1016/j.jmst.2018.05.020

Silber, J., Kramer, A., Labes, A., & Tasdemir, D. (2016). From discovery to production: Biotechnology of marine fungi for the production of new antibiotics. *Marine Drugs*, *14*(7), 137. DOI: 10.3390/md14070137 PMID: 27455283

Simões, M., Simões, L. C., & Vieira, M. J. (2010). A review of current and emergent biofilm control strategies. *Lebensmittel-Wissenschaft + Technologie*, *43*(4), 573–583. DOI: 10.1016/j.lwt.2009.12.008

Snider, R. M., Strycharz-Glaven, S. M., Tsoi, S. D., Erickson, J. S., & Tender, L. M. (2012). Long-range electron transport in *Geobacter sulfurreducens* biofilms is redox gradient-driven. *Proceedings of the National Academy of Sciences of the United States of America*, *109*(38), 15467–15472. DOI: 10.1073/pnas.1209829109 PMID: 22955881

Starosvetsky, D., Armon, R., Yahalom, J., & Starosvetsky, J. (2001). Pitting corrosion of carbon steel caused by iron bacteria. *International Biodeterioration & Biodegradation*, *47*(2), 79–87. DOI: 10.1016/S0964-8305(99)00081-5

Steidl, R. J., Lampa-Pastirk, S., & Reguera, G. (2016). Mechanistic stratification in electroactive biofilms of *Geobacter sulfurreducens* mediated by pilus nanowires. *Nature Communications*, *7*(1), 12217. DOI: 10.1038/ncomms12217 PMID: 27481214

Sun, D., Xu, D., Yang, C., Shahzad, M. B., Sun, Z., Xia, J., Zhao, J., Gu, T., Yang, K., & Wang, G. (2016). An investigation of the antibacterial ability and cytotoxicity of a novel Cu-bearing 317L stainless steel. *Scientific Reports*, *6*(1), 29244. DOI: 10.1038/srep29244 PMID: 27385507

Tang, H.-Y., Yang, C., Ueki, T., Pittman, C. C., Xu, D., Woodard, T. L., Holmes, D. E., Gu, T., Wang, F., & Lovley, D. R. (2021). Stainless steel corrosion via direct iron-to-microbe electron transfer by Geobacter species. *The ISME Journal, 15*(10), 3084–3093. DOI: 10.1038/s41396-021-00990-2 PMID: 33972726

Usher, K. M., Kaksonen, A. H., & MacLeod, I. D. (2014). Marine rust tubercles harbour iron corroding archaea and sulphate reducing bacteria. *Corrosion Science, 83*, 189–197. DOI: 10.1016/j.corsci.2014.02.014

Venzlaff, H., Enning, D., Srinivasan, J., Mayrhofer, K. J. J., Hassel, A. W., Widdel, F., & Stratmann, M. (2013). Accelerated cathodic reaction in microbial corrosion of iron due to direct electron uptake by sulfate-reducing bacteria. *Corrosion Science, 66*, 88–96. DOI: 10.1016/j.corsci.2012.09.006

Vigneron, A., Alsop, E. B., Chambers, B., Lomans, B. P., Head, I. M., & Tsesmetzis, N. (2016). Complementary microorganisms in highly corrosive biofilms from an offshore oil production facility. *Applied and Environmental Microbiology, 82*(8), 2545–2554. DOI: 10.1128/AEM.03842-15 PMID: 26896143

Wang, F., Gu, Y., O'Brien, J. P., Yi, S. M., Yalcin, S. E., Srikanth, V., Shen, C., Vu, D., Ing, N. L., Hochbaum, A. I., Egelman, E. H., & Malvankar, N. S. (2019). Structure of microbial nanowires reveals stacked hemes that transport electrons over micrometers. *Cell, 177*(2), 361–369.e10. DOI: 10.1016/j.cell.2019.03.029 PMID: 30951668

Wang, J., Du, M., Li, G., & Shi, P. (2022). Research progress on microbiological inhibition of corrosion: A review. *Journal of Cleaner Production, 373*, 133658. DOI: 10.1016/j.jclepro.2022.133658

Wu, J., Zhang, D., Wang, P., Cheng, Y., Sun, S., Sun, Y., & Chen, S. (2016). The influence of Desulfovibrio sp. and Pseudoalteromonas sp. on the corrosion of Q235 carbon steel in natural seawater. *Corrosion Science, 112*, 552–562. DOI: 10.1016/j.corsci.2016.04.047

Xu, D., & Gu, T. (2014). Carbon source starvation triggered more aggressive corrosion against carbon steel by the Desulfovibrio vulgaris biofilm. *International Biodeterioration & Biodegradation, 91*, 74–81. DOI: 10.1016/j.ibiod.2014.03.014

Xu, D., Gu, T., & Lovley, D. R. (2023). Microbially mediated metal corrosion. *Nature Reviews. Microbiology, 21*(11), 1–14. DOI: 10.1038/s41579-023-00920-3 PMID: 37344552

Xu, D., Li, Y., Song, F. & Gu, T. (2013). Laboratory investigation of microbiologically influenced corrosion of C1018 carbon steel by nitrate-reducing bacterium Bacillus licheniformis. *Corrosion Science*, *77*, 385–390. DOI: https://doi.org/DOI: 10.1016/j.corsci.2013.08.007

Yu, L., Duan, J., Du, X., Huang, Y., & Hou, B. (2013). Accelerated anaerobic corrosion of electroactive sulfate-reducing bacteria by electrochemical impedance spectroscopy and chronoamperometry. *Electrochemistry Communications*, *26*, 101–104. DOI: 10.1016/j.elecom.2012.10.022

Yuan, S., Liang, B., Zhao, Y., & Pehkonen, S. O. (2013). Surface chemistry and corrosion behaviour of 304 stainless steel in simulated seawater containing inorganic sulphide and sulphate-reducing bacteria. *Corrosion Science*, *75*, 171–181. DOI: 10.1016/j.corsci.2013.04.058

Zhang, P., Xu, D., Li, Y., Yang, K., & Gu, T. (2015). Electron mediators accelerate the microbiologically influenced corrosion of 304 stainless steel by the Desulfovibrio vulgaris biofilm. *Bioelectrochemistry (Amsterdam, Netherlands)*, *101*, 14–21. DOI: 10.1016/j.bioelechem.2014.06.010 PMID: 25023048

Zhang, W., Ding, W., Li, Y., Tam, C., Bougouffa, S., Wang, R., Pei, B., Chiang, H., Leung, P., Lu, Y., Sun, J., Fu, H., Bajic, V. B., Liu, H., Webster, N. S., & Qian, P. Y. (2019). Marine biofilms constitute a bank of hidden microbial diversity and functional potential. *Nature Communications*, *10*(1), 84–463. DOI: 10.1038/s41467-019-08463-z PMID: 30705275

Zhang, Y., Ma, Y., Duan, J., Li, X., Wang, J., & Hou, B. (2019). Analysis of marine microbial communities colonizing various metallic materials and rust layers. *Biofouling*, *35*(4), 429–442. DOI: 10.1080/08927014.2019.1610881 PMID: 31109195

Zhang, Y., Zhai, X., Guan, F., Dong, X., Sun, J., Zhang, R., Duan, J., Zhang, B., & Hou, B. (2022). Microbiologically influenced corrosion of steel in coastal surface seawater contaminated by crude oil. *NPJ Materials Degradation*, *6*(1), 35. DOI: 10.1038/s41529-022-00242-4

Zhang, Y. M., Ma, Y., Duan, J. Z., Li, X. H., Wang, J., & Hou, B. R. (2019). Analysis of marine microbial communities colonizing various metallic materials and rust layers. *Biofouling*, *35*(4), 429–442. DOI: 10.1080/08927014.2019.1610881 PMID: 31109195

Zhou, E., Lekbach, Y., Gu, T., & Xu, D. (2022). Bioenergetics and extracellular electron transfer in microbial fuel cells and microbial corrosion. *Current Opinion in Electrochemistry*, *31*, 100830. DOI: 10.1016/j.coelec.2021.100830

Zhou, E., Zhang, M., Huang, Y., Li, H., Wang, J., Jiang, G., Jiang, C., Xu, D., Wang, Q., & Wang, F. (2022). Accelerated biocorrosion of stainless steel in marine water via extracellular electron transfer encoding gene phzH of Pseudomonas aeruginosa. *Water Research, 220*, 118634. DOI: 10.1016/j.watres.2022.118634 PMID: 35691192

Zhou, E.-Z., Ren, G.-Y., Sun, Y.-L., Fan, Y.-Q., Yang, Y., Sun, M.-Y., Xu, D.-K., Lu, Y.-P., Wang, J.-J., Li, H.-B., & Wang, F.-H. (2022). Antibacterial activities of a novel Cu-bearing high-entropy alloy against multi-drug-resistant Acinetobacter baumannii and Staphylococcus aureus. *Rare Metals, 41*(2), 570–579. DOI: 10.1007/s12598-021-01837-7

Zhou, M., Wang, H., Hassett, D. J., & Gu, T. (2013). Recent advances in microbial fuel cells (MFCs) and microbial electrolysis cells (MECs) for wastewater treatment, bioenergy and bioproducts. *Journal of Chemical Technology and Biotechnology, 88*(4), 508–518. DOI: 10.1002/jctb.4004

Zuo, R. (2007). Biofilms: Strategies for metal corrosion inhibition employing microorganisms. *Applied Microbiology and Biotechnology, 76*(6), 1245–1253. DOI: 10.1007/s00253-007-1130-6 PMID: 17701408

Chapter 6
Sustainable Nanobiotechnological Approaches for Corrosion Inhibition in Water Purification

Surya Pratap Singh
https://orcid.org/0000-0001-7366-560X
Srinath University, Jamshedpur, India

Diwesh Kumar
https://orcid.org/0000-0002-6321-5602
Srinath University, Jamshedpur, India

Sudarshana Banerjee
https://orcid.org/0000-0002-5045-2684
Srinath University, Jamshedpur, India

ABSTRACT

Trace amounts of heavy metals pose serious threats to aquatic environments and corrosion inhibition in the water due to traditional purification technologies. It impacts human health and requires modern, sustainable purification technologies. The chapter discusses nanobiotechnological water filtration systems as a revolutionary technique for the elimination of heavy metals and other toxins. The approach emphasizes creating eco-friendly, adsorptive, natural-source-oriented nanomaterials and their surface functionalization using biological components in order to create selectivity and efficiency. Merging nanotechnology with biotechnology produces a

DOI: 10.4018/979-8-3373-2617-7.ch006

synergistic system rooted in adsorption, catalysis, and bioremediation. They help in the prevention of corrosion by surface coating or adsorbing harmful ions. Case studies demonstrate the real-world uses, performance standards, and scalability of these systems in actual application. The present chapter addresses the ecological and economic benefits of such systems and emphasizes the decentralized water treatment potential.

1. INTRODUCTION

Safe and clean water is a birthright for every human being and an essential element of public health, food security, and sustainable economic development. Ever-increasing global water scarcity and pollution have been caused by environmental degradation, industrial effluent; agricultural runoff, population explosion, and climate change (Gupta et al., 2021). More than 2 billion people worldwide drink water contaminated with either fecal matter or toxic chemicals, resulting in the spread of waterborne diseases and causing health-related problems in the long run, stated WHO. The situation presents a global challenge emphasizing the urgent need for water treatment technologies that are effective, cost-efficient, and sustainable and can keep pace with a wide variety of fluctuating quality problems (Saravanan et al., 2021).

Originating from conventional water treatment technologies, coagulation, flocculation, sedimentation, filtration, and disinfection have been applied to standard municipal and industrial water treatment. However, while the above processes are generally very effective at the removal of suspended solids and microbial pathogens, they are found insufficient for removing persistent and emerging contaminants, e.g., heavy metals, endocrine-disrupting chemicals or EDCs, pharmaceuticals, microplastics, and PFAS (Kumar et al., 2025a). Nevertheless, the good operation of such systems is greatly prevented by physicochemical water source heterogeneity, process complexities, and slow build-up of secondary polluters or sludge through time.

The limitations of traditional treatment systems have, however, laid the foundation for the initiation and development of novel water purification technologies such as membrane filtration (Aziz et al., 2024), ion exchange (Alkhadra et al., 2022), AOPs (Guo et al., 2022), and electrochemical treatment (dos Santos et al., 2021). These systems can perform better in various scenarios and requires more energy large infrastructure, and proper setup. Hence there is a need for several bio-integrated methods which can be more energy efficient and ecofriendly (Bigham et al., 2025).

By maintaining the parity between two different domains such as nanotechnology and biology; nanobiotechnology plays a pivotal role in the sector of water treatment

This emerging sector is favorably applicable to water cycle management (Sahu et al., 2021).

The base of the nanobiotechnological system are nanoparticles, nanofibers, and nanocomposites with the biological phase, including enzymes, microbial biofilms, polysaccharides, and plant-derived materials. These systems have various physicochemical properties which augment their efficiency in water treatment processes (Sunita & Chakrabarty, 2022).

Nanobiotechnological methods are most suitable for removal of microplastics and hence it has fast recovery of microplastic particles. By incorporating novelties, these techniques can increase the recycling process and decrease waste generation (López-Maldonado et al., 2024).

Among the different available applications Nanobiotechnological applications are among the convincing application areas for the extraction of heavy metals from contaminated waters. Environmental pollutants such as lead (Pb), cadmium (Cd), arsenic (As), mercury (Hg), chromium (Cr), and copper (Cu) are highly toxic and persist for long periods (Jagaba et al., 2023; Jomova et al., 2025). These are emitted from different anthropogenic sources like mining, electroplating, leather tanning, battery manufacturing, and disposal of electronic waste. Heavy metals do not get degraded once released into the aquatic environment and have the potential to bioaccumulate and bio-magnify in the food chain (Ray & Vashishth, 2024). Prolonged exposure to heavy metals has been associated with serious health issues, such as neurotoxicity, renal failure, carcinogenicity, and developmental disorders (Haidar et al., 2023).

Traditional methods of heavy metal removal—chemical precipitation, adsorption onto activated carbon, ion exchange, and membrane separation—are generally constrained by low selectivity, partial removal, operation costs, and toxic sludge generation (Razzak et al., 2022). By contrast, nanobiotechnological systems provide highly selective, efficient, and environmentally friendly alternatives. For instance, functional nanomaterials can selectively metal chelation process from water, whereas agricultural or microbial origin ecosorbents offer inexpensive and renewable matrices for metal adsorption (Hussain et al., 2024). Enzymatic network systems can further catalyze redox conversions of heavy metals into less toxic or separable forms.

Among several available functional gains, one of the key features of nanobiotechnological filtering systems is their ability to offer a wide range of applications in decentralized water treatment arrangements, especially in infrastructure-deficient areas.

These systems require low energy, and their self-regenerative techniques allow them to be installed in rural areas and disaster-prone areas. By upgrading the functional technology, the contamination detection capacity of a water treatment system can be increased (Kishore et al., 2024).

In order to study the advanced material engineering, methods like synthetic biology and green synthesis are eco-friendly initiatives (Soni et al., 2021). On the other hand bio-nanocomposites, which consist of natural polymers such as cellulose, chitosan, and alginate, can find out several water quality parameters. Nano-biotechnology blended with IoT and machine learning algorithms can pave a smart and real-time monitoring and provide water treatment solutions (Sharma et al., 2024b).

There are limitations of the nanobiotechnological systems, like exaggeration of their potential, at times of exceeding important constraints and different socio-economic variables (Pokrajac et al., 2021). These technologies can reduce the cost, challenges and can work in aquatic environments. There is social adequacy and low value of the materials at the end of their life cycle.

To requires rigorous research and risk assessment among regulatory ambiguity, ecotoxicological impacts of engineered nanomaterials, scalability, and long-term performance stability.

Additionally, it requires more analysis with ecological risk, regulatory vigilance, and socio-economic viability for the modification of the system at the small scale (Kolossváry et al., 2023; Padilla-Rivera et al., 2023).

The present chapter overviews the available modern water purification technologies giving special emphasis on nanobiotechnological filtration systems for heavy metal ion removal. The present chapter also discuses about the working principles, design criteria, strengths, and weaknesses along with the several application areas of the nanobiotechnological systems. Different available case studies were showcased to present successful applications and nascent trends (You, 2024).

Treatment of water is a big issue and causes socio-environmental problems. The integration of nanobiotechnological systems is a promising good result in the treatment of water and fulfillment of the United Nations Sustainable Development Goal 6: Ensure the availability and sustainable management of water and sanitation for all (Sharma et al., 2024a). We can create more intelligent and cleaner solutions by applying the strength of biology and nanoscience which restore and maintain the integrity of our world's most essential resource—water (Carofiglio et al., 2022). The chapter project ideas about the prospects of nanobiotechnology with an evaluation of future directions of research, policy implications, helping cleaner ecosystems and sustainable water security.

2. AVAILABLE TRADITIONAL PURIFICATION TECHNOLOGIES

2.1 Physical and Chemical Processes

Sedimentation and filtration are basic physical processes in the treatment of water and wastewater. They are the mainstay of traditional treatment plants, efficiently removing suspended solids and lowering turbidity before more sophisticated purification processes (Fig. 1).

Figure 1. Conventional purification technologies

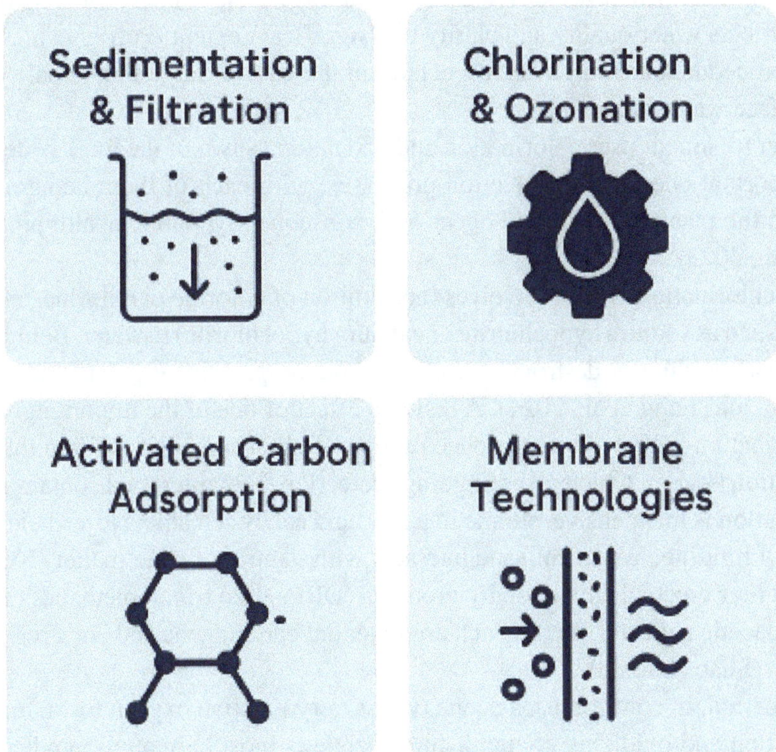

Sedimentation, or gravity settling, is the process of removing particulate matter from water by letting it settle undisturbed in a basin or clarifier (Garg, 2022). Sediment particles are heavier particles, like sand, silt, and flocculated organic or inorganic substances, which eventually settle at the bottom as a result of gravity. Sedimentation efficiency is affected by some of the factors, such as particle size and weight, temperature of water, velocity of flow, and the basin geometry of the sedimentation tank. In numerous systems, coagulants such as alum or ferric chloride are added before aggregating fine particles into larger flocs for increasing the settling process — a procedure popularly referred to as coagulation-flocculation (Pillai & Thombre, 2024).

After sedimentation, the partly filtered water is subject to filtration, where water is passed through permeable media and residual particulate is separated. Filtration media may include sand, gravel, anthracite, and now more advanced materials such as granular-activated carbon (GAC) (Singh et al., 2022; Xu et al., 2021). The operation may be divided into slow sand filtration, rapid gravity filtration, and pressure filtration, depending on use and flow rates.

Combining sedimentation and filtration not only reduces the physical strain on subsequent treatment processes (e.g., membrane systems, disinfection) but also positively affects water quality and clarity to a significant extent (Poirier et al., 2023). These procedures are still vital, especially in the case of large municipal systems and surface water treatment.

Water treatment uses chlorination and ozonation as two of the most widespread and important chemical disinfection processes, with each of them being a major factor in the inactivation of pathogens and, consequently, public health protection (He et al., 2023).

The chlorination process involves the addition of chlorine or chlorine-releasing agents (such as sodium hypochlorite or calcium hypochlorite) to water. Being a very strong oxidant, chlorine destroys or inactivates microorganisms, e.g., bacteria, viruses, and protozoa (Jiang et al., 2024). A residual effect of one of the important benefits of chlorination is that a minute concentration of chlorine will remain in the water distribution system, which gives ongoing protection from microbial contamination. Chlorination is inexpensive, simple to apply, and easily scalable. Nevertheless, it is also very limiting. When chlorine interacts with natural organic matter (NOM) in water, it may create disinfection by-products (DBPs) like trihalomethanes (THMs) and haloacetic acids (HAAs), which are potential carcinogens and are a regulatory problem (Shao et al., 2023).

Ozonation, in contrast, uses ozone (O_3), a very reactive oxygen form, for water disinfection and oxidizing contaminants. Ozone is most commonly produced on-site through the electrical discharge of oxygen or air. It quickly inactivates microorganisms by destroying their cell walls and disturbing nucleic acids. Ozonation works especially well against chlorine-resistant organisms like Cryptosporidium and Giardia, and it also aids in the degradation of micropollutants and taste- and odor-forming substances (Zhao et al., 2022). Ozone has no residual effect, is fairly unstable, and is more complicated to install concerning infrastructure and energy input than chlorine-based systems.

In recent water treatment, the cooperative or sequential application of ozonation and chlorination is becoming more common to achieve maximum disinfection with minimum harmful by-products (Li et al., 2024b).

Activated carbon is widely used in adsorption technology involved in various physicochemical processes of wastewater and water treatment, mainly for removing dissolved organic compounds, micropollutants, materials imparting taste or odor to water, and some heavy metals (Alver et al., 2022). Due to its extremely high surface area (between 500 and 1500 m^2/g), porous nature, and high extent of functional groups, activated carbon is among the most efficient adsorbents present.

Activated carbon is normally made by the activation and carbonization of high-carbon-content materials like peat, coconut shells, coal, or wood. Activation by physical (via steam or carbon dioxide at high temperature) or chemical (by using activating agents like phosphoric acid) creates a microporous matrix to improve adsorption capacity.

The two main types of activated carbon employed in water treatment are powdered activated carbon (PAC) and granular activated carbon (GAC) (Zhang et al., 2023). PAC is typically dosed as a slurry for short-contact-time adsorption, usually as a reaction to seasonal contamination events. GAC, on the other hand, is filled into fixed-bed filters and offers long-term adsorption with the possibility of regeneration and reuse.

The adsorption is through physical forces (Vander Waals forces) and, to a reduced degree, chemical connection. It is very well-organized in the removal of pesticides, pharmaceuticals, chlorinated solvents, and forerunners of disinfection by-products. Activated carbon can also adsorb assured inorganic pollutants, like mercury and lead, particularly when functionalized or saturated with other chemicals (Sharma et al., 2022).

Although very effective, activated carbon has to be substituted or redeveloped from time to time, and its usefulness is susceptible to the pH of water, temperature, and the availability of interfering solutes. However, it remains to be a pivotal part of refined water purification systems.

2.2 Membrane Technologies

Membrane separation technologies are currently an vital module of contemporary water and wastewater management recognitions to their capacity to undertake high contaminant removal with low chemical contributions and small system designs. Among all these, microfiltration (MF) and ultrafiltration (UF) are low-pressure membrane processes that are very efficient in the removal of suspended solids, bacteria, protozoa, and macromolecules and therefore play a critical role in both municipal and industrial use (Urošević & Trivunac, 2020).

Microfiltration (MF)

Microfiltration is a physical separation of water from impurities using semi-permeable membranes of pore sizes often between 0.1 and 1.0 micrometers (μm) (Sharma et al., 2023) (Fig. 2). The process is applied at comparatively low pressures, typically between 0.1 and 2 bar, and it essentially works by a sieving action, with the larger particles being blocked on the surface by the membrane pores.

MF is widely applied as a pretreatment process in drinking water treatment, particularly for surface water sources with a tendency towards high turbidity and biological contamination (Costa et al., 2024). It successfully removes suspended solids, bacteria (e.g., E. coli), Protozoan cysts (e.g., Giardia, Cryptosporidium), Algae and large organic particles.

However, because of its larger pore size, MF is ineffective against viruses, dissolved salts, and low-molecular-weight organics, and therefore, it is generally combined with other treatment processes, like disinfection or reverse osmosis, in order to attain complete purification.

MF membranes typically are made up of materials such as polypropylene (PP), polyvinylidene fluoride (PVDF), or cellulose acetate and may be found in configurations such as hollow fiber, flat sheet, and tubular modules (Moattari et al., 2021). MF systems are highly regarded for mechanical strength, back washability, and simplicity in being integrated in modular treatment facilities.

Ultrafiltration (UF)

Ultrafiltration works similarly to microfiltration, but with membranes having lower pore sizes, generally ranging from 0.01 to 0.1 μm, or having a molecular weight cut-off (MWCO) ranging from 1,000 to 100,000 Daltons. Due to this, UF is capable of rejecting a larger range of pollutants, including: Colloidal particles, Emulsified oils and fats, Proteins and polysaccharides, Viruses and bacteria.

UF membranes are usually run slightly above pressures (1–5 bar) compared to MF, and the filtration process employs both sieving and adsorption, depending on the characteristics of membrane surfaces and solute properties (Fig. 2). Because of its better separation efficiency, ultrafiltration is applied on a large scale in:

- Drinking water treatment for the removal of pathogens
- Reclamation and reuse of wastewater
- Industrial process water treatment, particularly in food, pharmaceutical, and biotechnology industries
- Reverse osmosis (RO) pretreatment, where UF considerably lowers membrane fouling and improves system lifespan

The materials for UF membranes are usually polysulfone (PS), polyethersulfone (PES), and ceramic materials, and their configurations are akin to MF (e.g., hollow fiber and spiral wound). Automated backwashing and chemical cleaning (CIP) are often found in UF systems to maintain flux rates and increase membrane life.

Figure 2. Microfiltration (MF) and Ultrafiltration (UF) Techniques

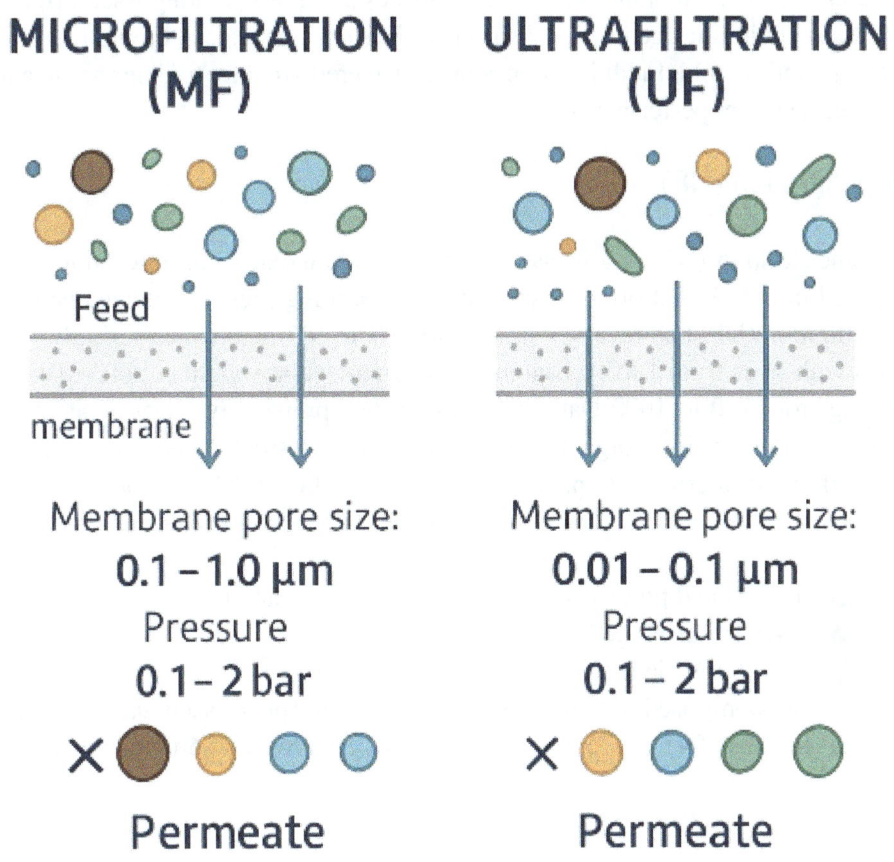

Operational Considerations

Both UF and MF systems need to be properly managed for operational parameters, including:

- Transmembrane pressure (TMP)
- Crossflow velocity to minimize fouling
- Cleaning cycles and chemical dosing to recover membrane permeability
- Pretreatment (e.g., screening or coagulation) to avoid clogging and prolong membrane life

Membrane fouling due to the deposition of organic matter, biofilms, or scaling minerals on the membrane surface continues as a major operating issue (Tong et al., 2023). Inhibiting agents and advanced surface designs (e.g., hydrophilic surface coatings, antimicrobial additives) are being considered to reduce fouling and enhance long-term system performance.

Nanofiltration (NF)

Nanofiltration (NF) is a pressure-driven membrane separation technology, as depicted in Fig. 3, that offers selectivity and operating pressure situated between ultrafiltration (UF) and reverse osmosis (RO). The NF membranes generally have pores with sizes from 1 to 10 nanometers and molecular weight cut-off (MWCO) ranging from 200 to 1000 Daltons such that they preferively reject divalent and higher monovalent ions, organic compounds, and very small pathogens.

Working at intermediate pressures (around 4–30 bar), NF membranes permit permeation of water and several monovalent ions (e.g., sodium, chloride), rejecting:

- Divalent and polyvalent salts (e.g., Ca^{2+}, Mg^{2+}, and Fe^{3+})
- Heavy metals (e.g., Pb, As)
- Natural organic matter
- Commonly used contaminants are pesticides and pharmaceutical compounds
- Hardness-forming ions, which make NF effective for softening operations

Figure 3. Process of nanofiltration membrane

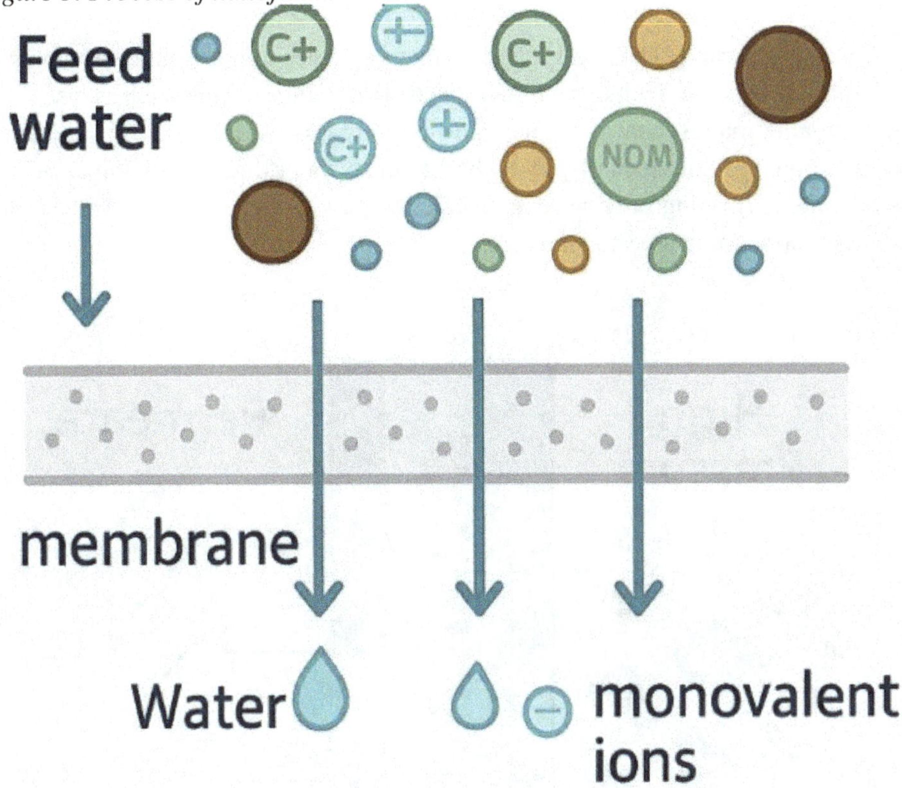

One of the key advantages of NF is its low energy consumption when compared to reverse osmosis (RO) because of its lower pressure requirement while maintaining high removal rates of a wide range of contaminants. It is most useful in cases like:

- Treatment of drinking water
- Reuse and recovery of water in industrial operations
- Treatment of wastewater from textile, dairy and dye

NF membranes are typically fabricated from thin-film composites (TFC) and may be either spiral-wound or hollow-fiber modules. As with other membrane technologies, however, NF systems are vulnerable to fouling and therefore require pretreatment and periodic maintenance in order to sustain performance and membrane life.

Reverse Osmosis (RO)

Reverse Osmosis (RO) is a very efficient membrane separation technique employed to desalinate water and remove a broad spectrum of contaminants such as dissolved salts, organic matter, heavy metals, and microorganisms. It works by subjecting water to high pressure (usually 30–80 bar) to push water through a semipermeable membrane, permitting only water molecules to pass and excluding dissolved ions and other contaminants, shown in Fig. 4.

Figure 4. Reverse Osmosis Treatment Process

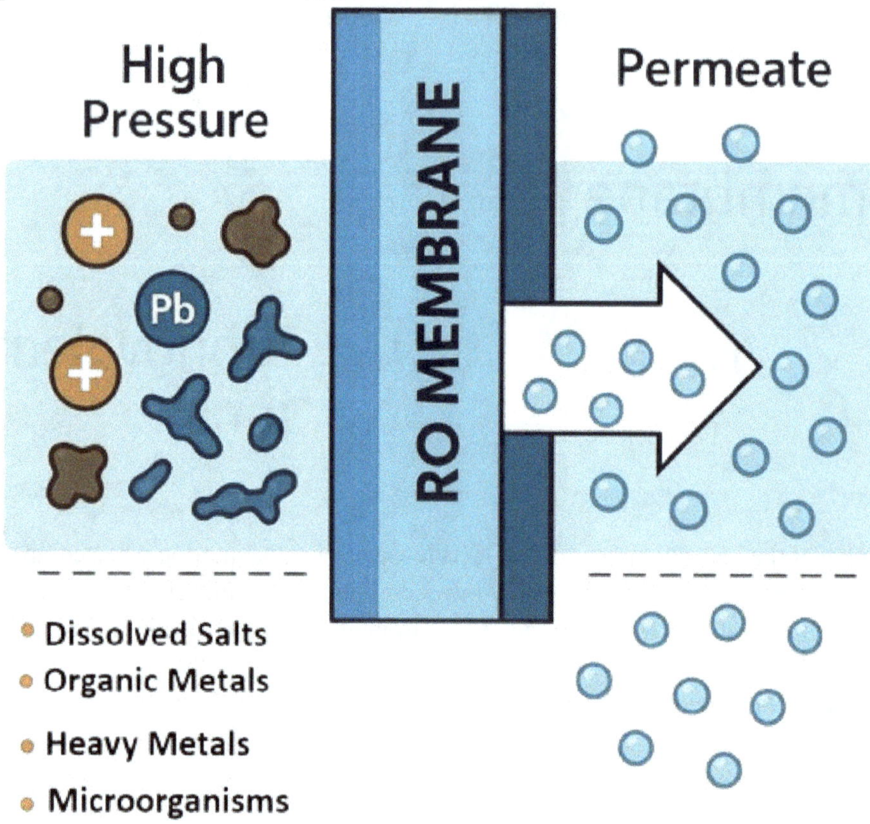

In contrast to microfiltration and ultrafiltration, which are pore-based separation technologies, RO membranes operate mainly by a solution-diffusion process in which impurities are rejected at the molecular level. RO membranes possess an extremely low molecular weight cut-off (~100 Da) and are capable of removing:

- Dissolved salts up to 99% (desalination)
- Heavy metals such as lead, arsenic, and mercury
- Microorganisms such as bacteria and viruses
- Organic pollutants such as pesticides and pharmaceutical residues
- RO is commonly applied in applications such as:
- Seawater and brackish water desalination
- Ultraclean water production for pharmaceuticals, electronics, and laboratories
- Industrial wastewater treatment and recycling
- Municipal drinking water treatment

RO membranes are generally fabricated from thin-film composite (TFC) polyamide materials and are designed in spiral-wound modules for compact, high-efficiency configurations (Rana et al., 2023). Although very effective, RO systems need extensive pretreatment (e.g., filtration, antiscalant dosing) to prevent membrane fouling and scaling, which can impair system performance and life.

While energy-consuming relative to other filtration processes, continuous improvement in membrane materials and energy recovery devices has progressively rendered RO more efficient and cost-effective for large- and small-scale water treatment.

3. NANOBIOTECHNOLOGICAL FILTRATION SYSTEMS

Nanobiotechnological filtration systems are a synergistic integration of nanotechnology and biotechnology that presents transformative potential in water purification. Nanoscale materials are integrated with bioactive substances—like peptides, enzymes, or microbial consortia—to design very specific, efficient, and sustainable nanobiotechnological filtration systems for the elimination of a wide variety of water-transmitted pathogens.

Nano-substances like carbon nanotubes (CNTs), metal-organic frameworks (MOFs), graphene oxide, silver nanoparticles, and nanofibers also possess unique characteristics, such as porosity tunability, high surface area, and antimicrobial properties. Being functionalized by bioactive compounds, such as metalloproteins towards heavy metal interaction or enzymes to degrade organics, they allow the bio-selective removal of pollutants like arsenic, lead, nitrates, endocrine-disrupting chemicals, or even pathogens (Thakur et al., 2025).

One of the leading applications is the construction of bio-functionalized membranes, wherein the nanomaterial acts as a support matrix and bioreceptor enables target-selective adsorption or catalytic degradation. For instance, enzyme-immobilized nanofilters would be capable of degrading pesticides or drugs upon contact, while nanocomposite filters have the potential to integrate adsorption, disinfection, and biodegradation into one platform (Najim et al., 2024).

In addition, nanobiotechnological systems can be engineered to be self-cleaning, anti-biofouling, and capable of real-time sensing, improving operational lifespan and minimizing maintenance needs. Their modularity also pre-adapts them for decentralized water treatment in rural or resource-poor areas.

While still in its nascent stages, these technologies hold great promise towards realizing next-generation water cleaning, with increased selectivity, low energy requirements, and environmental impact minimization, opening up the doors to cleaner and safer water sources across the globe.

3.1 Nanobiotechnological Filtration Systems: Advanced Mechanisms for Sustainable Water Purification

Nanobiotechnological filtration devices represent a remarkable alliance of biologically-based systems and engineered nanomaterials aimed at the provision of an extremely selective, effective, and eco-friendly water cleaning method (Joseph et al., 2023). The devices are made possible through exploiting the specific properties of nanomaterials, such as large surface area, great chemical reactivity, and customizable surface, which allow them to get rid of all kinds of water pollutants—molecular and ionic ones—with high efficiency (Fig. 5).

Microorganisms like fungi, bacteria, etc. and plants are less harmful systems and act as compatible ones. The impact on the environment is very low if these kinds of systems with the natural, nanostructures facilitated by microorganisms, are used in the water management process.

Biofilms consist of microbial communities that act as pioneering elements in nanobiotechnological filtration systems which helps in degradation of organic pollutants (Mahto & Das, 2022). These systems also restrict the motion of nanostructured membranes, which allows the integration of dynamic and regenerative filtration layers and creates sustainable treatment capacity (Salehi et al., 2022).

Figure 5. Nanobiotechnological-Biofilm Filtration Systems

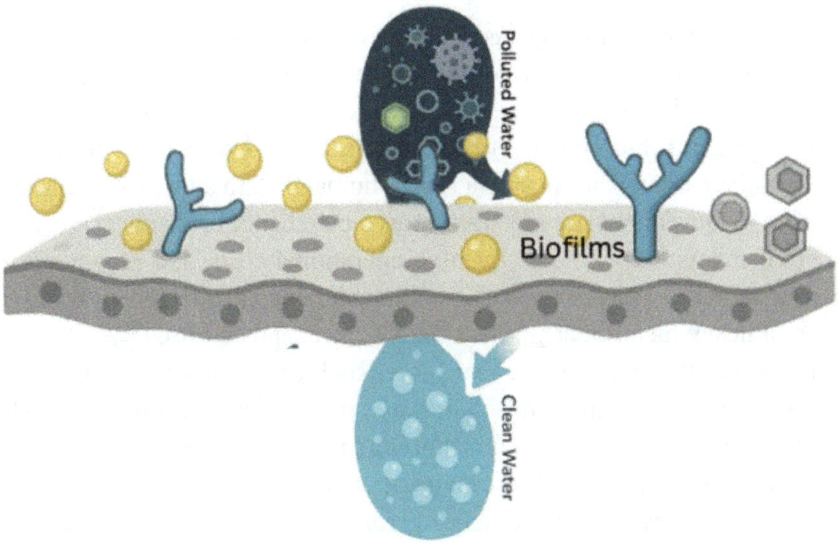

To adsorb and break the specific micropollutant characteristics of nanomaterials are being used. These structures not only respond to concentrated contaminants but also are used as sustained functional materials.

The scientific reason for the nanobiotechnological filtration process depends upon absorption, adsorption, and catalysis. Bioremediation is a process where biologically active materials metabolize or detoxify harmful substances in a natural process (Bhatt et al., 2022).

These processes taken together offer several vital benefits:

- Outstanding selectivity for target pollutants by biomolecular recognition and enzyme action;
- Increased reusability and regenerability of materials, leading to cost-effectiveness in operation;
- Smooth integration into decentralized, modular treatment systems, especially useful in remote or resource-poor areas without conventional water infrastructure.

In total, nanobiotechnological filtration systems represent a visionary solution for the future of water treatment, providing scalable, multifunctional, and sustainable solutions to address global water quality issues (Palit & Ranjit, 2024).

4. HEAVY METAL REMOVAL USING NANOBIOTECHNOLOGY

Nanobiotechnology has opened up a new door and given rise to extremely effective methods for getting rid of heavy metals in water sources that have been contaminated. Extremely toxic metals like lead (Pb), cadmium (Cd), arsenic (As), mercury (Hg), and chromium (Cr) can still cause harm even when they are present in very small amounts, thus posing a risk to human health and ecosystems (Angon et al., 2024). Conventional treatments such as ion exchange or chemical precipitation have problems with low selectivity, high operational costs, and generating waste. However, the case is different for nanobiotechnological methods, which ensure cleanup of heavy metals that is selective, effective, and eco-friendly.

Nanomaterials engineered at their core, such as magnetic nanoparticles, derivatives of graphene, carbon nanotubes, and metal-organic frameworks (MOFs), have a large surface area and customizable chemical properties for heavy metal ion adsorption and sequestration. These materials when made functional with biological compounds like enzymes, peptides, or microbial metabolites, show specific selectivity and binding affinity for the targeted metal ions (Li et al., 2024a).

Biogenic nanoparticles, made through microbial or plant-mediated processes, are the green and economically viable alternative to chemically synthesized nanomaterials. These particles that are most commonly made of metals or metal oxides, have the ability to adsorb heavy metals and also lower their bioavailability and power. For example, biosynthesized iron oxide or silver nanoparticles pass on to the next stage of water treatment in rural areas because they have been shown to have a very strong attraction for arsenic and lead (Yadav et al., 2024).

Heavy metal bioremediation can also be achieved through the use of biofilms and microbial consortia. A group of bacteria and fungi have the ability to produce naturally extracellular polymeric substances (EPS) that can combine with and make heavy metals unreactive.

In this way, biofilms can be used in filtration systems for constant removal of pollutants, provided they are attached to nanostructured surfaces (Lago et al., 2024).

Furthermore, enzyme-conjugated nanoparticles have turned into super specific catalysts for the process of converting heavy metals to less toxic forms and eliminating them from water (Tiwari et al., 2025).

4.1 Limitations in Conventional Technologies

The traditional methods are not only Dr. inefficient and but also create huge sludge which could still be a source of pollution. The conventional water treatment methods include sedimentation, chlorination, activated carbon filtration, coagulation-flocculation and ion exchange (Duggireddy & Pisharody, 2024). However, these methods are still limited in their capacity to deal with chemical water quality problems.

The majority of the conventional methods are not suitable and consequently, are not effective against new pollutants like drugs, EDCs and microplastics. In addition, some methods, such as chemical disinfecting, are capable of producing harmful disinfection byproducts (DBPs) which in turn can pose health problems (Malik & Kumar, 2024). Reverse osmosis and ion exchange methods are among the most energy consuming ones and, thus, economically not feasible for large-scale application in low-resource communities. Besides, traditional methods are slow in removing even trace amounts of heavy metals and organic micropollutants at environmentally relevant concentrations. Besides, sludge production, membrane fouling and frequent chemical dosing reduce process efficiency and sustainability even more. These problems show that there is an urgent need for green nanobiotechnological systems which are innovative, specific, scalable and environment-friendly to replace the existing ones (Prajapati et al, 2024). Nanobiotechnological innovations present the possibility of overcoming a number of the limitations of conventional techniques, with a few summarized in Table 1.

Table 1. Comparative Analysis of Traditional Treatment Limitations vs. Nanobiotechnological Solutions

Traditional Method Limitation	Nanobiotechnological Solution
Not suitable for emerging pollutants for example pharmaceuticals, microplastics, PFAS etc.	Enzyme membranes and magnetic nanoparticles help in degradation and recovery of contaminants
It requires high energy	It requires minimal external energy and operate at ambient conditions
It generates toxic sludge and secondary pollutants	It minimizes residual waste and facilitate circular usage
Unable to remove trace-level heavy metals fully	It offer high surface area and metal-binding affinity
It requires intensive infrastructure and is unsuitable for decentralized deployment	These type of systems are lightweight and suitable for rural and off-grid applications
Limited adaptability for water with varying pollutants	Bio-nanostructures are suitable for varying water conditions

4.2 Nanobiotechnological Methodologies

Nanobiotechnological Methodologies in water purification and wastewater treatment are techniques that merge nanotechnology and biocatalysis, such as enzymes, microbes, or biopolymers, to efficiently identify and remove or degrade pollutants from water.

Present water management technologies prevent the accumulation of contaminants, which creates corrosion on tanks, pipes, and membrane modules. Such technologies are very essential for the survival of life, but at the same time, they cause the loss of metals like Fe, Cu, and Zn through leaching. To tackle this problem, eco-friendly green decay preventers having plant-based origin like tannins, chitosan, and polyphenols are employed to stabilize the systems (Sesia et al., 2023).

In the case of use of bio-nanocomposite materials, these inhibitors provide better adhesion, self-healing, and long-term pitting resistance in saline and acidic environments. Besides that, metal oxide nanostructures such as ZnO, TiO_2, and CeO_2. They offer dual advantages to both the protection and the electrochemical stability of the microbial membranes (Mehrabi et al., 2023).

The inclusion of corrosion inhibition in the water treatment systems results in better water quality, reduced maintenance needs, and increased infrastructure life span. It paves the way for the transition to decentralized, durable, and sustainable purification technologies. Hence, corrosion inhibition is going to be the need of the day for futures, green water infrastructure.

4.2.1 Incorporation of Corrosion Inhibition in Water Purification

Magnetic nanoparticles modified with biosorbents are used for selective removal of heavy metals from water that has been polluted (Inobeme et al., 2023). The very methods conduct the synergistic effects of nanomaterials and bio-substances, which give high effectiveness, ecological compatibility, and the possibility of running in various water treatment environments.

Magnetic nanoparticles that are functioning with biosorbents such as chitosan, proteins, or polysaccharides have been reported to possess a tremendous affinity for heavy metal ions like lead, cadmium, and mercury (Bakhtiari et al., 2024).

Biosynthesized nanoparticles accelerate the separation and recovery by magnetic means after the treatment, which is a less expensive and more environmentally friendly method (S. Singh et al., 2024).

Metal-binding protein with enzyme-mediated systems isolates heavy metals into stable biosynthetic matrices (Mishra et al., 2024). These biological compounds can be used to increase operational stability and recyclability.

Bio-nanocomposite membranes have good permeability and adsorption capacity due to nanomaterials (Anwer et al., 2023). These membranes are capable of adsorbing heavy metals and retain their structural integrity. The incorporation of nanobiotechnological approaches is a next-generation solution.

4.3 Performance Metrics

Sorption capacity and biosorbent-functionalized nanomaterials are key indicators of nanobiotechnological filtration systems. Nanomaterials have demonstrative capacities of greater than 200 mg/g for Pb^{2+}, which shows their high efficiency in the removal of heavy metal ions (Salehi et al., 2022).

These systems require low energy, and they are suitable for small areas and off-grid operations. Most of the nanobiotechnological systems work under ambient conditions as compared to reverse osmosis. Thus operational cost and environmental concerns are minimized (Matin et al., 2021; Zhao et al., 2024). Such types of advanced technologies allow effective removal of contaminants. This can also work under influent conditions. These systems are lowering the environmental burden and resource use by maintaining excellent performance without the need for heavy chemical modification (Kumar et al., 2025b). Their on-site reliability and flexibility increase their value as green approaches for water quality management.

5. CASE STUDIES OVERVIEW: WATER TREATMENT USING NANOBIOTECHNOLOGY

The nanobiotechnological processes are crucial for characterization. To understand this several case studies have been studded that how nanomaterials degrade the pollutants. These case studies also highlight the large-scale industrial applications of nanomaterials. These case studies include community acceptability, regulatory restrictions, infrastructure needs, and cost-benefit feasibility. it would provide a better solution by considering these problems in a holistic, integrated way.

Case Study 1: Removal of Arsenic in Bangladesh

Severe public health issues have been caused due to the presence of Arsenic in groundwater in Bangladesh. Various Arsenic removal technologies are available which measures the presence of Arsenic acceptable level of groundwater.

Adsorption Techniques

This type of technique includes materials with extraordinary affinity towards arsenic. Dr. Abdul Hussam has invented the solo arsenic filter. This type of filter is low in cost and requires low maintenance (Mourdikoudis & Dominguez-Benetton, 2025). This method can treat 500 liters of water daily which is quite sufficient for 60 people. More than 20,000 units were installed throughout Bangladesh.

Coagulation Flocculation

In this method, coagulants like alum or other similar chemicals are used to precipitate the arsenic contaminated water. With the help of sedimentation and filtration process precipitates are taken out. However, this method needs strict control of chemical dosing and proper disposal of sludge (Hassan, 2023).

Ion Exchange and Membrane Filtration

This method includes the ion exchange resin and membrane filtration to remove the arsenic from the water. (Chauhan et al., 2023). These approaches require high operational costs and different maintenance challenges. This makes them less attractive for application on a vast scale in rural Bangladesh (Rana et al., 2024).

Subsurface Arsenic Removal (SAR)

This is one of the most suitable methods to reduce the Arsenic from groundwater. This method is favorable for on-site removal of arsenic. SAR technology has been proven to have good results in field tests in the villages of Bangladesh (Kundu et al., 2018).

Electrochemical Arsenic Remediation (ECAR)

ECAR is one of the modern arsenic removal techniques that employ electricity for the creation of iron hydroxides. These iron hydroxides are made up of such forms of iron that are to a great extent, water soluble and hence they can easily remove arsenic from water. Groundwater research has demonstrated that the ECAR system can reduce the arsenic content of water from 500 ppb to below 10 ppb, which means this technique can be adopted for water treatment where the cost of getting rid of arsenic is already high and especially in places where no water with acceptable arsenic levels is available (Ghosh et al., 2021).

Health hazards due to arsenic exposure are numerous. It hampers the immune system making an individual more prone to microbial infections and it has been correlated to more infant deaths (Selim et al., 2024; Woolf et al., 2023).

Decentralized systems are already in operation in places like Bangladesh to improve water and sanitation problem and to make the systems more efficient in the countries where such a scenario exists. For instance, incorporating DEWATS (decentralized wastewater treatment systems), renewable energy-based options, and even blockchain-supported food and water supply chains are some of the examples of the implementation of such systems (Kilthau et al., 2025).

The applicability of DEWATS systems in a decentralized manner for institutional, industrial, and community wastewater treatment is steadily increasing. Generally, the systems tend to apply a combination of physical and biological processes, such as sedimentation, filtration, and anaerobic digestion (Varma et al., 2022).

Physical and biological processes are constructed in a low-cost, low-maintenance, and local materials-friendly way. A good example to support this is a decentralised greywater treatment and reuse system in a rural Indian setting that achieved high removal efficacies for most of the pollutants.

Lastly, despite betterment in arsenic removal technology designs in Bangladesh, unified efforts need to now go towards applying cost-effective, actually long-lasting, and community-acceptable technologies to supply safe drinking water to the entire nation.

The biosynthesized iron oxide nanoparticle system was implemented within rural villages in efforts to treat arsenic-contaminated groundwater. > 95% removal that was low maintenance but with high community health benefit uplift was reported in the study (Bhagawati et al., 2025).

Case Study 2: Lead and Cadmium Adsorption Using Fungal Nano-Biosorbents

Indian researchers prepared a myco-nanocomposite of Aspergillus niger and iron nanoparticles. This system has excellent sorption and reusability abilities for industrial wastewater treatment.

To remove heavy metals such as lead (Pb^{2+}) and cadmium (Cd^{2+}), the fungal nano-biosorbents are an emerging and highly efficient option (Ali et al., 2024). By using fungi at a nanoscale, these biosorbents exploit the inherent metal-binding capacity, present a higher surface area, improve adsorption kinetics, and have denser functional groups (Jamir et al., 2024).

Some fungal isolates have demonstrated very strong biosorption potential (El-Mahdy et al., 2021). The cell walls contain chitin, glucans, and proteins. It also contains several functional groups like carboxyl, hydroxyl, amine, and phosphate. This provides them various free groups which are capable of complexing metal ions. The sorption efficiency can work greatly when fungal biomasses are treated into nanostructured forms. Various studies report shows that the adsorption capacities of fungal nano-biosorbents is varies from 150 to 200 mg/g for Pb^{2+} and Cd^{2+} ions respectively. Thus, these systems work in high pH values (Miyah et al., 2024). Their biodegradable and renewable nature further signifies their ability in sustainable green remediation technologies.

The key strengths of nano-biosorbents is their reusability and regenerability over several rounds of treatment with minimal reduction in adsorption effcency. This makes them ideal to be integrated into modular water treatment systems (Stoler et al., 2022).

Overall, nano-biosorbents derived from fungi provide an economically viable, high-performance, and eco-friendly solution for the removal of such toxic heavy metals as lead and cadmium from polluted water (Moghadas et al., 2022). Their scalability and versatility make them equally credible tools in future water purification and ecological regeneration processes.

Case Study 3: Bio-Inspired Membranes in Municipal Water Plants (USA)

California cities have experimented with hybrid nano-biotech membranes along with enzymes to take out pharmaceuticals. The whole setup gave off better flux, less fouling, and enhanced breakdown of the contaminant.

In the past few years, bio-inspired membranes have won great popularity in the municipal water treatment facilities across the U.S. due to their selectivity, resistance to foulings, and low-energy consumption (Bernardes et al., 2025). These bio-inspired membranes made us of aquaporins and cellular lipid bilayers as nature's models and used nanoscale engineering combined with the principles of biomimicry to deliver improved water purification performance.

Pilot plants in California and Texas have worked on the membranes that employed natural water-channel proteins as a cutting-edge technique for water purification. These proteins, found in living cells, are incorporated into a polymer or nanocomposite structure, forming a filter that is very selective. It lets water molecules pass while blocking salts, heavy metals, microbes, and other organic pollutants (Khan et al., 2023).

Companies such as Nano H_2O and other innovators have carried out trials in areas with water shortages and contamination, showing how these systems can work in real-life conditions.

In Orange countries like California has demonstrated bio-inspired membrane in the Groundwater Replenishment System (GWRS). As compared to traditional reverse osmosis the membranes provides better operational costs, as well as longevity of the membranes (OCWD, 2025).

These membranes could treat water with different pH and organic components (Ahmed et al., 2022). Finally, bio-inspired membranes require minimum chemical for cleaning, which reduces environmental effects (Singh et al., 2024).

6. TOWARD CLEANER ECOSYSTEMS: ENVIRONMENTAL IMPLICATIONS

Nanobiotechnological water filtration process brings revolution in the field of filtration process and this is becoming the first technology in the world. The traditional method of water treatment is unable to solve the current water pollution problems (Rathod et al., 2024). The advanced nanomaterials together with biological systems are applied in nanobiotechnological systems to not only increase the removal of contaminants but also to accomplish the goals of the environment.

This technique demonstrates a commendable result in the process of water treatment. Fresh water bodies restore their ecological balance and promote aquatic biodiversity, as well as improves the soil health and reduces the bioaccumulation of toxins through food webs. This system helps in the treatment of water in local areas such as rural and undeveloped communities. Whereas on the industrial side, nanobiotechnology platforms are flexible, scalable, and can work better to remove specific pollutants depending on the site. Thanks to the latest innovations, these features are coming to the drinking water purification process, which is becoming more advanced and self-sufficient tech-wise (Liu et al., 2021).

Large-scale implementation of this technique requires thorough contemplation of long-term environmental consequences. It also requires the modification of the policies as well and to modify this, there must be interdisciplinary efforts by scientists, engineers, policymakers, and industry players (da Silva & Blasimme, 2024).

Synthetic biology and ecological engineering might be the future technologies that nature-friendly and resilient ecosystems (Samson Prince et al., 2025). This kind of action is presumed to bring water security to the whole world and at the same time, keep the environment safe and make sustainability better. The raw materials used in these systems are living organisms, and they consume the least amount of energy by the normal standards of working. Moreover, they not only reduce the use of chemicals but also avoid pollution through chemical facilities (Feder-Kubis et al., 2024).

The flexible design of these systems is another plus point. They are small and compact, hence very suitable for rural areas to meet their needs for safe drinking water (Schillinger & Özerol, 2025). Being reusable and renewable, they produce less waste and conserve natural resources. Therefore, they can be considered a good long-term option for water security and healthy ecosystems.

Besides, they adhere to the principles of a circular economy (Al-Gethami et al., 2024). In conclusion, nanobiotechnology not only looks like a cleaner and more sustainable way but also a very innovative one in the area of water treatment and taking care of the planet.

7. FUTURE PERSPECTIVE

In the future, the nanobiotechnical water treatment processes might become more intelligent due to the use of digital tools. The filters can be installed with IoT devices that would allow for real-time monitoring of water quality through the sensors. Pollution levels could be revealed and the state of water displayed (Pal et al., 2025).

AI and machine learning could also be of great assistance. Rather than stopping the process and analysing the parts, such tools would be guessing all the changes in a material's reaction with a variety of impurities. This would mean making the system readily available and keeping the water quality constant (Cairone et al., 2024).

Nonetheless, technology by itself will not suffice. In order to use such systems on a global scale, we will have to set proper rules. Safety, risk testing, and the disposal of materials after their life cycle are among the issues that should be addressed by policy. Regulations that are explicit, that rely on truthful reporting, and that involve the local community will be the basis for establishing trust (Chithra & Bhambri, 2024).

8. CONCLUSION

In this chapter, several water purification processes and technologies are discussed, among them traditional treatments, the new, very little known scientific purification through the nanobiotechnological filtration, and others. The most common approaches, such as the use of sedimentation and chlorination along with activated carbon, have been the foundation of water treatment systems and have performed their job well for many decades. However, such technologies have some drawbacks when it comes to the removal of new and complicated contaminants linked to the drinking water sources and supply. Therefore, a wider range of methods and technologies can support water purification to the extent of making it more efficient and versatile.

Nanobiotechnology, which is a new framework for supplying clean drinking water, can be an instant solution. The synergistic effect of nanoparticles and other biological types of treatment will enable the removal of metals and the degradation of pollutants under milder and more eco-friendly conditions. It cuts down on production costs, and disposal is smaller.

The chapter justifies that the future of water purification will be the development of technology concerning the combination of improvements to water purification that harmonize with the science of water treatment. Scale, cost, and environmental impact are the three main factors that will continue to be of utmost importance in both market technologies and academic research. The contamination of rivers due to the latter increased occurrence and the conflict of water resources that may lead to a global catastrophe are some trends of water supply insecurity that are supplemented by the responsible use of neobiotechnological water resources and, consequently, the topic of clean water could be an important aspect to tackle in the quest of sustainable water resources and human health.

ACKNOWLEDGMENT

The authors acknowledge the use of AI-based tools, such as ChatGPT, for assistance in editing, grammar enhancement, and spelling checks during the preparation of this manuscript.

REFERENCES

Ahmed, S. F., Mehejabin, F., Momtahin, A., Tasannum, N., Faria, N. T., Mofijur, M., Hoang, A. T., Vo, D.-V. N., & Mahlia, T. M. I. (2022). Strategies to improve membrane performance in wastewater treatment. *Chemosphere*, *306*, 135527. DOI: 10.1016/j.chemosphere.2022.135527 PMID: 35780994

Al-Gethami, W., Qamar, M. A., Shariq, M., Alaghaz, A.-N. M. A., Farhan, A., Areshi, A. A., & Alnasir, M. H. (2024). Emerging environmentally friendly bio-based nanocomposites for the efficient removal of dyes and micropollutants from wastewater by adsorption: A comprehensive review. *RSC Advances*, *14*(4), 2804–2834. DOI: 10.1039/D3RA06501D PMID: 38234871

Ali, S., Naseer, S., Rehman, M., & Wei, Z. (2024). Recent trends and sources of lead toxicity: A review of state-of-the-art nano-remediation strategies. *Journal of Nanoparticle Research*, *26*(7), 168. DOI: 10.1007/s11051-024-06081-5

Alkhadra, M. A., Su, X., Suss, M. E., Tian, H., Guyes, E. N., Shocron, A. N., Conforti, K. M., de Souza, J. P., Kim, N., Tedesco, M., Khoiruddin, K., Wenten, I. G., Santiago, J. G., Hatton, T. A., & Bazant, M. Z. (2022). Electrochemical Methods for Water Purification, Ion Separations, and Energy Conversion. *Chemical Reviews*, *122*(16), 13547–13635. DOI: 10.1021/acs.chemrev.1c00396 PMID: 35904408

Alver, A., Baştürk, E., Altaş, L., & Işık, M. (2022). A solution of taste and odor problem with activated carbon adsorption in drinking water: Detailed kinetics and isotherms. *Desalination and Water Treatment*, *252*, 300–318. DOI: 10.5004/dwt.2022.28269

Angon, P. B., Islam, M. S., Kc, S., Das, A., Anjum, N., Poudel, A., & Suchi, S. A. (2024). Sources, effects and present perspectives of heavy metals contamination: Soil, plants and human food chain. *Heliyon*, *10*(7), e28357. DOI: 10.1016/j.heliyon.2024.e28357 PMID: 38590838

Anwer, A. H., Ahtesham, A., Shoeb, M., Mashkoor, F., Ansari, M. Z., Zhu, S., & Jeong, C. (2023). State-of-the-art advances in nanocomposite and bio-nanocomposite polymeric materials: A comprehensive review. *Advances in Colloid and Interface Science*, *318*, 102955. DOI: 10.1016/j.cis.2023.102955 PMID: 37467558

Aziz, S., Mazhar, A. R., Ubaid, A., Shah, S. M. H., Riaz, Y., Talha, T., & Jung, D.-W. (2024). A comprehensive review of membrane-based water filtration techniques. *Applied Water Science*, *14*(8), 169. DOI: 10.1007/s13201-024-02226-y

Bakhtiari, S., Salari, M., Shahrashoub, M., Zeidabadinejad, A., Sharma, G., & Sillanpää, M. (2024). A Comprehensive Review on Green and Eco-Friendly Nano-Adsorbents for the Removal of Heavy Metal Ions: Synthesis, Adsorption Mechanisms, and Applications. *Current Pollution Reports*, *10*(1), 1–39. DOI: 10.1007/s40726-023-00290-7

Bernardes, A. F., Meng, Z., Campos, L. C., & Coppens, M.-O. (2025). Bio-inspired anti-fouling strategies for membrane-based separations. *Chemical Communications*, *61*(27), 5064–5071. DOI: 10.1039/D4CC05149A PMID: 40084840

Bhagawati, M., Datta, B. K., Newar, R., Sonowal, S., Buragohain, K., Tamuly, D., & Nath, R. (2025). Nanoparticle-Assisted Microbial Removal of Arsenic (As) from Drinking Water Sources. In *Nano-Bioremediation for Wastewater Treatment* (pp. 249–289). Wiley. DOI: 10.1002/9781394271641.ch10

Bhatt, P., Pandey, S. C., Joshi, S., Chaudhary, P., Pathak, V. M., Huang, Y., Wu, X., Zhou, Z., & Chen, S. (2022). Nanobioremediation: A sustainable approach for the removal of toxic pollutants from the environment. *Journal of Hazardous Materials*, *427*, 128033. DOI: 10.1016/j.jhazmat.2021.128033 PMID: 34999406

Bigham, A., Zarepour, A., Safarkhani, M., Huh, Y., Khosravi, A., Rabiee, N., Iravani, S., & Zarrabi, A. (2025). Inspired by nature: Bioinspired and biomimetic photocatalysts for biomedical applications. *Nano Materials Science*, *7*(1), 1–23. DOI: 10.1016/j.nanoms.2024.02.006

Cairone, S., Hasan, S. W., Choo, K.-H., Li, C.-W., Zarra, T., Belgiorno, V., & Naddeo, V. (2024). Integrating artificial intelligence modeling and membrane technologies for advanced wastewater treatment: Research progress and future perspectives. *The Science of the Total Environment*, *944*, 173999. DOI: 10.1016/j.scitotenv.2024.173999 PMID: 38879019

Carofiglio, M., Laurenti, M., & Cauda, V. (2022). *The World Big Challenges Faced by Nanoscience: Examples of How Integrated Science Can Save the World* (pp. 153–181). DOI: 10.1007/978-3-030-94651-7_8

Chauhan, M. S., Rahul, A. K., Shekhar, S., & Kumar, S. (2023). Removal of heavy metal from wastewater using ion exchange with membrane filtration from Swarnamukhi river in Tirupati. *Materials Today: Proceedings*, *78*, 1–6. DOI: 10.1016/j.matpr.2022.08.280

Chithra, N., & Bhambri, P. (2024). Ethics in Sustainable Technology. In *Handbook of Technological Sustainability* (pp. 245–256). CRC Press., DOI: 10.1201/9781003475989-21

Costa, F. C. R., Moreira, V. R., Guimarães, R. N., Moser, P. B., Santos, L. V. S., de Paula, E. C., & Amaral, M. C. S. (2024). Pre-oxidation and coagulation-flocculation as a pretreatment to UF-RO applied for surface water treatment and arsenic removal. *Desalination*, *586*, 117855. DOI: 10.1016/j.desal.2024.117855

da Silva, R. G. L., & Blasimme, A. (2024). From lab to society: Fostering clinical translation of molecular systems engineering. *Bioengineering & Translational Medicine*, *9*(1), e10564. Advance online publication. DOI: 10.1002/btm2.10564 PMID: 38193130

dos Santos, A. J., Kronka, M. S., Fortunato, G. V., & Lanza, M. R. V. (2021). Recent advances in electrochemical water technologies for the treatment of antibiotics: A short review. *Current Opinion in Electrochemistry*, *26*, 100674. DOI: 10.1016/j.coelec.2020.100674

Duggireddy, R. P., & Pisharody, L. (2024). Traditional methods of water purification in rural areas. In *Water Resources Management for Rural Development* (pp. 55–64). Elsevier., DOI: 10.1016/B978-0-443-18778-0.00003-9

El-Mahdy, O. M., Mohamed, H. I., & Mogazy, A. M. (2021). Biosorption effect of Aspergillus niger and Penicillium chrysosporium for Cd- and Pb-contaminated soil and their physiological effects on Vicia faba L. *Environmental Science and Pollution Research International*, *28*(47), 67608–67631. DOI: 10.1007/s11356-021-15382-4 PMID: 34258698

Feder-Kubis, J., Wirwis, A., Policht, M., Singh, J., & Kim, K.-H. (2024). Principles and practice of greener ionic liquid–nanoparticles biosystem. *Green Chemistry*, *26*(6), 3072–3124. DOI: 10.1039/D3GC04387H

Garg, S. (2022). *Industrial Wastewater: Characteristics*. Treatment Techniques and Reclamation of Water., DOI: 10.1007/978-3-030-83811-9_1

Geetha Varma, V., Jha, S., Himesh Karthik Raju, L., Lalith Kishore, R., & Ranjith, V. (2022). A review on decentralized wastewater treatment systems in India. *Chemosphere*, *300*, 134462. DOI: 10.1016/j.chemosphere.2022.134462 PMID: 35390407

Ghosh, S., Debsarkar, A., Dutta, A., Bhandary, S., & Chopra, D. (2021). Delivering Arsenic-free Drinking Water-Made Practically Possible: Continuous Scale Electrochemical Arsenic Remediation Process Furnished, based on Experimental Studies and ANN Simulation. *Environment, Development and Sustainability*, *23*(9), 13087–13112. DOI: 10.1007/s10668-020-01200-3

Guo, K., Wu, Z., Chen, C., & Fang, J. (2022). UV/Chlorine Process: An Efficient Advanced Oxidation Process with Multiple Radicals and Functions in Water Treatment. *Accounts of Chemical Research*, *55*(3), 286–297. DOI: 10.1021/acs.accounts.1c00269 PMID: 35025201

Gupta, P., Singh, J., Verma, S., Chandel, A. S., & Bhatla, R. (2021). Impact of climate change and water quality degradation on food security and agriculture. In *Water Conservation in the Era of Global Climate Change* (pp. 1–22). Elsevier., DOI: 10.1016/B978-0-12-820200-5.00006-3

Haidar, Z., Fatema, K., Shoily, S. S., & Sajib, A. A. (2023). Disease-associated metabolic pathways affected by heavy metals and metalloid. *Toxicology Reports*, *10*, 554–570. DOI: 10.1016/j.toxrep.2023.04.010 PMID: 37396849

He, Z., Fan, X., Jin, W., Gao, S., Yan, B., Chen, C., Ding, W., Yin, S., Zhou, X., Liu, H., Li, X., & Wang, Q. (2023). Chlorine-resistant bacteria in drinking water: Generation, identification and inactivation using ozone-based technologies. *Journal of Water Process Engineering*, *53*, 103772. DOI: 10.1016/j.jwpe.2023.103772

Hussain, M. K., Khatoon, S., Nizami, G., Fatma, U. K., Ali, M., Singh, B., Quraishi, A., Assiri, M. A., Ahamad, S., & Saquib, M. (2024). Unleashing the power of bio-adsorbents: Efficient heavy metal removal for sustainable water purification. *Journal of Water Process Engineering*, *64*, 105705. DOI: 10.1016/j.jwpe.2024.105705

Inobeme, A., Mathew, J. T., Adetunji, C. O., Ajai, A. I., Inobeme, J., Maliki, M., Okonkwo, S., Adekoya, M. A., Bamigboye, M. O., Jacob, J. O., & Eziukwu, C. A. (2023). Recent advances in nanotechnology for remediation of heavy metals. *Environmental Monitoring and Assessment*, *195*(1), 111. DOI: 10.1007/s10661-022-10614-7 PMID: 36378336

Jagaba, A. H., Lawal, I. M., Birniwa, A. H., Affam, A. C., Usman, A. K., Soja, U. B., Saleh, D., Hussaini, A., Noor, A., & Yaro, N. S. A. (2023). Sources of Water Contamination by Heavy Metals. In *Membrane Technologies for Heavy Metal Removal from Water* (pp. 3–27). CRC Press., DOI: 10.1201/9781003326281-2

Jamir, I., Ezung, L. Y., Merry, L., Tikendra, L., Devi, R. S., & Nongdam, P. (2024). Heavy Metals Clean Up: The Application of Fungi for Biosorption. *Geomicrobiology Journal*, *41*(3), 201–212. DOI: 10.1080/01490451.2024.2307899

Jiang, Y., Qiao, Y., Jin, R., Jia, M., Liu, J., He, Z., & Liu, Z. (2024). Application of chlorine dioxide and its disinfection mechanism. *Archives of Microbiology*, *206*(10), 400. DOI: 10.1007/s00203-024-04137-7 PMID: 39256286

Jomova, K., Alomar, S. Y., Nepovimova, E., Kuca, K., & Valko, M. (2025). Heavy metals: Toxicity and human health effects. *Archives of Toxicology*, *99*(1), 153–209. DOI: 10.1007/s00204-024-03903-2 PMID: 39567405

Joseph, T. M., Al-Hazmi, H. E., Śniatała, B., Esmaeili, A., & Habibzadeh, S. (2023). Nanoparticles and nanofiltration for wastewater treatment: From polluted to fresh water. *Environmental Research*, *238*, 117114. DOI: 10.1016/j.envres.2023.117114 PMID: 37716387

Khan, N. A., Singh, S., López-Maldonado, E. A., N, P., Méndez-Herrera, P. F., López-López, J. R., Baig, U., Ramamurthy, P. C., Mubarak, N. M., Karri, R. R., & Aljundi, I. H. (2023). Emerging membrane technology and hybrid treatment systems for the removal of micropollutants from wastewater. *Desalination*, *565*, 116873. DOI: 10.1016/j.desal.2023.116873

Kilthau, M., Henkel, V., Wagner, L. P., Gehlhoff, F., & Fay, A. (2025). A decentralized optimization approach for scalable agent-based energy dispatch and congestion management. *Applied Energy*, *377*, 124606. DOI: 10.1016/j.apenergy.2024.124606

Kishore, S., Malik, S., Shah, M. P., Bora, J., Chaudhary, V., Kumar, L., Sayyed, R. Z., & Ranjan, A. (2024). A comprehensive review on removal of pollutants from wastewater through microbial nanobiotechnology -based solutions. *Biotechnology & Genetic Engineering Reviews*, *40*(4), 3087–3112. DOI: 10.1080/02648725.2022.2106014 PMID: 35923085

Kolossváry, E., Farkas, K., Karahan, O., Golledge, J., Schernthaner, G.-H., Karplus, T., Bernardo, J. J., Marschang, S., Abola, M. T., Heinzmann, M., Edmonds, M., & Catalano, M. (2023). The importance of socio-economic determinants of health in the care of patients with peripheral artery disease: A narrative review from VAS. *Vascular Medicine*, *28*(3), 241–253. DOI: 10.1177/1358863X231169316 PMID: 37154387

Kumar, A., Kumar, A., Seema, K., & Kumari, K. (2025a). *Detection and Remediation Techniques for Emerging Contaminants*., DOI: 10.1007/978-3-031-82579-8_10

Kumar, V., Verma, P., de Freitas, F. A., Srivastava, P. K., Vashishth, A., & Américo-Pinheiro, J. H. P. (2025b). A critical review on biofuels generation from pulp-paper mill sludge with emphasis on pretreatment methods: Renewable energy for environmental sustainability. *BMC Environmental Science*, *2*(1), 2. DOI: 10.1186/s44329-024-00016-0

Kundu, D. K., Gupta, A., Mol, A. P. J., Rahman, M. M., & van Halem, D. (2018). Experimenting with a novel technology for provision of safe drinking water in rural Bangladesh: The case of sub-surface arsenic removal (SAR). *Technology in Society*, *53*, 161–172. DOI: 10.1016/j.techsoc.2018.01.010

Lago, A., Rocha, V., Barros, O., Silva, B., & Tavares, T. (2024). Bacterial biofilm attachment to sustainable carriers as a clean-up strategy for wastewater treatment: A review. *Journal of Water Process Engineering*, *63*, 105368. DOI: 10.1016/j.jwpe.2024.105368

Li, L., Ma, C., Yang, Y., Wang, B., Liu, X., Wang, Y., Bian, X., Zhang, G., & Zhang, N. (2024a). Exploring the potential of plant-derived metal ion binding peptides: Preparation, structure-activity relationship, and biological activities. *Trends in Food Science & Technology*, *152*, 104650. DOI: 10.1016/j.tifs.2024.104650

Li, X., Li, Z., Yang, D., Li, S., Yang, L., Yan, W., Xiang, W., Pei, L., Ye, Z., & Xu, H. (2024b). Electrochemical disinfection for human health protection: Disinfection mechanisms, system innovations, applications. *Journal of Environmental Chemical Engineering*, *12*(5), 114073. DOI: 10.1016/j.jece.2024.114073

Liu, L., Guo, X., Liu, W., & Lee, C. (2021). Recent Progress in the Energy Harvesting Technology—From Self-Powered Sensors to Self-Sustained IoT, and New Applications. *Nanomaterials (Basel, Switzerland)*, *11*(11), 2975. DOI: 10.3390/nano11112975 PMID: 34835739

López-Maldonado, E. A., Khan, N. A., Singh, S., Ramamurthy, P. C., Kabak, B., Baudrit, J. R. V., Alkahtani, M. Q., Álvarez-Torrellas, S., Varshney, R., Serra-Pérez, E., García, J., Gkika, D. A., Kyzas, G. Z., Kadier, A., Singh, R., Periyasamy, S., Gizaw, D. G., Hossain, N., & Zahmatkesh, S. (2024). Magnetic polymeric composites: Potential for separating and degrading micro/nano plastics. *Desalination and Water Treatment*, *317*, 100198. DOI: 10.1016/j.dwt.2024.100198

Mahto, K. U., & Das, S. (2022). Bacterial biofilm and extracellular polymeric substances in the moving bed biofilm reactor for wastewater treatment: A review. *Bioresource Technology*, *345*, 126476. DOI: 10.1016/j.biortech.2021.126476 PMID: 34864174

Malik, N., & Kumar, R. (2024). A Comparative Study of Different Disinfection Processes, and Management Practices to Control the Formation of Disinfection by-Products (DBPs). In *Drinking Water Disinfection By-products* (pp. 59–83). Springer Nature Switzerland., DOI: 10.1007/978-3-031-49047-7_4

Matin, A., Laoui, T., Falath, W., & Farooque, M. (2021). Fouling control in reverse osmosis for water desalination & reuse: Current practices & emerging environment-friendly technologies. *The Science of the Total Environment, 765*, 142721. DOI: 10.1016/j.scitotenv.2020.142721 PMID: 33129530

Mehrabi, M. R., Soltani, M., Chiani, M., Raahemifar, K., & Farhangi, A. (2023). Nanomedicine: New Frontiers in Fighting Microbial Infections. *Nanomaterials (Basel, Switzerland)*, *13*(3), 483. DOI: 10.3390/nano13030483 PMID: 36770443

Mishra, S., Patel, A., Bhatt, P., Chen, S., & Srivastava, P. K. (2024). Perspective Evaluation of Synthetic Biology Approaches for Effective Mitigation of Heavy Metal Pollution. *Reviews of Environmental Contamination and Toxicology*, *262*(1), 21. DOI: 10.1007/s44169-024-00072-2

Miyah, Y., El Messaoudi, N., Benjelloun, M., Georgin, J., Franco, D. S. P., Acikbas, Y., Kusuma, H. S., & Sillanpää, M. (2024). MOF-derived magnetic nanocomposites as potential formulations for the efficient removal of organic pollutants from water via adsorption and advanced oxidation processes: A review. *Materials Today Sustainability*, *28*, 100985. DOI: 10.1016/j.mtsust.2024.100985

Moattari, R. M., Mohammadi, T., Rajabzadeh, S., Dabiryan, H., & Matsuyama, H. (2021). Reinforced hollow fiber membranes: A comprehensive review. *Journal of the Taiwan Institute of Chemical Engineers*, *122*, 284–310. DOI: 10.1016/j.jtice.2021.04.052

Moghadas, B. K., Esmaeili, H., Tamjidi, S., & Geramifard, A. (2022). Advantages of nanoadsorbents, biosorbents, and nanobiosorbents for contaminant removal. In *Nano-Biosorbents for Decontamination of Water* (pp. 105–133). Air, and Soil Pollution., DOI: 10.1016/B978-0-323-90912-9.00006-X

Mourdikoudis, S., & Dominguez-Benetton, X. (2025). Physicochemical vs Electrochemical Technologies for Metal Recovery – Main Insights, Comparison, Complementarity and Challenges. *Chemistry Methods*, *5*(3), e202400046. Advance online publication. DOI: 10.1002/cmtd.202400046

Najim, A. A., Radeef, A. Y., Al-Doori, I., & Jabbar, Z. H. (2024). Immobilization: The promising technique to protect and increase the efficiency of microorganisms to remove contaminants. *Journal of Chemical Technology and Biotechnology*, *99*(8), 1707–1733. DOI: 10.1002/jctb.7638

OCWD. (2025). *ORANGE COUNTY WATER DISTRICT*. https://www.ocwd.com/

Padilla-Rivera, A., Hannouf, M., Assefa, G., & Gates, I. (2023). A systematic literature review on current application of life cycle sustainability assessment: A focus on economic dimension and emerging technologies. *Environmental Impact Assessment Review*, *103*, 107268. DOI: 10.1016/j.eiar.2023.107268

Pal, P., Pandey, B., Prakash, A., Sarangi, P. K., Rathore, S. S., Agarwal, V., Gupta, A. K., & Singh, A. K. (2025). *Exploration of Biosensors in Innovative Applications and Sustainable Solutions for Environmental Monitoring and Management.*, DOI: 10.1007/978-981-97-7221-6_19

Palit, S., & Ranjit, P. S. (2024). Large Scale Applications of Nanomaterials for Water Treatment. In *Sustainable Production Innovations* (pp. 137–162). DOI: 10.1002/9781119792888.ch4

Pillai, S. B., & Thombre, N. V. (2024). Coagulation, Flocculation, and Precipitation in Water and Used Water Purification. In *Handbook of Water and Used Water Purification* (pp. 3–27). DOI: 10.1007/978-3-319-78000-9_63

Poirier, K., Lotfi, M., Garg, K., Patchigolla, K., Anthony, E. J., Faisal, N. H., Mulgundmath, V., Sahith, J. K., Jadhawar, P., Koh, L., Morosuk, T., & Al Mhanna, N. (2023). A comprehensive review of pre- and post-treatment approaches to achieve sustainable desalination for different water streams. *Desalination*, *566*, 116944. DOI: 10.1016/j.desal.2023.116944

Pokrajac, L., Abbas, A., Chrzanowski, W., Dias, G. M., Eggleton, B. J., Maguire, S., Maine, E., Malloy, T., Nathwani, J., Nazar, L., Sips, A., Sone, J., van den Berg, A., Weiss, P. S., & Mitra, S. (2021). Nanotechnology for a Sustainable Future: Addressing Global Challenges with the International Network4Sustainable Nanotechnology. *ACS Nano*, *15*(12), 18608–18623. DOI: 10.1021/acsnano.1c10919 PMID: 34910476

Prajapati, D., Jabborova, D., Saharan, B. S., Singh, N., Patani, A., Singh, S., & Joshi, C. (2024). Bionanotechnology: A Paradigm for Advancing Environmental Sustainability. *Indian Journal of Microbiology*. Advance online publication. DOI: 10.1007/s12088-024-01389-1 PMID: 40371027

Rahidul Hassan, H. (2023). A review on different arsenic removal techniques used for decontamination of drinking water. *Environmental Pollutants and Bioavailability*, *35*(1), 2165964. Advance online publication. DOI: 10.1080/26395940.2023.2165964

Rana, D., Matsuura, T., Kassim, M. A., & Ismail, A. F. (2023). Reverse Osmosis Membrane. In *Handbook of Membrane Separations* (pp. 6–32). CRC Press., DOI: 10.1201/9781003285656-2

Rana, V., Sheokand, M., Jain, K., Dhaka, S., Godara, S. K., Urmaliya, D. N., Madhav, S., Singh, K. P., & Dhaka, R. K. (2024). Effective and affordable water purification technologies for rural development. In *Water Resources Management for Rural Development* (pp. 91–106). Elsevier., DOI: 10.1016/B978-0-443-18778-0.00016-7

Rathod, P. B., Singh, M. P., Taware, A. S., Deshmukh, S. U., Tagad, C. K., Kulkarni, A., & Kanagare, A. B. (2024). Comprehensive insights into water remediation: Chemical, biotechnological, and nanotechnological perspectives. *Environmental Pollutants and Bioavailability*, *36*(1), 2329660. Advance online publication. DOI: 10.1080/26395940.2024.2329660

Ray, S., & Vashishth, R. (2024). From water to plate: Reviewing the bioaccumulation of heavy metals in fish and unraveling human health risks in the food chain. *Emerging Contaminants*, *10*(4), 100358. DOI: 10.1016/j.emcon.2024.100358

Razzak, S. A., Faruque, M. O., Alsheikh, Z., Alsheikhmohamad, L., Alkuroud, D., Alfayez, A., Hossain, S. M. Z., & Hossain, M. M. (2022). A comprehensive review on conventional and biological-driven heavy metals removal from industrial wastewater. *Environmental Advances*, *7*, 100168. DOI: 10.1016/j.envadv.2022.100168

Sahu, J. N., Karri, R. R., Zabed, H. M., Shams, S., & Qi, X. (2021). Current Perspectives and Future Prospects of Nano-Biotechnology in Wastewater Treatment. *Separation and Purification Reviews*, *50*(2), 139–158. DOI: 10.1080/15422119.2019.1630430

Salehi, E., Khajavian, M., Sahebjamee, N., Mahmoudi, M., Drioli, E., & Matsuura, T. (2022). Advances in nanocomposite and nanostructured chitosan membrane adsorbents for environmental remediation: A review. *Desalination*, *527*, 115565. DOI: 10.1016/j.desal.2022.115565

Samson Prince, H., Macarius, N. M. N., Chandrashekhar Walkikar, S., Das, S., Rajan, A. P., & Deepankumar, K. (2025). *Harmonizing Nature and Innovation: Sustainable Environmental Solutions Through Nano-biotech Synergy* (pp. 35–83). DOI: 10.1007/978-3-031-78483-5_2

Saravanan, A., Senthil Kumar, P., Jeevanantham, S., Karishma, S., Tajsabreen, B., Yaashikaa, P. R., & Reshma, B. (2021). Effective water/wastewater treatment methodologies for toxic pollutants removal: Processes and applications towards sustainable development. *Chemosphere*, *280*, 130595. DOI: 10.1016/j.chemosphere.2021.130595 PMID: 33940449

Schillinger, J., & Özerol, G. (2025). For better or worse: The influence of conflict-driven decentralization on the resilience of urban water supply infrastructure in the Middle East. *Sustainable and Resilient Infrastructure*, *10*(1), 1–21. DOI: 10.1080/23789689.2024.2355768

Selim, S., Ahmad, B., & Uddin, M. M. (2024). Guardians of the Depths: Managing Groundwater Contamination in Developing Countries. In *Water Crises and Sustainable Management in the Global South* (pp. 595–623). Springer Nature Singapore., DOI: 10.1007/978-981-97-4966-9_19

Sesia, R., Spriano, S., Sangermano, M., & Ferraris, S. (2023). Natural Polyphenols and the Corrosion Protection of Steel: Recent Advances and Future Perspectives for Green and Promising Strategies. *Metals*, *13*(6), 1070. DOI: 10.3390/met13061070

Shao, B., Shen, L., Liu, Z., Tang, L., Tan, X., Wang, D., Zeng, W., Wu, T., Pan, Y., Zhang, X., Ge, L., & He, M. (2023). Disinfection byproducts formation from emerging organic micropollutants during chlorine-based disinfection processes. *Chemical Engineering Journal, 455*, 140476. DOI: 10.1016/j.cej.2022.140476

Sharma, A., Goel, H., Sharma, S., Rathore, H. S., Jamir, I., Kumar, A., Thimmappa, S. C., Kesari, K. K., & Kashyap, B. K. (2024a). Cutting edge technology for wastewater treatment using smart nanomaterials: Recent trends and futuristic advancements. *Environmental Science and Pollution Research International, 31*(48), 58263–58293. DOI: 10.1007/s11356-024-34977-1 PMID: 39298031

Sharma, A., Kumar, N., Sillanpää, M., Makgwane, P. R., Kumar, S., & Kumari, K. (2022). Carbon nano-structures and functionalized associates: Adsorptive detoxification of organic and inorganic water pollutants. *Inorganic Chemistry Communications, 141*, 109579. DOI: 10.1016/j.inoche.2022.109579

Sharma, S., Gupta, S., Kaur, S., Kumar, D., Banerjee, P., & Nadda, A. K. (2023). *Polymeric Membranes for Water Treatment*., DOI: 10.1007/978-981-19-9176-9_1

Sharma, S., Sharma, K., & Grover, S. (2024b). Real-Time Data Analysis with Smart Sensors. In *Application of Artificial Intelligence in Wastewater Treatment* (pp. 127–153). DOI: 10.1007/978-3-031-69433-2_5

Singh, S., Tiwari, H., Verma, A., Gupta, P., Chattopadhaya, A., Singh, A., Singh, S., Kumar, B., Mandal, A., Kumar, R., Yadav, A. K., Gautam, H. K., & Gautam, V. (2024). Sustainable Synthesis of Novel Green-Based Nanoparticles for Therapeutic Interventions and Environmental Remediation. *ACS Synthetic Biology*, *13*(7), 1994–2007. DOI: 10.1021/acssynbio.4c00206 PMID: 38899943

Singh, S. P., Sharma, M. K., Kumar, C., & Gaur, R. C. (2022). Effect of Filter Media on Performance of Biofilm Reactor for Treatment of Domestic Wastewater. *Journal of Hazardous, Toxic and Radioactive Waste*, *26*(3), 04022014. DOI: 10.1061/(ASCE)HZ.2153-5515.0000700

Soni, V., Raizada, P., Singh, P., Cuong, H. N., S, R., Saini, A., Saini, R. V., Le, Q. V., Nadda, A. K., Le, T.-T., & Nguyen, V.-H. (2021). Sustainable and green trends in using plant extracts for the synthesis of biogenic metal nanoparticles toward environmental and pharmaceutical advances: A review. *Environmental Research, 202*, 111622. DOI: 10.1016/j.envres.2021.111622 PMID: 34245729

Stoler, J., Jepson, W., Wutich, A., Velasco, C. A., Thomson, P., Staddon, C., & Westerhoff, P. (2022). Modular, adaptive, and decentralised water infrastructure: Promises and perils for water justice. *Current Opinion in Environmental Sustainability*, *57*, 101202. DOI: 10.1016/j.cosust.2022.101202

Sunita, S., & Chakrabarty, J. (2022). Size-Dependent Catalytic Properties of Nanomaterials, Their Suitability in Terms of Efficiency, Cost-Effectiveness and Sustainability. In *Advanced Nanocatalysts for Biodiesel Production* (pp. 199–215). CRC Press., DOI: 10.1201/9781003120858-9

Thakur, S., Chandra, A., Kumar, V., & Bharti, S. (2025). *Environmental Pollutants: Endocrine Disruptors/Pesticides/Reactive Dyes and Inorganic Toxic Compounds Metals*. Radionuclides, and Metalloids and Their Impact on the Ecosystem., DOI: 10.1007/978-981-97-7221-6_3

Tiwari, A., Tiwari, V., & Banik, B. K. (2025). Waste Management: Nano Photocatalysis as an Efficient Future Pathway. In *Emerging Trends in Photoredox Synthetic Transformation* (pp. 305–339). DOI: 10.1007/978-981-97-8205-5_11

Tong, T., Liu, X., Li, T., Park, S., & Anger, B. (2023). A Tale of Two Foulants: The Coupling of Organic Fouling and Mineral Scaling in Membrane Desalination. *Environmental Science & Technology*, *57*(18), 7129–7149. DOI: 10.1021/acs.est.3c00414 PMID: 37104038

Urošević, T., & Trivunac, K. (2020). Achievements in low-pressure membrane processes microfiltration (MF) and ultrafiltration (UF) for wastewater and water treatment. In *Current Trends and Future Developments on (Bio-)* (pp. 67–107). Membranes., DOI: 10.1016/B978-0-12-817378-7.00003-3

Woolf, A. D., Stierman, B. D., Barnett, E. D., Byron, L. G., Woolf, A. D., Stierman, B. D., Barnett, E. D., Byron, L. G., Bole, A., Balk, S. J., Huerta-Montañez, G. M., Landrigan, P. J., Marcus, S. M., Nerlinger, A. L., Patel, L. H., Philipsborn, R., Zajac, L., Gray, K. A., Briskin, J., & Frantz, J. M. (2023). Drinking Water From Private Wells and Risks to Children. *Pediatrics*, *151*(2), e2022060645. Advance online publication. DOI: 10.1542/peds.2022-060645 PMID: 36995188

Xu, L., Campos, L. C., Li, J., Karu, K., & Ciric, L. (2021). Removal of antibiotics in sand, GAC, GAC sandwich and anthracite/sand biofiltration systems. *Chemosphere, 275*, 130004. DOI: 10.1016/j.chemosphere.2021.130004 PMID: 33640744

Yadav, J., Rani, M., Shanker, U., & Sillanpaa, M. (2024). Forging the advances of iron-based nanomaterials by functionalizing charge carriers regions for eradication of heavy metal ion contamination. *Inorganic Chemistry Communications, 164*, 112440. DOI: 10.1016/j.inoche.2024.112440

You, S. (2024). Cosmopolitan pathways from the Global South: How non-middle-class students become desirable Fulbright applicants. *Global Networks, 24*(2), e12458. Advance online publication. DOI: 10.1111/glob.12458

Zhang, J., Yu, S., Wang, J., Zhao, Z.-P., & Cai, W. (2023). Advanced water treatment process by simultaneous coupling granular activated carbon (GAC) and powdered carbon with ultrafiltration: Role of GAC particle shape and powdered carbon type. *Water Research, 231*, 119606. DOI: 10.1016/j.watres.2023.119606 PMID: 36680821

Zhao, H.-X., Zhang, T.-Y., Wang, H., Hu, C.-Y., Tang, Y.-L., & Xu, B. (2022). Occurrence of fungal spores in drinking water: A review of pathogenicity, odor, chlorine resistance and control strategies. *The Science of the Total Environment, 853*, 158626. DOI: 10.1016/j.scitotenv.2022.158626 PMID: 36087680

Zhao, S., Zhao, Y., Cui, Z., Zhang, H., & Zhang, J. (2024). Effect of pH, Temperature, and Salinity Levels on Heavy Metal Fraction in Lake Sediments. *Toxics, 12*(7), 494. DOI: 10.3390/toxics12070494 PMID: 39058146

ADDITIONAL READING

Ahmad, A., & Azam, T. (2019). Water Purification Technologies. In *Bottled and Packaged Water* (pp. 83–120). Elsevier., DOI: 10.1016/B978-0-12-815272-0.00004-0

Malviya, A., & Jaspal, D. (2021). Artificial intelligence as an upcoming technology in wastewater treatment: A comprehensive review. *Environmental Technology Reviews, 10*(1), 177–187. DOI: 10.1080/21622515.2021.1913242

Metcalf & Eddy. (2003). *Wastewater Engineering: Treatment and Reuse - Edition 4*.

Nourani, V., Elkiran, G., & Abba, S. I. (2018). Wastewater treatment plant performance analysis using artificial intelligence – an ensemble approach. *Water Science and Technology, 78*(10), 2064–2076. DOI: 10.2166/wst.2018.477 PMID: 30629534

Vanitha, S. K. R., & Boopathi, S. (2023). *Artificial Intelligence Techniques in Water Purification and Utilization.*, DOI: 10.4018/978-1-6684-4118-3.ch010

KEY TERMS AND DEFINITIONS

Activated Carbon: Efficient carbon absorbent generated from wood, coal, coconut cells, etc.

Agricultural Runoff: Water from farms that carries fertilizers, pesticides, and soil into nearby rivers and lakes, often polluting them.

Bionanocomposite: Eco-friendly materials made from natural fibers and nanoparticles, used to clean water and kill harmful microbes.

Climate Change: Shifts in weather patterns caused by global warming, affecting water cycles, rainfall, and freshwater availability.

Coagulation: A chemical process in which some coagulants are used that help to promote the aggregation of colloidal and suspended particles.

Decentralized Water Treatment: Community-level water treatment plants, society-level water treatment plants, and other small-scale water treatment infrastructure.

Disinfection: Removal of biological contaminants.

Endocrine-Disrupting Chemicals (EDCs): Pollutants that disturb the body's hormonal balance.

Environmental Degradation: Damage of nature by anthropogenic activities.

Flocculation: The settling process of floc particles after the coagulation process.

Heavy Metals: Toxic elements like Pb, Hg, and As that can contaminate water and harm health over time.

Industrial Effluent: Wastewater from factories that may contain chemicals, oils, and metals needing proper treatment before discharge.

Life Cycle Assessment (LCA): A process to measure a product's environmental impact—from raw material to disposal.

Membrane Filtration: Semipermeable barrier with tiny pores that block fine and micro particles, bacteria, and even viruses from water.

Microplastics: Tiny plastic fragments polluting surface and subsurface water

Nanobiotechnology: The merger of biology and nanotech.

PFAs (Per- and Polyfluoroalkyl Substances): "Forever chemicals" used in non-stick and waterproof products.

Sustainable Economic Development: Growing the economy while protecting nature—fulfilling today's needs without compromising the future.

Waterborne Diseases: Illnesses caused by poor water and sanitation facilities.

Chapter 7
Biodegradable Nanomaterials in Textiles

Dipan Kumar Das
https://orcid.org/0000-0003-0224-3295
Centurion University of Technology and Management

Padmaja Patnaik
https://orcid.org/0000-0003-4468-7871
Centurion University of Technology and Management, Odisha, India

Nibedita Nayak
Centurion University of Technology and Management, Odisha, India

Sudip Kumar Das
https://orcid.org/0009-0003-0814-0884
Dr. C.V. Raman University, India

ABSTRACT

Biodegradable nanomaterials have emerged as a revolutionary class of materials in the textile industry, offering sustainable solutions to combat environmental pollution and enhance fabric functionalities. These nanomaterials, derived from natural polymers like cellulose, chitosan, and starch, exhibit remarkable biodegradability, biocompatibility, and antimicrobial properties. Recent advancements in nanotechnology have enabled the development of smart textiles with improved mechanical strength, thermal stability, and responsive behaviors. The integration of biodegradable nanomaterials into textiles not only mitigates the impact of synthetic microplastics but also enhances the durability and functionality of fabrics. This chapter explores the synthesis, properties, and applications of biodegradable nanomaterials in textiles, drawing insights from recent research on biopolymer-based

DOI: 10.4018/979-8-3373-2617-7.ch007

nanocomposites, antimicrobial coatings, and sustainable production techniques. Additionally, challenges such as scalability, cost-effectiveness, and environmental safety are discussed.

1. INTRODUCTION

Biodegradable nanomaterials are emerging strong contenders in textile engineering owing to their sustainability in the environment and unique functionalities. These nature-based nanomaterials, when they reach the end of their life, can degrade harmlessly and reduce ecological footprint. Their nano-sized dimension offers better mechanical, chemical, and thermal properties. On application in textiles, they improve performance without compromising on sustainability. With growing concerns over synthetic textile pollution, biodegradable nanomaterials are a good substitute. Their application comes into the transition towards greener production and circular economy principles globally. In order to build clothing that protects people from environmental challenges, man has dominated the creation of fibres and textiles for ages. As a result, progressively disparate cultural and ethnic identification qualities have emerged. Natural materials like vegetable fibres, animal leather, and synthetic fabrics combine to create materials that are incredibly resilient and long-lasting for our clothing. However, population expansion will restrict the use of natural resources, including land and water, just for food, which makes the textile business non-sustainable. At the same time, because of their contribution to global climate change and their accumulation of toxins in the local environment, materials generated from petrochemicals will eventually be replaced by more biodegradable alternatives. By using biotechnological techniques to replicate nature's biosynthetic machinery, human ingenuity is making it possible to replace clothing. To satisfy consumer needs, this new vision must incorporate new features while maintaining the functionality and appearance of clothing. The textile sector may now offer bio-based materials with cognitive functions thanks to advancements in nanotechnology, which can give the desired qualities. This reveals nano-biotechnological techniques to address future societies' material consciousness and dressing up future societies (Fuentes et al., 2022). Because of their nanoscale size, organic nanomaterials have special qualities like improved biodegradability, reduced toxicity, and improved functionality in functional fabrics. The increasing demand for eco-friendly clothing and textiles has led to consideration of these biodegradable materials. They are being included into the production of multipurpose textiles and have energy-harvesting, antimicrobial, and self-cleaning properties. Utilizing organic nanomaterials as raw materials has several advantages for the economy and ecology, such as a decreased reliance on non-renewable resources, decreased waste, and possible maintenance cost reductions. The goal is to give producers and decision-makers a thorough understanding of the benefits and difficulties of incorporating organic nanoparticles into multipurpose textiles (Khan et al., 2025). To solve environmental concerns and improve functioning, the textile industry is putting more and more effort into creating sustainable and eco-friendly nano textile materials. The stain resistance, durability, and antibacterial properties of textiles have all been enhanced by nanotechnology. The incorporation of biobased and biodegradable nanoparticles improves textile performance and promotes sustainable production methods. The synthesis of nanomaterials will be improved, production will be scaled up, and full lifecycle sustainability will be guaranteed. It is anticipated that new developments like smart features and circular economy strategies will further transform the sector (Rahaman & Khan, 2024). Fig. 1 shows various applications.

Figure 1. Applications of organic nanomaterials in textiles (Rahaman & Khan, 2024).

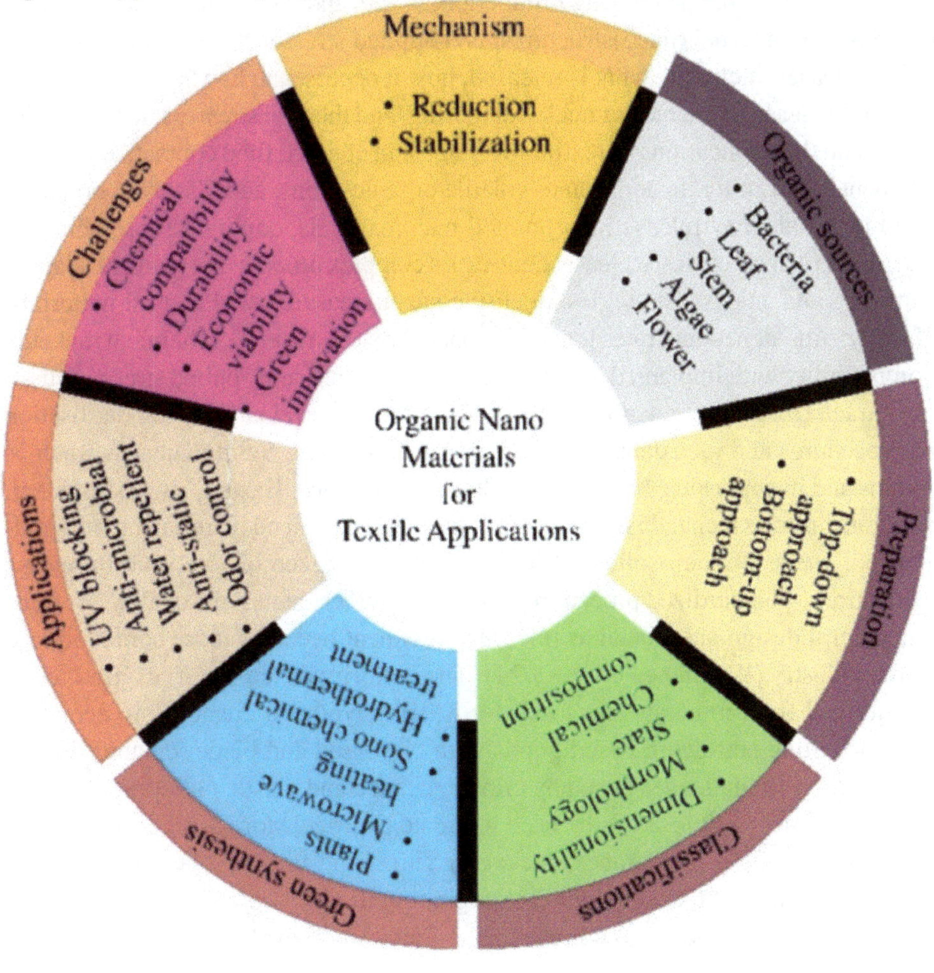

The rising production and consumption of textile items is closely related to the world's expanding population. The use of textiles and clothing materials, which is predicted to rise, is one of the main causes of microfiber production. The invisible pollution caused by textile microfibers, which have been found in marine species and sediments, is the result of the textile industry. shows that a significant percentage of the microfibers released from functionalized fabrics have hazardous qualities and are not biodegradable. The biodegradability of textiles is influenced by their material functionalization, with microfibers containing dyes, toxic chemicals, and nanomaterials potentially posing health risks to humans and other organisms. Preventative measures for reduction, covering sustainable production, consumer, end of life, domestic washing, and wastewater treatment phases was discussed (Periyasamy, 2023). It is necessary to evaluate the type and level of exposure in order to guarantee the safe usage of materials. Throughout their life cycle, solid materials like plastics, metals, coatings, and cements undergo some degree of degradation, and releases may happen during production, usage, and end-of-life. The material's composition, internal (nano)structures, and applied stresses during its lifecycle all affect releases, including what is released, how it occurs, and how much material is released. releases from mechanical, weathering, and thermal stressors in detail, with a focus on the applications of textile washing, skin contact, food contact, and fused-filament 3D printing. In addition to volatile organics, ions, and dissolved organics, solid materials may release incorporated nanomaterials, composite fragments, or micro- and Nano plastics. Release identity is a complex process that requires adapted sampling and analysis strategies, quality control measures, and control materials. Usually, time-dependent rates that are impacted by the material's chemical makeup, compatibility with implanted engineered nanomaterials, and applied stress are used to characterise release quantities. To produce reliable evaluations, standardisation of procedures and metadata documentation are essential. Solid materials, such as micro- and nanoplastics, have methodological synergies. Hazard, risk assessment, and transfer to associated issues such as tyre wear, advanced materials, biodegradable polymers, and non-solid matrices should all be taken into account in release evaluations. Standardized procedures, agreed-upon test materials, quality controls, and careful design are crucial as release assessment becomes more commonplace in the industry (Wohlleben et al., 2024). The majority of the materials used today are derived from minerals and fossil feedstocks. We must create a bio economy by substituting petroleum-based goods with bio-based and biodegradable ones in response to the challenges posed by climate change and plastic pollution. A versatile biopolymer, cellulose can be used to create hydrogels for absorbents, aerogels for insulation, filter membranes, packaging films, and fibres for textiles and reinforcements. Relevant stakeholders are beginning to view wood-based cellulose as sustainable, biodegradable, and renewable. Some study explores the potential and challenges of using cellulose-based materials as a sustainable alternative to synthetic materials, analysing their mechanical properties and tensile properties, and comparing them to other materials, aiming to provide insights into their potential applications (Wang et al., 2021).

Researchers worldwide are working to improve the efficiency and dependability of bio based materials due to growing concerns about protecting our ecosystem and environment from the excessive use of dangerous compounds derived from plastic. It is thought that bio based materials could be a good substitute for synthetic materials, fulfilling their function while being incredibly environmentally friendly. The replacement of synthetic materials is still a long way off, despite the fact that bio based materials are currently widely employed in many industrial and domestic applications. To fully utilise these environmentally friendly materials, more advancements are needed in a number of industries, including the energy sector, green chemistry, medicinal, and food packaging. Researchers are exploring bio based materials like bio-fibers, biopolymers, biofilms, and bio composites to prevent overuse issues and explore their applications and potential prospects in the field (Tyagi et al., 2023).

2. LITERATURE REVIEW

Historical Context and Conceptual Clarifications

Integration of nanotechnology in the textile industry began in the late 1990s and early 2000s when the first applications were aimed at enhancing stain resistance and durability. Gradually, this changed to biomedical and eco-friendly applications, leading to smart textiles and biodegradable nanomaterials. With technological revolution and the awareness of the environment, the application of biodegradable nanomaterials has grown tremendously.

For establishing a firm conceptual basis, the following terms are defined:

Biodegradability is the property of a material to biodegrade naturally by biological entities into harmless constituents.

Nano-finishing is the technique of depositing nanomaterials on fabrics to enhance or provide new qualities like antimicrobial or UV resistance.

Smart textiles are fabrics embedded with nanotechnology or electronics to provide functionalities like sensing, actuation, or responsiveness to environmental inputs.

Thematic Literature Review

1. Sources and Types of Nanomaterials: Studies have explored a wide array of biodegradable nanomaterials sourced from cellulose, lignin, chitosan, starch, and bacterial polymers. Each has distinct advantages; for example, cellulose nanocrystals exhibit high mechanical strength, while chitosan offers natural antimicrobial properties.

Table 1. Advantages and Disadvantages of Nanomaterials in the Textile Industry

Nanomaterial Source	Advantages	Disadvantages
Cellulose	High mechanical strength- Renewable and abundant Biodegradable	Hydrophilic nature limits water, resistant applications, Requires modification (He et al., 2021)
Chitosan	Antimicrobial, Biocompatible, Biodegradable	Poor solubility in neutral/alkaline conditions, Needs chemical enhancement (Li et al., 2021)
Lignin	Antioxidant properties, Abundant by-product, Renewable	Dark Color limits use in light-coloured fabrics, Variability in composition (Raman et al., 2022)
Starch	Biodegradable, Cost-effective, Easy to process	Low mechanical strength, Moisture sensitivity (Moore, 1967)
Bacterial Polymers	High purity and strength, Excellent water retention, Biocompatible	High production cost, Scale-up limitations (Fernandes et al., 2021)
Algae/Microbial	Sustainable sourcing, Biodegradable, Novel functionalities	Less industrial maturity, Requires consistent cultivation conditions (Arun et al., 2020)

2. Processing and Functionalization Processes: Acid hydrolysis, enzymatic treatment, and mechanical fibrillation are typical extraction processes. Plasma treatment, silanization, or grafting have been used as functionalization processes to improve compatibility and performance. Methods of green synthesis are given more importance in terms of scalability and environmental compatibility.

Table 2. Processing and Functionalization Techniques of Biodegradable Nanomaterials in Textiles

Category	Technique	Description	Purpose/Benefits	Limitations
Extraction (Processing)	Acid Hydrolysis (Costa et al., 2022)	Uses strong acids (e.g., sulfuric or hydrochloric acid) to break down biomass into nanoscale cellulose.	Produces highly crystalline cellulose nanocrystals (CNCs) with uniform particle size.	Corrosive chemicals; requires neutralization and safe disposal.
	Enzymatic Treatment (Besegatto et al., 2018)	Employs enzymes like cellulase to selectively degrade amorphous regions of cellulose.	Eco-friendly and mild process; maintains structural integrity of nanofibers.	Slower process; enzyme cost and specificity are challenges.
	Mechanical Fibrillation (Ma et al., 2021)	High-shear or high-pressure mechanical force used to isolate nanofibers from biomass.	No chemical residues; scalable and clean method.	High energy consumption; wear and tear on equipment.
Functionalization	Plasma Treatment (Jelil, 2015)	Surface treatment with ionized gases to activate functional groups on nanomaterials.	Improves surface energy, bonding with textile substrates, and reactivity.	Equipment cost; may not penetrate deeper fiber layers.
	Silanization (Maia et al., 2022)	Use of silane coupling agents to chemically modify surfaces with hydrophobic or reactive groups.	Enhances compatibility with hydrophobic textiles; improves dispersion in composites.	May involve solvents; not always biodegradable.
	Grafting (Sun & Sun, 2002)	Covalent attachment of functional polymers onto nanomaterial surfaces.	Customizes functionality (e.g., antimicrobial, UV-resistance); improves stability.	Multi-step process; requires precise control and characterization.
Synthesis Strategy	Green Synthesis (Nambela et al., 2020)	Biological or eco-friendly chemical methods using plant extracts, enzymes, or microbes.	Reduces toxic waste; aligns with circular economy and sustainability goals.	Often limited to lab scale; variable consistency and yield.

3. Biomedical and Biodegradable Applications: Biodegradable nanomaterials have found applications in wound dressings, antimicrobial textiles, UV-blocking apparel, and self-cleaning surfaces. Smart textiles with biosensors and drug-delivery systems are being actively pursued.

Critical Evaluation and Comparative Insights

Several comparative studies indicate that cellulose nanocrystals have better mechanical reinforcement properties, whereas chitosan nanocomposites have better antimicrobial use. Yet, chitosan is limited by solubility, and chemical modification is needed for more general applicability. Similarly, despite nano-lignin being rich and renewable, its darkness constrains its application in aesthetic products. Scalability is another important challenge: most green synthesis approaches are currently limited to laboratory scale, and commercialization is the pacing item.

Summary of Literature Gaps

i. Though considerable advances have been achieved, certain research gaps still exist:
ii. Inadequate standardized testing of biodegradability under various environmental conditions.
iii. Inadequate lifecycle analysis of smart biodegradable textiles.
iv. Limited scalability and cost-effectiveness of the synthesis routes of green nanomaterials.
v. Lack of adequate data on long-term human and environmental effects of bio-degradable nano-finished textiles.

This chapter aims to fill some of these gaps, particularly by placing emphasis on material selection, synthesis routes, and performance characteristics of interest in practical textile use.

3. SOURCES AND TYPES OF BIODEGRADABLE NANOMATERIALS

Biodegradable nanomaterials from textiles are obtained mainly from renewable resources such as plants, algae, and microorganisms. Popular types include Nano cellulose, Nano lignin, chitosan nanoparticles, and starch nanocomposites. These materials show superior biodegradability, non-toxicity, and compatibility with the substrate of the textile. Their applicability enables them to be viable for uses from reinforcement agents to active functional coatings. Further research continues to widen sources and increase efficiency in extraction. Their categorization now also includes polymeric nanoparticles, hybrid nanostructures, and bio-nanocomposites as interest continues to increase. These days, nanomaterials are being studied because of their possible use in pharmaceutical sciences, cosmeceuticals, foods, textiles, and gene and tissue engineering. Because of their abundance in nature, low immunogenicity, biodegradability, and biocompatibility, biopolymers are favoured over synthetic polymers in this situation. Amylose and amylopectin combine to form the well-known biopolymer known as starch. It has recently been considered as a possible option for creating nanomaterials for use in biomedicine. There was discussion of many starch-based nanoparticles specifically created for medication delivery uses. (Ali et al., 2021). A non-toxic, biocompatible, and biodegradable polymer derived from chitin is called chitosan. Chitosan has good mechanical and thermal qualities, forms films, and has antibacterial activity. As a potential hybrid material, bio nanocomposites provide a renewable use. Bio nanocomposites based on chitosan offer promise as raw materials derived from natural polymers. Bio nanocomposites based on chitosan improve therapeutic properties like nontoxicity. There have been several initiatives in recent years to substitute renewable and biodegradable materials for petrochemical goods. The hardest aspect of this strategy is producing bio-based materials with qualities and capabilities comparable to those of manufactured goods. Numerous naturally occurring polymers, including chitin, collagen, alginate, cellulose, and starch, are appealing options because they may lessen reliance on artificial goods and have a favourable environmental impact. Another special bio-based polymer with outstanding inherent qualities is chitosan. It has excellent mechanical strength, strong thermal stability, and antibacterial and film-forming qualities. Nanotechnology has utilized chitosan-based materials, resulting in eco-friendly and economically enhanced bio nanocomposites with enhanced physical and chemical properties (Azmana et al.,2021). Biodegradable nanomaterials from biopolymers like cellulose, lignin, chitin, and starch are proposed for eco-friendly functional Nano systems. Nanolignin was extracted from coconut fibers through controlled hydrolysis and compared to nanolignins prepared through high shear homogenization and ultra-sonication processes. The resulting Nano lignin was analysed by FTIR for its chemical nature, DSC for its glass transition temperature, XRD for its crystallite size, AFM and FE-SEM for its shape, and the DLS particle size analyser for its size. While the homogenization and ultra-sonication methods produced 81.4 and 64.3 percent of the Nano lignin, respectively, the microbial process produced 58.4 percent. Thus, it is possible to extract Nano lignin, a valuable chemical, from coconut fibres for use in textile, medicinal, and environmental applications, among other uses (Juikar & Vigneshwaran, 2017). The need to investigate cutting-edge, environmentally friendly materials with remarkable physicochemical qualities for sustainable development is growing daily. All forms of Nano cellulose, including bacterial cellulose, cellulose nanofibers, and cellulose nanocrystals, have recently attracted a lot of interest in both the scientific and industrial communities because of their alluring intrinsic qualities, which include high surface area, biodegradability, high mechanical qualities, and a high concentration of hydroxyl groups for modification. Although Nano cellulose's hydrophilic characteristic limited its use in many applications, the presence of a functional group on its surface offers a platform for surface modification using a variety of methods. The search for environmentally friendly materials with exceptional physicochemical properties is increasing. Nano cellulose, including bacterial cellulose, cellulose nanofibers, and cellulose nanocrystals, is gaining interest due to its high surface area, biodegradability, mechanical properties, and high concentration of hydroxyl groups for modification. Despite its hydrophilic nature, its surface's functional group provides a platform for surface modification (Kumar et al.,2021). Fig. 2 shows sources and applications of biodegradable nanomaterials.

Figure 2. Sources and applications of Biodegradable Nanomaterials

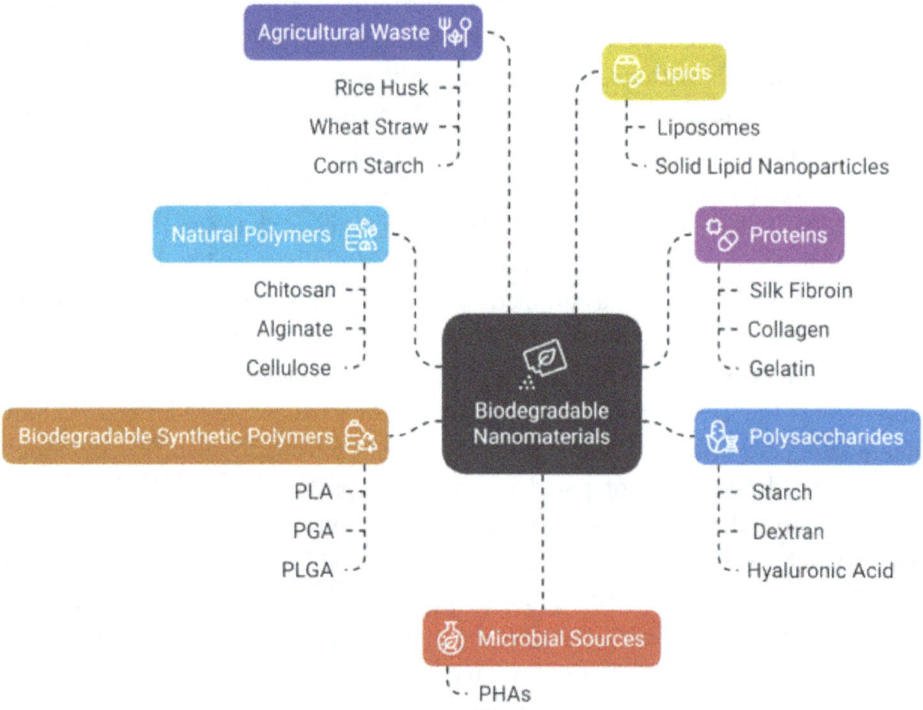

3.1. Extraction and Processing Methods of Nano Cellulose and Nano Lignin

Nano cellulose can generally be extracted from plant fibers by mechanical, chemical, or enzymatic treatment, e.g., acid hydrolysis or high-pressure homogenization. Nano lignin is a by-product of paper manufacturing that is purified using solvent extraction or hydrothermal treatments. The techniques seek to retain the nanoscale structure of the materials with maximal yield and purity. The final nanoparticles provide excellent surface area and mechanical strength. The processing methods are being optimized to minimize energy consumption and scalability. Their eco-friendly extraction favours the sustainability aims of textile manufacturing industries. Nano cellulose (NC) has been the subject of much research because of its many uses, natural availability, biodegradability, and non-toxicity. A promising renewable resource for creating sustainable textiles is biomass-derived NC, which may be used in place of non-biodegradable petrochemical chemicals. Nonetheless, the textile industry needs to be able to sufficiently investigate its possible uses inside the textile industry. The textile business creates diverse, worldwide goods with unique physicochemical characteristics that are used in many aspects of our life. Understanding the structure-function link between textile substrates and NC is therefore crucial, as is adapting NC-based renewable techniques for sustainable materials to satisfy the expanding demand and particular end applications of textile goods. An overview of recent research on nanomaterials (NC) applications in textiles, including enhancing wet processing, reinforcing fibers, and improving dye adsorption from wastewater was discussed. It also highlights challenges and opportunities for developing functionalized NC to enhance cost-effectiveness and sustainability, thereby expanding green and sustainable textile applications (Biranje et al., 2024).

Bio-nanomaterials are becoming more and more popular because of their environmentally benign and renewable qualities. One of the most cutting-edge materials for use in composite manufacturing, food, medicine, and other fields is nanocrystalline cellulose (NCC). In this work, NCC was effectively recovered from textile waste made of cotton by combining mechanical and chemical techniques. Following hydrolysis of the cellulose fibres with a diluted HCl solution, neutralisation, and drying, microcrystalline cellulose (MCC) with diameters ranging from 7 to 15 μm and lengths up to 300 μm was produced (as observed via optical microscopy and scanning electron microscopy, SEM). The extraction of nanocrystalline carbon (NCC) from undyed cotton textile wastes was achieved using a NaOH/urea solution with mechanical treatment, resulting in nanoscale dimensions. DLS and AFM analysis confirmed the presence of nanostructures, while FTIR spectroscopy confirmed the conversion of cellulose I to cellulose II, confirming the successful transformation into NCC. This marks the first time NCC was obtained from textile wastes (Işıtan et al., 2025).

3.2. Biodegradable Nanocomposites: Properties and Classification

Biodegradable nanocomposites are natural or bio-based matrices enhanced with nanoscale components like clay or nanocellulose. These composites are designed for specific textile applications, including antimicrobial clothing, flame retardants, and breathable membranes. Their biodegradability is their biggest advantage over traditional composites. Long-term innovation is making it feasible to create textile materials with many uses and minimal environmental effect. Cellulose is an abundant and renewable biopolymer that has recently attracted attention from nanotechnology. With special attention to the sources of cellulose, methods for isolating and purifying it, and the various procedures used to produce cellulose nanocrystals and cellulose nanofibers, this article examines the synthesis, functionalization, and industrial applications of cellulose-derived nanoparticles. Biodegradable materials are essential for reducing global pollution and conserving non-renewable resources. A thermoplastic poly-3-hydroxyalkanoates (PHA) storage material called BIOPOLTM was recently created by microorganisms. A degradation test was conducted on them to demonstrate the involvement of fungus and marine conditions in the deteriorating process (Matavulj & Molitoris, 2009).

4. FUNCTIONALIZATION AND FABRICATION TECHNIQUES

For improved compatibility with textiles, biodegradable nanomaterials are typically chemically treated or blended with other chemicals. These types of functionalization techniques offer higher dispersion, adhesion, and strength in fiber or fabric applications. Fabrication procedures such as electrospinning, solution casting, or in-situ polymerization are utilized for integrating these materials into textile matrices. The aim is to maintain both functional behaviour like strength, hydrophobicity, or bioactivity and biodegradability. Control of fabrication at the nanoscale ensures uniformity and reproducibility. Such progress makes possible tailor-made design of high-performance, sustainable textiles.

4.1. Surface Modification and Hybridization for Enhanced Textile Compatibility

Surface modification processes such as grafting, plasma treatment, or silanization are applied to design the interface of nanomaterials to textile fibers. These treatments increase compatibility, stability, and dispersion of nanoparticles in fabrics. Hybridization with polymers or other nanomaterials may produce synergistic effects, leading to multifunctional textiles. Surface modification also enhances interactions with dyes, coatings, or other functional agents. These modifications are necessary for durability against washing and wear. These treatments open the door to new applications in smart and technical textiles.

The majority of commercial polymers and nanomaterials are inert, which limits their use in a variety of industrial domains. Surface alterations that enhance physicochemical and biological characteristics, including adhesion, printability, wetting, and biocompatibility, can address this. Grafting certain moieties and conjugate compounds that enhance material performance is made possible by polymer functionalization. Several methods for grafting functional groups onto surfaces have been developed in the last several decades by both industry and academics. The need for more knowledge of cell-polymer interactions and nano safety was brought to light by the surface decorating of polymers and nanomaterials with an emphasis on commercial applications in food packaging, textile, medicinal, and water treatment (Makvandi et al., 2021). Every industrial sector has undergone a transformation due to the development of nanotechnology. By 2020, it is anticipated that the combined market share of all nanomaterial kinds would surpass \$10 billion. The majority of nanomaterial production is dependent on energy-, labor-, and cost-intensive processes such as lithography, milling, etching, vapour deposition, sol-gel, precipitation, pyrolysis, solvothermal, mechanochemistry, ultrasound, microwave irradiation, and chemical reduction. Environmentally friendly techniques for creating new nanomaterials, referred to as "greener nanotechnology," are driving current advancements in nanoscience. Greener nanotechnology uses the 12 green chemistry principles to synthesise nanoparticles in a less hazardous, safer, and energy-efficient way. For the creation of nanomaterials, the more environmentally friendly techniques make advantage of biological characteristics found in nature and obtain energy from the photochemistry and metabolic processes of bacteria, fungus, viruses, plants, and algae. Microorganisms, proteins, enzymes, and biomineralizing organisms are essential for the biological/greener synthesis of nanomaterials. These organisms allow minerals to accumulate as nanoparticles and control extracellular formation. The advanced method also allows for the formulation of nanomaterials with vitamins, sugars, plant polyphenols, and agricultural waste residues (Thangadurai et al., 2020).

4.2. Nanocoating and Nanofinishing Processes

Nanocoating involves the application of a thin film of nanoparticles on fabric surfaces to provide them with preferred properties like water repellency, UV resistance, or antimicrobial activity. Techniques involve dip-coating, spray-coating, and layer-by-layer assembly. Nano finishing enables the permanent introduction of nanomaterials into fabrics without inducing significant changes in fabric texture and permeability. The treatments enable balance between function and comfort. Biodegradable Nano coatings are gaining traction as green alternatives to fluorinated or heavy metal-containing finishes. They extend the life of products but reduce harm to the environment. Many sectors, including textile, pulp and paper, printing, leather, and others, employ dyes, which are complex, poisonous, and non-biodegradable aromatic chemicals, as colourants. One of the main concerns for environmental safety is the removal of dyes from contaminated wastewater. Using ultrasonic synthesis to create nanoscale zero valent iron (nZVI) particles and using them to extract methylene blue (MB) from aqueous solutions is the goal of the current investigation (Kamali et al., 2018). The intrinsic flammability of biomass-based fibres severely restricts the uses of bio-derived nanoparticles, despite the fact that they are attractive candidates for spinning high-performance sustainable textiles. Enhancing fibre flame retardancy while preserving superior mechanical performance is still urgently needed. Here, we provide a unique and effective method for creating robust, extremely durable, and flame-retardant nanocellulose/clay/sodium alginate (CRS) macrofibers using wet-spinning and directed drying, which is motivated by the structural characteristics of layered nanoclay. A novel inorganic structure with superior mechanical and thermal characteristics is produced by precisely orienting and arranging nanoclay within macrofibers. High flame retardancy and durability are attributed to the anisotropic structure. Temperature tolerance, thermal conductivity, and flame retardancy are all enhanced by rectorite nanosheets. High-strength, extremely durable, and flame-retardant fibres for biodegradable next-generation artificial fibres might be produced using this highly efficient method (Xie et al., 2023).

Nanotechnology has been used extensively in the textile industry recently, particularly in the creation of cosmetotextiles. This chapter examines how nanocoatings might improve these materials' sustainable and practical qualities. The chapter provides a thorough review of contemporary nanocoating methods, materials, and their effects on the environment by incorporating knowledge from materials science, sustainability, and cosmetic technology. Additionally, it included methods for creating environmentally friendly nanocoatings and carrying out lifetime studies, as well as a critical evaluation of the potential and problems associated with applying sustainable nanocoatings in cosmetotextiles. Nanocoatings are therefore crucial for enhancing effective, environmentally friendly cosmetotextiles and present opportunities for further study and industrial use. The development of biodegradable nanomaterials, toxicological research to guarantee user safety, and the use of circular economy concepts in the cosmetic textile sector are important avenues for future research (Yingngam, 2024). Over the past few decades, nanotechnology has helped build sophisticated and useful textile materials. Natural fiber-based textiles are now finished with a wide range of nanomaterials, both organic and inorganic, to provide properties like antibacterial, UV-resistant, flame-retardant, self-cleaning, super-absorbent, easy-care, and many more. It's never easy to incorporate nanoparticles onto natural fibres without compromising their native fibre qualities. To apply or synthesise nanoparticles inside the architecture of natural fibres, methods such as sol-gel, chemical and physical vapour deposition, nanoencapsulation, and plasma treatment are employed. The application of nanotechnology has opened up several opportunities for the textile finishing industry to produce textiles composed of natural fibres with multifunctional qualities (Paul & Ghosh, 2024).

4.3. Bio Functional Textile Design Using Nanoparticles and Biopolymers

Bio-functional textiles are engineered using the fusion of natural polymers and bioactive nanoparticles to achieve therapeutic or protective functionality. Examples are nanocellulose for moisture management or chitosan nanoparticles for antimicrobial purposes. They suit medical textiles, sportswear, and personal care. Integration is done using coating, embedding, or fibre-spinning techniques. Such fabrics provide features such as odor management, wound care, or sun protection but remain biodegradable. The convergence of nanotechnology and biology is revolutionizing the way we approach sustainable performance apparel. Multifunctional Using a bioinspired method, a polyhydroxybutyrate/poly-3-caprolactone (PHB/PCL) wound dressing was created. Biocompatible, antibacterial Melanin-TiO2 nanostructures were added to biodegradable electrospun PCL/PHB mats. The water absorption of PCL/PHB/Melanin-TiO2 fibres was enormous. PCL/PHB mats were given significant hydrophilicity, in-vitro bioactivity, and pertinent antibacterial activity by the surface melanin-TiO2 coating. Creating cutting-edge materials for wound dressings is a difficult yet unfinished endeavour. These systems are intended to serve as temporary skin replacements, carrying out a variety of tasks such as fluid absorption, antimicrobial activity, and facilitating cell migration and proliferation to aid in the process of skin regeneration. A multifunctional textile for wound dressing applications using biodegradable polyhydroxybutyrate/poly-3-caprolactone (PHB/PCL) mats fabricated by electrospinning was developed. The mats mimic the extracellular matrix, providing structural and biochemical support for tissue regeneration. The mats were modified with hybrid Melanin-TiO2 nanostructures, enhancing their antimicrobial activity against both Gram(+) and Gram(−) strains. The study provides a new paradigm for designing active wound dressings for regenerative medicine (Avossa et al., 2021).

Natural polymers have been utilised in biomedical applications in recent years; their nontoxic and biocompatible qualities make them the most favoured option. Chitosan, the second most readily accessible and abundant natural polysaccharide polymer after cellulose, has been effectively tested and described in nanopractical applications for the production of industrial and medical textiles. It is biodegradable and possesses a variety of natural qualities. The reactivity of the primary amino group and the primary and secondary hydroxyl groups in the different application areas is the basis for the synthesis of chitosan. It has improved thermophysiological qualities, functions as a bacteriostatic fabric adherence agent, and may be used as an antibacterial finish. The recycling of chitosan through precipitation, flocculation, or ultrafiltration processes, focusing on its biocomposite nanomaterials' applications in the biomedical field was explored (Singha et al., 2021). Recent advancements in surface nano-structured textiles have shown potential in biomedical applications. Coatings made of biopolymer films and nanoparticles on textile substrates have reduced hospital-acquired infections in the past decade. The combination of metal and oxide nanoparticles with biopolymers enhances textile properties. Few review papers provide a thorough understanding of how textiles may be surface tailored using nanoparticles and biopolymers, which can be used as an alternative to modify textile surfaces and provide them biocidal properties. The need for further research in biomedical textiles, focusing on antimicrobial properties was emphasized. It highlights the advantages of simultaneous nanoparticles-biopolymer deposition, which enhances wash-fastness and biocompatibility, stabilizes colloidal dispersions of nanoparticles, and uses common metal and metal oxide nanoparticles and biopolymers for antibacterial textile applications. The study also highlights the potential of these materials for long-lasting antibacterial effects (Vrinceanu et al., 2022).

5. APPLICATIONS IN TEXTILE ENGINEERING

Biodegradable nanomaterials have opened new doors in textile engineering, which offer environmentally friendly solutions to conventional materials. They are used to create intelligent textiles with built-in attributes like self-cleaning or deodorizing. Their biocompatibility also renders them suitable for medical and sanitary uses. Apart from that, they enable sustainable coloration, coating, and finishing treatments. The ecological safety and adaptability of these materials are transforming the textile industry. From apparel to medicine, their applications are increasing exponentially.

5.1. Smart and Functional Textiles: Antimicrobial, UV-Resistant, self-Cleaning Properties

Biodegradable nanoparticles are employed to design textiles that proactively react to climatic conditions or provide enhanced protection. Chitosan, zinc oxide, and silver-based biodegradable nanomaterials give antimicrobial functionality. Nano cellulose and titanium dioxide can offer UV-blocking and self-cleaning features. These properties enhance the functional properties of textiles in sportswear, outdoor gear, and healthcare. What is most important is that these functionalities are achieved without compromising the eco-friendliness of the final product. Intelligent biodegradable fabrics combine innovation and sustainability to serve modern needs. Chitosan is a naturally occurring organic polymer that has been utilised as an antibacterial agent and has good biocompatibility, biodegradability, and antimicrobial qualities. However, chitosan's extensive use is hampered by the ease with which its antibacterial qualities can be affected by both internal and external causes. Thus, the focus of research has shifted to enhancing the antibacterial action and encouraging the use of chitosan. A synergistic complementing effect between organic and inorganic components may be obtained when chitosan is added to inorganic nanoparticles to create composite materials. This will significantly increase the materials' antibacterial activity. Chitosan/inorganic nanocomposites have been widely employed in the fields of water treatment, food, textiles, cosmetics, and medicine due to their advantageous mechanical strength and biocompatibility, which has piqued the interest of many researchers. They study about the antimicrobial properties of chitosan, its mechanisms, and the application of nanocomposites made from chitosan and other inorganic materials, highlighting their potential for development in the field (Li et al., 2018). The primary constituent of crab shells, chitin polysaccharide, is the source of chitosan. Chitosan is a polymer that dissolves in acidic liquids and is biocompatible, nontoxic, and biodegradable. The pharmaceutical and medical industries make extensive use of it. One of chitosan's appealing qualities is its antimicrobial activity against various bacterial, fungal, and viral diseases. This makes chitosan useful for biological applications such as textile, food, tissue engineering, agriculture, and environmental protection. Furthermore, chitosan can boost immunity, increase feed conversion, and encourage growth in fish, crustaceans, poultry, and animals. The limitations of chitosan's antimicrobial capabilities due to its water solubility was discussed. It also highlights the preparation, factors affecting activity, structure, mechanism, and application of chitosan derivatives, providing an update on their potential (Lin et al., 2023).

Over the past thirty years, the field of nanomaterials research has advanced significantly. Nanomaterials are now easier and more accessible to synthesise, and it is also becoming more practical to integrate them into current technology. Because of their special qualities that are crammed into their tiny sizes, nanomaterials provide remarkable prospects that are being utilised to enhance a variety of applications. Nanotechnology is also helping with studies on textiles and clothing. Textiles have been enhanced with nanomaterials including graphene, carbon nanotubes, metal nanoparticles (Ag/Au, TiO2, ZnO, and CuO), and bio-based nanomaterials. Conductivity, water resistance, microbiological control, strength, anti-static qualities, colour blocking, etc. are a few examples of these functions. The introduction of nanomaterials has not only improved the goods that textile manufacturers produce, but it has also helped them because of the fact that these materials may be utilised to break down pollutants in textile effluent streams, which have long presented significant environmental problems. Therefore, it appears that the intelligent integration of nanomaterials is a key component of the textile industry's future. It should be noted that despite the amazing potential that nanomaterials provide the textile sector, there are still some issues that need to be resolved, such cost and scalability (Chakrabarty & Jasuja, 2022). Despite the enormous potential for a wide range of applications, the negative effects of nanofinished medical textiles on living cells have caused real concerns following their use in the customer and human services sectors. Because they are too small and can easily pass through living cell walls, they can have a very negative impact on the environment and the populace. Apart from the significance of engineered nanomaterials and all of their different applications, there is a knowledge gap between the cutting-edge developments in nanotechnology and nanosafety, which has led some organisations to emphasise the need for studies pertaining to occupational safety and health (Osman, 2020).

5.2. Biodegradable Nanomaterials for Wound Dressings and Biomedical Textiles

Biocompatibility, breathability, and controlled degradation are offered by biodegradable nanomaterial-based medical textiles. Chitosan and nanocellulose are utilized in wound dressings due to their antimicrobial properties and moisture holding ability. They can be made to release bioactive molecules or drugs slowly. Application reduces the infection risk and accelerates healing promptly. Removal of biodegradable components is not required, which makes them suitable for delicate medical uses. Their importance in personalized healthcare textiles is increasing tremendously.

Due to its significant socioeconomic influence on clinical practises, wound healing and ulcer treatment are broad topics that have sparked clinical study. With their advantages and disadvantages, materials at the microscale have greatly aided in the treatment of ulcers and wound healing throughout the past ten years of clinical science study. The present study covers the novel materials at nanoscale that have been created and made to solve the existing difficulties in wound care but significantly less familiar to doctors for tissue healing and ulcer control. Better materials than conventional wound dressings that imitate tissue in vivo were made possible by advancements in tissue engineering techniques that ranged from micro to nanoscale dimensions. Recent developments in polymer chemistry and material sciences have produced a whole class of new nanomaterials, from hydrogel and innovative electrospun polymers for wound care and tissue healing to bioactive tissue scaffolds. In specialised wound care, nanoscale materials with biodegradable chemistries and molecular compositions act as adhesives and bandages to promote wound healing. In vitro tissue substrate engineering for bio-responsive nanomedicine for wound healing is made possible by new materials. The design of soft nanomaterials, such as hydrogel and bio-scaffold, improves tissue regeneration and wound care. Additionally, nanoscale tissue engineering presents potential and problems for the use of synthetic biomaterials in tissue morphogenesis and remodelling (Singh et al., 2010).

Wound management is now a crucial component of healthcare systems as wound healing remains a problem for the medical community. One area that may provide several novel strategies for regenerative medicine is nanotechnology. It is important to note the significance of nanoparticles, which can induce certain characteristics when incorporated in biomaterials, making them attractive for use as wound dressing materials. Molecular engineering techniques for various self-assembling biocompatible nanoparticles have been developed in recent years by nano research. Nanomaterials are proven to enhance both the delayed wound healing process and burn therapy. Bioactive nanomaterials-based dressing categories in clinical practice, including semi-permeable films, foam dressings, hydrogel dressings, hydrocolloid dressings, alginate dressings, non-adherent contact layer dressings, and multilayer dressings was discussed. It highlights the lack of high-quality evidence and the need for advanced trials due to current wound healing therapies not providing excellent clinical outcomes (Stoica et al., 2020). Fig. 3 shows applications of nanomaterials for wound treatment.

Figure 3. Nanomaterials used for wound treatment (Stoica et al.,2020).

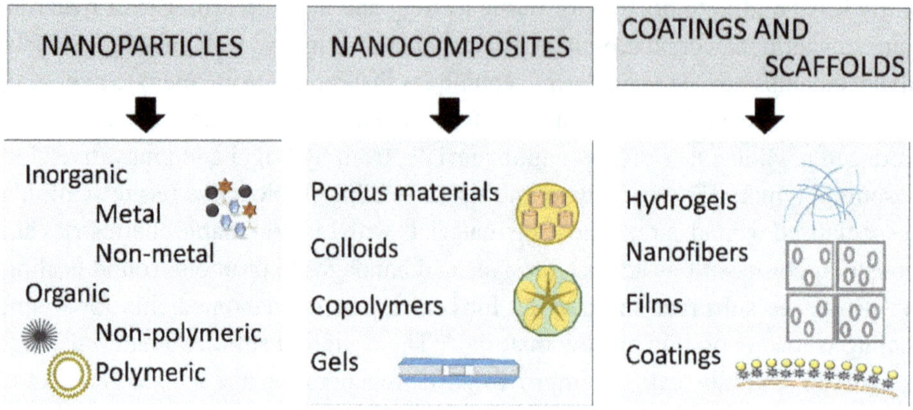

The newest field of study that draws researchers is smart textiles with sophisticated features that enhance human well-being and offer comfort in day-to-day living. Smart textiles have the ability to do a variety of jobs and have important uses in the healthcare industry, such as using sensors to monitor the wearer's physiological demands and health state in real time and integrating nanotechnology for improved performance. The incorporation of sensors and nanomaterials into textiles has forced the textile industry to address both physiological and technological issues, such as the fundamental need for a textile to retain its lifespan, washability, sweat resistance, and actual comfort. They explore the growing field of smart textiles, focusing on their applications in human health and welfare, the role of nanomaterials in antimicrobial textile development, and the challenges they face in societal acceptance (Barman et al., 2022).

In order to preserve the skin's physiological functioning, wound healing is essential. Using a bandage to cover the wound and lower the risk of infection and other injuries is the most popular therapy. Modern wound dressings are the best option for treating different kinds of wounds because of their exceptional biodegradability and biocompatibility. They also help relieve pain, keep the atmosphere wet and warm, and enhance hypoxic conditions to promote wound healing. We will present information on the clinical features of the wound, the qualities of common current dressings, and the in vitro, in vivo, and clinical trials on their efficacy because of the variety of wound types and advanced wound dressing products. The most often employed varieties in the production of contemporary dressings include films, alginates, hydrogels, hydrocolloids, and foams. The development of polymer materials for dressing applications, their optimal function, and the selection of dressings for wound treatment, while also estimating the current trend in new materials for wound healing was discussed (Nguyen et al., 2023). For almost fifty years, fiber-based implants (BTFIs) and biomedical textiles have been used in clinical settings to promote healing. Out of all the many types of biomaterials, silk-based biomaterials (SBBs) have been used extensively in clinical settings, such as sutures, for millennia. They are also becoming more and more acknowledged as a potential material for biomedical textiles. The application of SBBs for different BTFIs for extracorporeal implants, soft tissue repair, healthcare/hygiene goods, and related demands has been spurred by their exceptional mechanical qualities, biocompatibility, regulated degradability, and simplicity of processing. BTFIs from the standpoint of their kinds, physical characteristics, and biological attributes. The benefits and drawbacks of BTFIs from SBBs are then examined. discusses advancements in surface coatings, the physical and chemical changes made to SBBs for BTFIs, and the prospects and needs for future BTFI development employing SBBs (Li et al., 2015).

5.3. Sustainable Alternatives for Synthetic Dyes and Coatings

Biodegradable nanomaterials present eco-friendly options in place of synthetic dyes and coatings made of toxic chemicals. Plant pigments can be carried in nanoparticles with better adhesion and color strength. Biopolymer-based protective coatings using starch or PLA remove harmful residues. Eco-efficient alternatives reduce effluent contamination. They also promote closed-loop production systems because of simpler degradability and recyclability. Embracing such innovations will be central to a greener friendly textile sector.

One branch of computer science that enables the study of organic and inorganic molecules' three-dimensional structures at the nanoscale is computational chemistry (CC). CC is separated into ab-initio approaches, quantum mechanics, and molecular mechanics, respectively. Molecular energy, geometric optimization, entropy, enthalpy, dipole moments, kinetics, Fourier transform infrared spectroscopy (FTIR), UV-vis, nuclear magnetic resonance, and molecular dynamics are among the structural characteristics of molecules or macromolecules that can be ascertained using various theoretical models. In order to create nanomaterials with special qualities for a range of uses, such as nanomedicine, solar cells, electronics, and the environment, computational chemistry is becoming more and more crucial. Utilizing materials with superior adsorption capabilities, such as graphene, carbon nanotubes, biodegradable polymers, clays, and lignocellulosic materials, polymeric nanocomposites have been created to enhance the adsorption processes of contaminants in water. Because of its exceptional surface area, chemical characteristics, and electrical dispersion, graphene is perfect for maximising the adsorption of contaminants such as pharmaceutical waste and textile colours. The most frequent colourants in water are azo dyes, which are also the most extensively employed in industry. Nevertheless, no computer models have been used to analyse mixes of two or more azo dyes. The purpose of this chapter is to acquire computations such as Gibbs energy, partition coefficient, FTIR analysis, and electrostatic potential maps both before and after azo dye adsorption (de Jesús-González et al.,2022). Dye adsorption on graphite oxides generated using various techniques was compared. UV-visible spectroscopy was used to monitor the adsorption process, and physicochemical and spectroscopic methods were used to evaluate the adsorbents both before to and during adsorption. Remarkably, we discovered that adsorption efficiency for methylene blue (MB) increased with increasing graphite oxide oxidation degree, but adsorption efficiency was reduced by ultrasonic exfoliation of graphite oxide. The graphite oxide made using the modified Hummers' technique (HmGO) prior to ultrasonic exfoliation demonstrated the best MB adsorption capacity out of all the samples examined. When compared to other reported absorbents for MB, it is rather high. Adsorption kinetics and isotherms were also examined, as well as the effects of experiment parameters on MB adsorption on graphite oxide, including pH, adsorbent dose, stirring speed, and adsorption duration. According to the Langmuir model, graphite oxide has a maximum adsorption capacity of 751.9 mg · g−1 (He et al., 2013).

Water that is safe and clean to drink is essential. The anthropogenic and ongoing release of harmful and non-biodegradable organic pollutants into the aquatic environment has tragically resulted in the degradation of this quality. Because dye-containing effluent is frequently dumped into natural water bodies, the textile industry has emerged as one of the main causes of water pollution. A significant amount (about 20%) of dyes are lost during synthesis and processing and wind up in wastewater, according to studies. Textile dyes are considered pollutants of great concern because of their extensive industrial use, which makes them a constant hazard on a global scale. Dyes must be eliminated at their source since their release into the aquatic environment, together with the byproducts of their breakdown, poses major health concerns to aquatic plants, creatures, and people. is to introduce and talk about the latest cutting-edge scientific and technical advancements in the degradation and removal of colours from textile effluent. highlights physical and chemical methods while talking about the development of the newest nanomaterials, current research, and the superior efficiency of the state-of-the-art materials. The future outlook for treatment methods can be compared for cost and efficiency, providing insights into selection options, knowledge gaps, and improving efficiency. Ozonation and photo-Fenton are promising independent processes, combined with filtration methods, providing economically feasible and time-efficient solutions. However, further improvement is needed in developing models and processes that target dyes and secondary factors while remaining cost-effective and affordable globally (Sahu & Poler, 2024). When Dyers switched from natural dyes to synthetic dyes, they did it with skill and reason. Synthetic dyes produced wasted waste materials that needed to be disposed of, had no insoluble contaminants, and were more dependable in terms of strength and shade. In addition to offering rapid brilliant hues and, for the first time, fast black colours, they mostly eliminated the need for mordants. They were applied in less complicated procedures that used significantly less water and energy, and they could be combined to provide precise colour matching. Even though the amount of fibre used annually has increased 25 times, with at least 60% of it being synthetic, synthetic dyes still dominate textile colours. Environmental concerns prompted a review of colours and dyeing methods: several synthetic dyes that were deemed potentially harmful were banned and have since vanished from any dye house that complies with regulations and wants to work with big-box stores. Natural dyes have recently been the subject of a large body of study that (re)examines them due to "sustainability" and the false presumptions that natural = safe and synthetic = harmful. The literature overlooks the reasons for their abandonment, and natural dyes cannot produce enough colorant for fiber production. Synthetic dyes and their processes become more efficient, clean, and cost-effective for textile coloration (Bide, 2014) (Fig. 4).

Figure 4. Sustainable alternatives for synthetic dyes and coatings

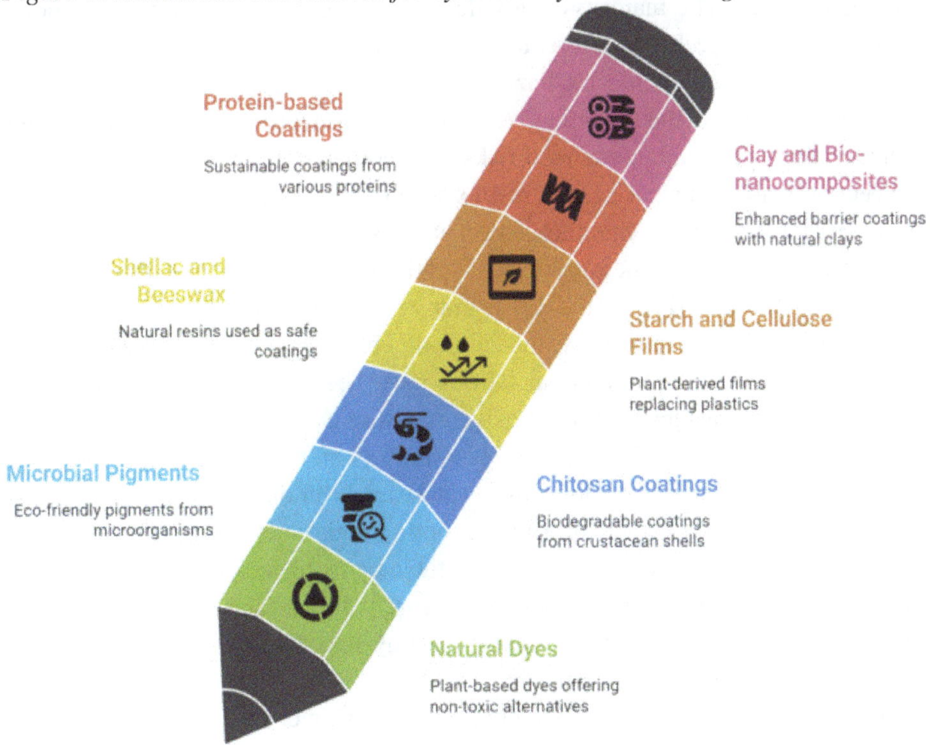

6. PERFORMANCE, BIODEGRADABILITY, AND ENVIRONMENTAL IMPACT

The function of biodegradable nanomaterials in clothing is usually comparable to or better than conventional materials. They are strong, flexible, and resistant with the added benefit of being eco-safe. Their biodegradability means they will degrade into non-toxic elements in their natural ecosystems, minimizing landfill buildup. Life cycle studies find drastically lower water and carbon footprints. Yet, having equalized testing protocols is still paramount to wider usage. Overall, they are a complete solution to sustainable fabric manufacturing and waste disposal.

Nanotechnology is now a top research priority in many countries because of its enormous potential and economic effect. But because of the uncertainties and irregularities in size, shape, and chemical composition, the presence of certain nanomaterials might have harmful impacts on the environment and human health. This overview covers the various cutting-edge uses of nanoparticles in the textile industry, along with the health and environmental hazards that go along with them. Nanomaterials, nanofinishing, nanocoatings, nanofibers, and nanocomposites—the four primary textile industry sectors are examined. various nanomaterial-based functional fabrics. The majority of textile materials come into close, continuous touch with human flesh. Therefore, it is imperative to conduct a thorough analysis of the impact of carcinogenic and hazardous compounds found in textiles. In order to further its growth in the upcoming years, it is important to properly recognise the potential advantages and unintended risks of nanomaterials to our environment. It is expected that the current paper's results would raise awareness of the potential impact of textile wastes containing nanomaterials and the importance of stronger restrictions with regard to the final disposal of these wastes (Saleem & Zaidi, 2020). Newly developed materials for the textile industry are called nanotextiles. However, there are dangers to sustainability, the environment, and human health when these materials are exposed. Commercial fabrics have been shown to emit nanosized compounds after washing and/or abrasive exposure, which leads to their buildup in soil and water bodies. The literature also reports on the production of biocidal nanotextiles, which, when released into waterways, can damage the aquatic echo system. It is crucial to gather information on the emissions and ambient concentrations of the nanoparticles (NPs) found in the nanotextiles. Analytical techniques and sustainable separation processes to measure nanomaterial concentrations are lacking. The measurement of nanotextile destiny is particularly crucial since, at the moment, sewage sludge, wastewater, air, and soil are some of the main sinks of nanotextiles, which degrades the environment. This chapter discusses the use of nanoparticles in textiles, their release from textiles, analytical techniques for identifying NPs in water, their toxicity, and the risks to the environment and public health. (Mazari et al.,2021). The prospective uses of smart textiles in a variety of domains, including biomedical and environmental monitoring, have attracted a lot of interest lately. Smart textiles may be made more sustainable and useful by using green nanoparticles. will describe the latest developments in smart textiles that use green nanomaterials for biomedical and environmental purposes. Synthesis, properties, and uses of green nanomaterials in the creation of smart textiles. We go over the difficulties and restrictions of utilising green nanoparticles in smart textiles as well as potential future developments for biocompatible and ecologically friendly smart textiles (Popescu & Ungureanu, 2023).

Figure 5. Different applications of Green Nanomaterials (Popescu & Ungureanu, 2023)

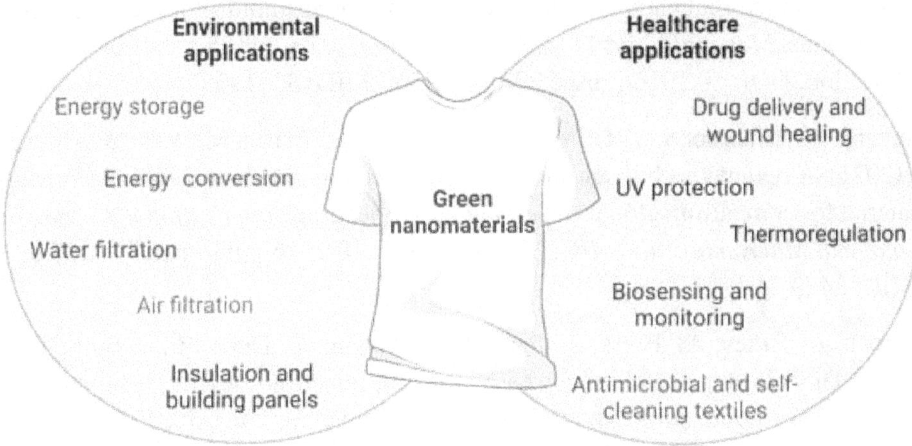

7. CONCLUSION

The use of biodegradable nanomaterials in textiles is a major step towards sustainable and eco-friendly production. Not only do these materials decrease the environmental impact of textiles but also improve the performance of fabric by imparting antimicrobial action, UV protection, and higher durability. Based on their inherent nature, toxicity-free degradation, and general applicability on the substrate of various textiles, there are immense chances of applying such materials for the next-generation use in textiles. As research progresses, more effective extraction, functionalization, and fabrication methods will expand their potential even further. Instead of synthetic and non-degradable materials, biodegradable nanomaterials

REFERENCES

Ali, K. A., Roy, P. K., Hossain, C. M., Dutta, D., Vichare, R., & Biswal, M. R. (2021). Starch-based nanomaterials in drug delivery applications. In Biopolymer-Based Nanomaterials in Drug Delivery and Biomedical Applications. https://doi .org/DOI: 10.1016/B978-0-12-820874-8.00023-3

Arun, S., Sinharoy, A., Pakshirajan, K., & Lens, P. N. (2020). Algae based microbial fuel cells for wastewater treatment and recovery of value-added products. *Renewable & Sustainable Energy Reviews*, *132*, 110041. DOI: 10.1016/j.rser.2020.110041

Avossa, J., Pota, G., Vitiello, G., Macagnano, A., Zanfardino, A., di Napoli, M., Pezzella, A., D'Errico, G., Varcamonti, M., & Luciani, G. (2021). Multifunctional mats by antimicrobial nanoparticles decoration for bioinspired smart wound dressing solutions. *Materials Science and Engineering C*, *123*, 111954. Advance online publication. DOI: 10.1016/j.msec.2021.111954 PMID: 33812582

Azmana, M., Mahmood, S., Hilles, A. R., Rahman, A., Arifin, M. A. B., & Ahmed, S. (2021). A review on chitosan and chitosan-based bionanocomposites: Promising material for combatting global issues and its applications. *International Journal of Biological Macromolecules*, *185*, 832–848. DOI: 10.1016/j.ijbiomac.2021.07.023 PMID: 34237361

Barman, J., Tirkey, A., Batra, S., Paul, A. A., Panda, K., Deka, R., & Babu, P. J. (2022). The role of nanotechnology based wearable electronic textiles in biomedical and healthcare applications. *Materials Today. Communications*, *32*, 104055. DOI: 10.1016/j.mtcomm.2022.104055

Besegatto, S. V., Costa, F. N., Damas, M. S. P., Colombi, B. L., De Rossi, A. C., de Aguiar, C. R. L., & Immich, A. P. S. (2018). Enzyme treatment at different stages of textile processing: A review. *Industrial Biotechnology (New Rochelle, N.Y.)*, *14*(6), 298–307. DOI: 10.1089/ind.2018.0018

Bide, M. (2014). Sustainable dyeing with synthetic dyes. Roadmap to sustainable textiles and clothing: eco-friendly raw Materials, technologies, and processing methods, 81-107.

Biranje, S. S., Kaushik, S., Marewad, D., Yadav, A., Vankundre, V., Panse, M., Joshi, I., Goli, A., Shahid, M., Kulkarni, K., Liu, J., & Adivarekar, R. (2024). Applications of nanocellulose and its derivatives in developing sustainable textiles. *Cellulose (London, England)*, *31*(9), 5343–5379. DOI: 10.1007/s10570-024-05935-4

Chakrabarty, S., & Jasuja, K. (2022). Applications of nanomaterials in the textile industry. In *Nanoscale Engineering of Biomaterials: Properties and Applications* (pp. 567–587). Springer Nature Singapore. DOI: 10.1007/978-981-16-3667-7_20

Costa, C., Viana, A., Silva, C., Marques, E. F., & Azoia, N. G. (2022). Recycling of textile wastes, by acid hydrolysis, into new cellulosic raw materials. *Waste Management (New York, N.Y.)*, *153*, 99–109. DOI: 10.1016/j.wasman.2022.08.019 PMID: 36067549

de Jesús-González, A., Rangel-Vázquez, N.-A., & de Velasco-Maldonado, P.-S. (2022). Thermodynamic and structural analysis of graphene/polymer nanocomposites for the adsorption of azo dyes present in binary mixed. In *Novel Materials for Environmental Remediation Applications*. Adsorption and Beyond., DOI: 10.1016/B978-0-323-91894-7.00015-3

Fernandes, M., Souto, A. P., Dourado, F., & Gama, M. (2021). Application of bacterial cellulose in the textile and shoe industry: Development of biocomposites. *Polysaccharides*, *2*(3), 566–581. DOI: 10.3390/polysaccharides2030034

Fuentes, K. M., Gómez, M., Rebolledo, H., Figueroa, J. M., Zamora, P., & Naranjo-Briceño, L. (2022). Nanomaterials in the future biotextile industry: A new cosmovision to obtain smart biotextiles. *Frontiers in Nanotechnology*, *4*, 1056498. Advance online publication. DOI: 10.3389/fnano.2022.1056498

He, B., Liu, X., Qi, S., Zheng, R., Chang, M., Lin, Q., & Ren, J. (2021). A review of water-resistant cellulose-based materials in pharmaceutical and biomedical application. *Current Medicinal Chemistry*, *28*(40), 8296–8318. DOI: 10.2174/0929867328666210208113354 PMID: 33557729

He, G., Zhang, J., Zhang, Y., Chen, H., & Wang, X. (2013). Fast and efficient removal of cationic dye using graphite oxide, adsorption, and kinetics studies. *Journal of Dispersion Science and Technology*, *34*(9), 1223–1229. DOI: 10.1080/01932691.2012.739946

Işıtan, A., Pasquardini, L., Bersani, M., Gök, C., Fioravanti, S., Lunelli, L., Çağlarer, E., & Koluman, A. (2025). Sustainable Production of Microcrystalline and Nanocrystalline Cellulose from Textile Waste Using HCl and NaOH/Urea Treatment. *Polymers*, *17*(1), 48. Advance online publication. DOI: 10.3390/polym17010048 PMID: 39795454

Jelil, R. A. (2015). A review of low-temperature plasma treatment of textile materials. *Journal of Materials Science*, *50*(18), 5913–5943. DOI: 10.1007/s10853-015-9152-4

Juikar, S. J., & Vigneshwaran, N. (2017). Extraction of nanolignin from coconut fibers by controlled microbial hydrolysis. *Industrial Crops and Products*, *109*, 420–425. DOI: 10.1016/j.indcrop.2017.08.067

Kamali, M., Capela, I., & Costa, M. E. (2018). Ultrasonic Synthesis of Zero Valent Iron Nanoparticles for the Efficient Discoloration of Aqueous Solutions Containing Methylene Blue Dye. In Nanomaterials in the Wet Processing of Textiles. DOI: 10.1002/9781119459804.ch8

Khan, M. S. H., Rahaman, M. T., Pranta, A. D., & Hasan, M. K. (2025). Eco-friendly organic nanomaterials for multifunctional textiles: Sources, applications, recent advances and future prospects towards sustainability. *International Journal of Environmental Science and Technology*, *22*(8), 7353–7410. Advance online publication. DOI: 10.1007/s13762-024-06299-9

Kumar, R., Rai, B., Gahlyan, S., & Kumar, G. (2021). A comprehensive review on production, surface modification and characterization of nanocellulose derived from biomass and its commercial applications. *Express Polymer Letters*, *15*(2), 104–120. DOI: 10.3144/expresspolymlett.2021.11

Li, G., Li, Y., Chen, G., He, J., Han, Y., Wang, X., & Kaplan, D. L. (2015). Silk-based biomaterials in biomedical textiles and fiber-based implants. *Advanced Healthcare Materials*, *4*(8), 1134–1151. DOI: 10.1002/adhm.201500002 PMID: 25772248

Li, J., Tian, X., Hua, T., Fu, J., Koo, M., Chan, W., & Poon, T. (2021). Chitosan natural polymer material for improving antibacterial properties of textiles. *ACS Applied Bio Materials*, *4*(5), 4014–4038. DOI: 10.1021/acsabm.1c00078 PMID: 35006820

Li, X., Che, Y., Lyu, Y., Liu, F., Wang, Y., & Zhao, C. (2018). Antimicrobial Properties and Applications of Chitosan/Inorganic Nanocomposites: An Overview | / . Cailiao Daobao. *Cailiao Daobao*, *32*(11), 3823–3830. DOI: 10.11896/j.issn.1005-023X.2018.21.020

Lin, X., Gong, X., Ruan, Q., Xu, W., Zhang, C., & Zhao, K. (2023). Antimicrobial Application of Chitosan Derivatives and their Nanocom-posites. *Current Medicinal Chemistry*, *30*(15), 1736–1755. DOI: 10.2174/0929867329666220803114729 PMID: 35927801

Ma, Y., Rissanen, M., You, X., Moriam, K., Hummel, M., & Sixta, H. (2021). New method for determining the degree of fibrillation of regenerated cellulose fibres. *Cellulose (London, England)*, *28*(1), 31–44. DOI: 10.1007/s10570-020-03513-y

Maia, M. T., Noronha, V. T., Oliveira, N. C., Alves, A. C., Faria, A. F., Martinez, D. T., Ferreira, O. P., & Paula, A. J. (2022). Silica nanoparticles and surface silanization for the fabrication of water-repellent cotton fibers. *ACS Applied Nano Materials*, *5*(4), 4634–4647. DOI: 10.1021/acsanm.1c03346

Makvandi, P., Iftekhar, S., Pizzetti, F., Zarepour, A., Zare, E. N., Ashrafizadeh, M., Agarwal, T., Padil, V. V. T., Mohammadinejad, R., Sillanpaa, M., Zarrabi, A., & Rossi, F. (2021). Functionalization of polymers and nanomaterials for water treatment, food packaging, textile and biomedical applications: A review. *Environmental Chemistry Letters*, *19*(1), 583–611. DOI: 10.1007/s10311-020-01089-4

Matavulj, M., & Molitoris, H. P. (2009). Marine fungi: Degraders of poly-3-hydroxyalkanoate based plastic materials. *Zbornik Matice Srpske za Prirodne Nauke*, (116), 253–265. DOI: 10.2298/ZMSPN0916253M

Mazari, S. A., Mubarak, N. M., Jatoi, A. S., Abro, R., Shah, A., Shah, A. K., & Lghari, Z. (2021). Environmental impact of using nanomaterials in textiles. In *Nanosensors and nanodevices for smart multifunctional textiles* (pp. 321–342). Elsevier. DOI: 10.1016/B978-0-12-820777-2.00018-2

Moore, C. A. (1967). An Economic Evaluation of Starch Use in the Textile Industry.

Nambela, L., Haule, L. V., & Mgani, Q. (2020). A review on source, chemistry, green synthesis and application of textile colorants. *Journal of Cleaner Production*, *246*, 119036. DOI: 10.1016/j.jclepro.2019.119036

Nguyen, H. M., Le, T. T. N., Nguyen, A. T., Le, H. N. T., & Pham, T. T. (2023). Biomedical materials for wound dressing: Recent advances and applications. *RSC Advances*, *13*(8), 5509–5528. DOI: 10.1039/D2RA07673J PMID: 36793301

Osman, E. (2020). Nanofinished medical textiles and their potential impact to health and environment. Nanoparticles and their biomedical applications, 127-145.

Paul, S., & Ghosh, S. (2024). Nano-finishing of Natural Fibres. In *Nanotechnology in Textile Finishing: Advancements and Applications* (pp. 313–352). Springer Nature Singapore. DOI: 10.1007/978-981-97-2696-7_11

Periyasamy, A. P. (2023). Microfiber Emissions from Functionalized Textiles: Potential Threat for Human Health and Environmental Risks. *Toxics*, *11*(5), 406. Advance online publication. DOI: 10.3390/toxics11050406 PMID: 37235219

Popescu, M., & Ungureanu, C. (2023). Green nanomaterials for smart textiles dedicated to environmental and biomedical applications. *Materials (Basel)*, *16*(11), 4075. DOI: 10.3390/ma16114075 PMID: 37297209

Rahaman, M. T., & Khan, M. S. H. (2024). Applications of green nano textile materials for environmental sustainability and functional performance: Past, present and future perspectives. *Nano-Structures and Nano-Objects*, *40*, 101332. Advance online publication. DOI: 10.1016/j.nanoso.2024.101332

Raman, A., & Sankar, A., SD, A., Anilkumar, A., & Saritha, A. (2022). Insights into the sustainable development of lignin-based textiles for functional applications. *Macromolecular Materials and Engineering*, *307*(8), 2200114. DOI: 10.1002/mame.202200114

Sahu, A., & Poler, J. C. (2024). Removal and degradation of dyes from textile industry wastewater: Benchmarking recent advancements, toxicity assessment and cost analysis of treatment processes. *Journal of Environmental Chemical Engineering*, *12*(5), 113754. Advance online publication. DOI: 10.1016/j.jece.2024.113754

Saleem, H., & Zaidi, S. J. (2020). Sustainable use of nanomaterials in textiles and their environmental impact. *Materials (Basel)*, *13*(22), 5134. DOI: 10.3390/ma13225134 PMID: 33203051

Singh, A. V., Aditi, A. S., Gade, W. N., Vats, T., Lenardi, C., & Milani, P. (2010). Nanomaterials: New generation therapeutics in wound healing and tissue repair. *Current Nanoscience*, *6*(6), 577–586. DOI: 10.2174/157341310793348632

Singha, K., Ray, A., Maity, S., & Pandit, P. (2021). Chitosan-based bionanocomposites in medical textile. In Bionanocomposites in Tissue Engineering and Regenerative Medicine. DOI: 10.1016/B978-0-12-821280-6.00027-1

Stoica, A. E., Chircov, C., & Grumezescu, A. M. (2020). Nanomaterials for wound dressings: An up-to-date overview. *Molecules (Basel, Switzerland)*, *25*(11), 2699. DOI: 10.3390/molecules25112699 PMID: 32532089

Sun, Y., & Sun, G. (2002). Durable and regenerable antimicrobial textile materials prepared by a continuous grafting process. *Journal of Applied Polymer Science*, *84*(8), 1592–1599. DOI: 10.1002/app.10456

Thangadurai, D., Prakash, L., Sangeetha, J., Al-Tawaha, A. R. M. S., David, M., Islam, S., . . . Adetunji, J. B. (2020). Sustainable synthesis of greener nanomaterials: Principles, processes, and products. Handbook of Nanomaterials and Nanocomposites for Energy and Environmental Applications, 1-23.

Tyagi, P., Ahmed, S., & Lai, C. W. (2023). Future prospects of biobased materials. In *Advanced Applications of Biobased Materials* (pp. 749–764). Elsevier. DOI: 10.1016/B978-0-323-91677-6.00013-1

Vrinceanu, N., Bucur, S., Rimbu, C. M., Neculai-Valeanu, S., Ferrandiz Bou, S., & Suchea, M. P. (2022). Nanoparticle/biopolymer-based coatings for functionalization of textiles: Recent developments (a minireview). *Textile Research Journal, 92*(19-20), 3889–3902. DOI: 10.1177/00405175211070613

Wang, J., Wang, L., Gardner, D. J., Shaler, S. M., & Cai, Z. (2021). Towards a cellulose-based society: Opportunities and challenges. *Cellulose (London, England), 28*(8), 4511–4543. DOI: 10.1007/s10570-021-03771-4

Wohlleben, W., Bossa, N., Mitrano, D. M., & Scott, K. (2024). Everything falls apart: How solids degrade and release nanomaterials, composite fragments, and microplastics. *NanoImpact, 34*, 100510. Advance online publication. DOI: 10.1016/j.impact.2024.100510 PMID: 38759729

Xie, W., Liang, X., Wang, H., Zhao, X., Tang, Y., Wu, M., & Yang, H. (2023). Structurally Tailoring Clay Nanosheets to Design Emerging Macrofibers with Tunable Mechanical Properties and Thermal Behavior. *ACS Applied Materials & Interfaces, 15*(2), 3141–3151. DOI: 10.1021/acsami.2c19295 PMID: 36598369

Yingngam, B. (2024). Nanocoatings for cosmetotextiles. In Sustainable Approach to Protective Nanocoatings. https://doi.org/DOI: 10.4018/979-8-3693-3136-1.ch006

KEY TERMS AND DEFINITIONS

Biodegradable Nanomaterials: Nano-sized materials derived from renewable biological sources that can naturally decompose without harming the environment. They are used to enhance textile properties while reducing ecological impact.

Nano Cellulose: A form of cellulose processed to the nanoscale, exhibiting high mechanical strength, biodegradability, and surface area, making it suitable for reinforcing fibers and eco-friendly textiles.

Chitosan Nanoparticles: Biodegradable and antimicrobial nanoparticles derived from chitin, used in textiles for wound care, odour control, and antimicrobial finishes.

Bio-Nanocomposites: Materials composed of biopolymers (e.g., starch, chitosan) and Nano fillers (e.g., clay, Nano cellulose) to enhance mechanical, thermal, or functional properties in textiles while maintaining biodegradability.

Functionalization: Chemical modification of nanomaterials to enhance their interaction with textile substrates, improving dispersion, adhesion, durability, and bioactivity.

Green Nanotechnology: Environmentally sustainable synthesis and application of nanomaterials using biological organisms or processes to reduce toxicity, waste, and energy usage.

Nano Coating: Application of a nanoscale film onto textile surfaces to provide functional properties like water repellency, UV resistance, or antibacterial effects, often without altering fabric texture.

Smart Textiles: Textile systems integrated with functional nanomaterials that respond to external stimuli, providing features such as antimicrobial action, UV-blocking, or self-cleaning capabilities.

Sustainable Dyeing and Coating: Use of biodegradable Nano carriers and biopolymers for coloration and fabric finishing, minimizing toxic effluents and enabling closed-loop recycling.

Circular Economy in Textiles: A design and production approach emphasizing reuse, biodegradability, and minimal waste through the integration of biodegradable nanomaterials, promoting a sustainable lifecycle for textiles.

Chapter 8
Introduction to Water Purification Technologies

Janani Karthikeyan
Sacred Heart College, Tirupattur, India

Revathi Anbarasu
Sacred Heart College, Tirupattur, India

Kaviya Ganesan
Sacred Heart College, Tirupattur, India

Amudhanila Padmanaban
https://orcid.org/0009-0003-1970-6150
Sacred Heart College, Tirupattur, India

Amala Kangasalam
https://orcid.org/0000-0003-4210-9664
Sacred Heart College, Tirupattur, India

ABSTRACT

Water purification is essential for ensuring safe drinking water, as natural sources often contain physical, chemical, and microbial contaminants that pose health risks. Organizations like the EPA and WHO set quality standards, promoting technologies such as electrodeionization, reverse osmosis, ultrafiltration, sedimentation, biosorption, ion exchange, and UV treatment. These methods improve water quality by removing pollutants and pathogens. However, rising pollution and water scarcity demand innovative approaches. Nanobiotechnology offers promise by integrating nanomaterials and biological systems for efficient treatment. Techniques like nanofiltration, nano-adsorption, and nano-catalytic degradation remove bacteria, viruses,

DOI: 10.4018/979-8-3373-2617-7.ch008

heavy metals, and pesticides effectively. Despite their benefits, challenges like cost, scalability, and regulation limit widespread use. Research, investment, and sustainable strategies are key to expanding access and ensuring global water security.

1. INTRODUCTION TO WATER PURIFICATION TECHNOLOGIES

The ongoing rise of industrialization and urban expansion, our environment is increasingly at risk. Water scarcity has become one of the most pressing environmental concerns worldwide. The rapid population increase in developing regions is anticipated to further heighten the demand for clean water across residential, agricultural, industrial, and energy sectors in the coming years. (Savage & Diallo, 2005). In recent years, nanotechnology has become a prominent applied technology in numerous sectors. This progress stems from the convergence of different scientific fields, allowing for manipulation at the atomic scale and the development of innovative structures.Nanotechnology encompasses the creation of materials and devices at the nanoscale, along with the ability to manipulate them to leverage their distinct properties. This understanding highlights the vast potential of nanotechnology across various fields, including life sciences, engineering, and medicine (Bayda & Sutliff, 2020; Safaei *et al.,* 2019; Imani & Safaei, 2019). The application of nanotechnology is poised to improve the effectiveness of clean energy production. For example, the incorporation of nanomaterials in solar panels, wind turbines, marine energy systems, and geothermal sources can enhance energy capture, facilitating a shift from fossil fuels to renewable energy alternatives. Furthermore, nanotechnology allows for the development of materials that significantly reduce environmental pollutants typically generated by human activities. As industries adapt to the integration of nanotechnology, it becomes possible to create eco-friendly materials that can be repurposed into other valuable products. A deliberate transition towards specific industries will aid in minimizing environmental damage (Ahmadi *et al.,* 2019; Babatunde *et al.,* 2020). As a result of evolving lifestyles and the growing reliance on technology, the global industry is encountering numerous environmental challenges. Modern nanotechnology is the design, characterization, production, and application of structures, devices, and systems by manipulating their shape and size at the nanometer scale. Presently, considerable attention has been diverted to using nanostructured materials as adsorbents or catalysts for the removal of toxic and harmful contaminants from wastewater (Savage & Diallo, 2005).

1.1. Importance of Water Purification

Access to safe drinking water is a fundamental human necessity and a crucial human right, particularly for the preservation of good health. Around 1 billion people in the world is facing a shortage of access to safe drinking water, despite the fact that around 70% of the surface of the Earth is covered with water in the form of oceans, seas, rivers, lakes, and underground aquifers. Research indicates that nearly half of these individuals suffer from health problems linked to the scarcity of clean drinking water. In underdeveloped regions and economically disadvantaged countries, many people resort to drinking directly from rivers and ponds, which can lead to severe illnesses such as typhoid, diarrhoea, cholera, hepatitis, polio, dysentery, and tapeworm infections (Werber *et al.,* 2016). Data from various countries regarding enhanced water resources concerning life expectancy reveals significant disparities between developing and developed nations, highlighting key indicators related to water quality, health, and sanitation.

It is a fundamental prerequisite for better children's health conditions to provide adequate sanitation and safe drinking water. In developed countries, access to clean drinking water correlates closely with improved health outcomes; for instance, almost 99% of a country's population has access to clean drinking water in a country like Spain or Norway. In contrast, nations such as Mozambique and Niger report access to clean drinking water at less than 45%, resulting in lower birth success rates. It is a stark reality that around 1.5 million children die each year due to a lack of hygienic water, sanitary conditions, and sanitation facilities. Furthermore, over 30% of child fatalities stem from preventable diseases associated with water, with nearly 90% of deaths among children under five attributed to diarrheal diseases in developing countries (Bain *et al.,* 2014; Mehmood *et al.,* 2013). Growing awareness of this critical issue in recent years has underscored the necessity for safe and hygienic drinking water, prompting global attention towards potential solutions, as highlighted in various international declarations, including the 21 Program and the UN Millennium Declaration of 2000 (Bain *et al.,* 2014).

Water sanitation is an eminent issue needing further scrutiny which has been somewhat overlooked, as indicated by the reports "State of the World's Children," published in 2005, and "Human Development Report 2006."They emphasize that to achieve the Millennium Development Goals related to water sanitation in several developing nations, the number of individuals lacking access to safe, clean, and hygienic drinking water must be reduced by half by 2015. Addressing the water crisis in these countries is essential to mitigate significant health risks. To this end, there is a necessity for innovative, cost-effective, sustainable, and efficient technologies to ensure the provision of safe drinking water. A high level of expertise is crucial in this field to restore hygienic living conditions and improve the environment in underdeveloped regions (Khalid *et al.,* 2014; Michalak *et al.,* 2013). It is very important to focus on water sources that are polluted through human and animal waste and do whatever possible to treat them for reducing the susceptibility to waterborne diseases.

Engineering solutions must be designed to maintain safe levels of various physicochemical and microbiological parameters in the available water resources. Various methods for water treatment, including coagulation and flocculation, sedimentation, and chemical disinfection, can be employed to purify these water sources. However, the application of these methods in developing countries can be inconsistent due to the limited availability of chemicals and resources (Bain *et al.,* 2014). Many individuals still opt to consume untreated water from natural sources, often utilizing simple methods for treating household water, such as chlorination and boiling, which are both effective and economical.

1.2. Global Water Contamination Challenges

A major focus of water conservation initiatives should be on the restoration of leaking distribution systems, as mentioned earlier. In some cities in developing nations, it is estimated that between 40% and 70% of water is lost to transit (Hinrichsen *et al.,* 1997). Water preservation at the city level can be enhanced through the application of efficient plumbing fixtures. Low flush, dual flush, and vacuum flush toilets can achieve substantial savings compared with traditional flush toilets, which constitute 20% to 40% of per capita water use in developed countries. In the United States, uniform water efficiency standards for nearly all toilets, urinals, showerheads, and faucets manufactured after January 1994 were established by federal Energy Policy Act of 1992. More so, many states and local governments in the U.S. require water-efficient products for use in any new construction and renovation projects. Further precipitation collected from rain measures include rainwater harvesting systems designed for toilet flushing and gardening, recycling of greywater for landscaping, and the use of native plants with low water requirements in arid and semi-arid regions. Water conservation initiatives also encompass industrial water usage. Many industries are heavily reliant on water; however, in some developed countries, industrial water consumption is decreasing as new production methods emerge that focus on water recycling and enhancing water productivity (Gleick 2003; Hinrichsen *et al.,* 1997).

1.3. Standards for Safe Drinking Water (WHO, EPA)

The initial standard for drinking water, aimed at safeguarding the public from acute bacterial infections, was established in 1914 by the U.S. Public Health Service. This standard has undergone several revisions over the years to incorporate source protection, chemical criteria, and a standard for radioactivity (Safe Drinking Water Act, 1986). The act of Congress in 1974 called the Safe Drinking Water Act protects groundwater against pollution by organic and inorganic chemicals, radionuclides, and microorganisms. The act was amended in 1986. This Act has had a considerable amount of support as nearly 50 percent of the U.S. population relies on groundwater as the principal source of drinking water (Cotruvo, 1987). The national initiative over the past twenty years to enhance water quality has yielded positive outcomes; however, new challenges persist. Few regions in the country are free from the quality of water issues. Polluted groundwater is often found near industrial and densely populated areas, increasing the likelihood of exposure. Also, while more point sources of pollution are being controlled, nonpoint sources still pose threats to groundwater. Sources potentially responsible for contaminants include agricultural runoff (pesticide, fertilizer), landfills, underground storage tanks, septic systems, or inadequate information as well (Espey & Shaw, 1997). Thanks in part to the efforts of the U.S. Environmental Protection Agency, the inventory and monitoring of specific contamination sources are improving. Monitoring will become increasingly vital as we assess the scope of groundwater contamination. The regulation of chemical carcinogens in groundwater is a topic of considerable debate. These regulations must balance the goal of reducing cancer incidence with the uncertainties surrounding causation, especially regarding low-dose levels of suspected carcinogens in water supplies (U.S. Environmental Protection Agency, 2011).

A robust Safe Drinking Water Act should encompass several essential elements:

1. It is crucial to monitor and regulate a wide array of compounds, acknowledging the significant variety of contaminants that threaten our groundwater. Currently, numerous substances that are present in groundwater remain unregulated (Mishel & Shierholz, 2013).
2. We must establish a comprehensive information management system to assess the extensive data that has been collected. Such a system would be instrumental in identifying trends and implementing preventive measures when necessary. For optimal effectiveness, this information system should recognize that pollutants released into the environment can disperse widely (Cory & Rahman, 2009) Contaminants found in the air or soil frequently infiltrate groundwater sources.
3. States are required to formulate thorough groundwater management plans and regulations. These plans should prioritize the safeguarding of groundwater quality from contamination and promote coordinated management of shared aquifers across state lines (Kuhn & Pfister, 1990).
4. Greater emphasis must be placed on preventing groundwater contamination. In addition to existing initiatives, we should explore alternatives to harmful activities, lower waste hazard levels through the imposition of taxes or fees,

and impose restrictions on the production, distribution, and use of contaminating substances. Only through proactive prevention can we reduce the costly and often distressing cleanup and remediation efforts currently faced by many communitie (Trevors, 1989).

2. TRADITIONAL WATER PURIFICATION METHODS

Membrane technologies, including nanofiltration (NF), are becoming essential elements in the field of water purification (US Bureau of Reclamation and Sandia National Laboratories, 2003). Nanofiltration membranes (NF membranes) are extensively utilized in the treatment of both drinking water and wastewater. This membrane process utilizes low pressure to separate substances effectively in the range of 0.001 to 0.1 micrometer. NF membranes are pressure-driven membranes, which bridge characteristics between reverse osmosis and ultrafiltration membranes, with pore sizes ranging from 0.2 to 4 nanometers. Research has exhibited that NF membranes can effectively eliminate turbidity, microorganisms, and inorganic ions such as calcium and sodium. Their applications encompass the softening of groundwater to lessen hardness, elimination of dissolved organic materials and trace pollutants from surface water, wastewater treatment to get rid of both organic and inorganic pollutants plus organic carbon, and pretreatment of seawater for desalination processes. (Van der Bruggen and Vandercasteele et al.,2003) explored the effectiveness of nanofiltration in removing cations, natural organic matter, biological pollutants, organic contaminants, nitrates, and arsenic from both surface water and groundwater (Van der Bruggen et al., 2003). Carbon nanotube filters are also being better known in water treatment, as the latest reports demonstrate the effective development of carbon nanotube filters (Srivastava et al., 2004).

Noticeably, such new membranes are fabricated as tubular hollow membranes with carbon nanotube walls having radial alignment. The study showed that the filters efficiently remove bacteria, including Escherichia coli and Staphylococcus aureus, from polluted water sources. The carbon nanotube act as a filters are easily purified through ultrasonication and autoclaving methods. Nanoceramic filters, on the other hand, consist of a mixture of nano alumina fibbers and micro glass that have a very positive charge that is effective in trapping negatively charged particles. Nanoceramic filters are highly effective in removing viruses and bacteria, as well as having a high particulate capacity, low clogging, and the capability to chemisorb dissolved heavy metals (Shah & Ahmed, 2011). Iron nanoparticles are a useful element in the area of nano remediation.

Nanoscale zero-valent iron particles with a size of 10 to 100 nm and generally having a core-shell structure are synthesized from Fe (II) and Fe (III) employing borohydride as the reducing agent. The centerpiece is represented chiefly by zero valent or metallic iron, while upon oxidation of metallic iron, the shell gets formed comprising mixed valence iron oxides (Fe (II) and Fe (III)). Nanoscale zero valent iron is preferred in nano-remediation due to its superior surface area and higher reactivity-in comparison to larger-sized particles-with the added abilities of reduction and adsorption. Nanoscale Zerovalent Iron (nZVI) has recently become the most investigated groundwater remediation technique in environmental nanotechnology. Experiments have shown that nanoscale metal iron is very effective for destroying a wide array of common pollutants such as chlorinated methane's, brominated methane's, trihalomethanes, chlorinated ethenes, chlorinated benzenes, other poly-chlorinated hydrocarbons, as well as pesticides and dyes (Zhang, 2003). Nanoscale zerovalent iron (nZVI) could be used efficiently for reducing organic contaminants organic and inorganic anions such as nitrate, perchlorate, selenate, arsenate, arsenite, and chromate. The rates of reaction of nZVI are very much higher in value and its sorption capacity is very high as compared to the traditional granular iron. The nZVI can also remove dissolved metals from solutions, i.e., lead (Pb) and nickel (Ni), by either reducing them to zerovalent species or to lower oxidation states (USEPA, 2013). Nanoscale zerovalent iron has been a hopeful candidate for this technology in recent years.

These responsive substances are strategically positioned in subterranean channels situated downstream of the contaminated area, compelling the contaminated water to pass through them. This process allows for the treatment of contaminants without the need for water excavation. Overall, this economical removal technique tends to inflict less environmental damage compared to alternative methods.

The incorporation of nanoparticles utilized in environmental applications will inevitably result in their release into the ecosystem. In order to make an assessment of the impending risks, an understanding of the mobility, bioavailability, toxicity, and persistence of the pollutants is essential. The knowledge of exposure to aquatic and terrestrial organisms constituted from nanoparticles suspended in water and soil is limited. The increasing utilization of engineered nanoparticles across various industrial applications, along with their potential roles in wastewater treatment and drinking water purification, raises critical questions about their removal within the urban water cycle. The source for environmental openness for the seismic study is a Close Observatory Nature Park (CONP) project there. However, due to the diminutive size of nanoparticles, their sedimentation velocities are relatively low, preventing significant sedimentation unless larger aggregates are formed (Frimmel & Niessner, 2010). The majority of nanoparticles utilized in contemporary technical applications are being functionalized, which suggests that research conducted with unmodified nanoparticles may not accurately reflect the behaviour of the particles currently in use. Functionalization is commonly employed to reduce agglomeration, thereby enhancing the mobility of the particles. Regrettably, there is still limited understanding regarding the impact of functionalization on the environmental behaviour of nanoparticles.

2.1. Reverse Osmosis

Reverse osmosis (RO) is increasingly recognized globally for its effectiveness in water treatment and desalination processes. This method operates under pressure, utilizing a semipermeable membrane that effectively filters out dissolved substances from the feed water. The rejection of these substances occurs through mechanisms such as size exclusion, charge exclusion, and various physical-chemical interactions among the solute, solvent, and membrane (Radjenović et al., 2008; Bellona & Drewes, 2005). The effectiveness of the procedure is influenced by operational specifications as well as the features of the membrane and feed water. The most commonly used modules in the market are spiral-wound and hollow fiber configurations. Their development in terms of hollow fiber module is that it possesses a very high packing density and thus can achieve even greater production rates of permeate. However, they are more susceptible to fouling. Membranes are classified as either asymmetric, comprising a single polymer layer, or composite, which consists of two or more layers. The introduction of functional groups into the polymer matrix plays a crucial role in determining the charge valence and strength of the membrane (Gabelich et al., 2005), while the extent of adsorption of dissolved species is influenced by the membrane's hydrophobicity, charge, and surface roughness (Mondal *et al.,*2008).

2.2. Electrode Ionization (EDI)

Electrode ionization is an advanced hybrid technology that integrates two distinct processes, electrodialysis and ion exchange. The dynamic interaction between these two methods mitigates the limitations associated with each, such as concentration polarization and the need for chemical regeneration. This synergy enables the technology to treat solutions with low ionic strength far more effectively than either process could achieve independently. An EDI cell is structured with alternating cationic and anionic membranes positioned between the anode and cathode, creating both dilute and concentrated compartments. Ion exchange resins are placed within the dilute sections, where the spheres symbolize the resin. The anionic exchange membrane (AEM) and cationic exchange membrane (CEM) facilitate electrical activity, which helps to avert concentration polarization. The applied electric field enhances ionic movement across the two compartments through the resin bed, which captures and releases ionic species, thus initiating mass transport through the ion exchange materials (DiMascio & Ganzi, 1999). Concurrently, water dissociates at the interfaces of the anionic and cationic exchange materials, generating protons and hydroxyl ions that contribute to the in-situ regeneration of the resin. The purpose of ion exchange resin is to lower the resistance of the cell, which increases with dilution in the solution. Ionogenic sites can achieve loadings between 1,000- and 100,000-times higher concentrations of resin ions than ion concentration in solution; therefore, increasing conductivity of the resin bed when electrified. When the electric field is applied to the bath solution, it affects the dissolved ionic species and ions obtained from the solid (mobile ions). However, in those cases when such solids contain a higher number of ions than their respective dissolved ions, the solid contribution ion transport toward the membrane becomes further because of solution being much poorer towards it. This occurs because, although ion mobility within the solid is approximately 20 times lower than in the solution, the transport rate is influenced by the mobility and concentration of the products, resulting in a transport rate that can be 50 to 5,000 times greater within the solid (Mir, 2001).

2.3. Sedimentation and Filtration

Sedimentation refers to the movement of molecules or colloids within a centrifugal or gravitational field. By analysing the migration rate and the eventual sedimentation or the equilibrium of suspended diffusion materials, one can gather information regarding particle size and their interactions. For colloids typically exceeding 100 nm in size, gravitational settling is a prevalent occurrence. To conduct an analytical observation, a simple graded vessel maintained at a constant temperature is sufficient to track the descent of the suspension/solvent interface or the accumulation of sediment (Yeoman & Lester *et al.,* 1992). This straightforward experiment, as this report aims to demonstrate, can yield significant insights. In contrast, for particles of much smaller dimensions, a centrifuge is required. The resurgence of interest in analytical sedimentation can be attributed to advancements in ultracentrifuge technology(Alon & Adin, 1994). This renewed focus, particularly in the biochemical realm, is understandable given the historical context of ultracentrifugation, which began with Svedberg's demonstration of the monodispersed of proteins. However, within the field of colloid science, the application of ultracentrifuge sedimentation as a research method remains relatively underrepresented, and literature on ultracentrifugation seldom addresses colloids (Graham, 1988).

It is regrettably, simple-shaped colloids, like rigid rods or spheres, can be made to simulate much of the more complicated sedimentation phenomena of interacting macromolecules. Additionally, studying the migration rates of well-defined colloids is pertinent to the materials processing of powders (Kaye, 1981; Lum et al., 1996) and colloids. Understanding the settling behaviour is crucial, as it can be either beneficial (in applications like filtration and centrifugal consolidation (Brinker &

Scherer, 1990) or detrimental (as seen in paints and ceramic suspensions). It is important to highlight that many colloids currently under investigation are sufficiently small for ultracentrifuge analysis. For instance, there is a rapidly expanding body of research focused on small colloids for solgel processing and ultrafine materials that exhibit unique mechanical and optical properties (Edelstein & Cammarata, 1996). For these nano colloids, centrifugal sedimentation is likely to become a significant technique for characterization.

2.4. Ultrafiltration and Ion Exchange

When the ion-exchange–ultrafiltration system was compared to the ultrafiltration process alone, it proved more effective for surfactants. In the hybrid system, the surfactant separation was increased by the ion exchange resin throughout the filtration time (Nirali Kansara *et al.,* 2016) It was noted that the type of resin significantly impacted the hybrid process. Among the various ion-exchange resins evaluated, the magnetic ion exchange resin (MIEX) proved to be the most effective. The hybrid system, with a minimal resin dosage of 5 cm^3/dm^3, achieved near-complete (almost 100%) anionic surfactant reduction in the case of a model solution at a concentration of 0.25 CMC using a 30 kD poly ether sulfone membrane (Sharma et al., 2018). However, the hybrid system's effectiveness in separation was primarily attributed to the ion-exchange resin, while the membrane's role was limited to separating the saturated resin from the solution. With a lower limit of 5 cm_3/dm^3 for resin consumption in the hybrid system, the one-parameter operating condition for the model solution at 0.25 CMC almost fully (i.e. almost 100%) reduced anionic surfactants by using a 30 kD poly ether sulfone membrane. In contrast, when the processes operated independently, surfactant removal was 25% for ultrafiltration and 90% for ion exchange. The hybrid system with A100 resin yielded the lowest surfactant separation from the model solutions, achieving only 47% (Alshameri *et al.,* 2019).

2.5. Biosorption and UV Treatment

Biosorption refers to the ability of particular biomolecules or types of biomasses to attach to and concentrate particular ions or other substances from water solutions. Unlike the more intricate process of bioaccumulation, which relies on active metabolic transport, biosorption involving non-living biomass or certain molecules and their functional groups is a passive process (Bal & Thakur, 2022). It primarily depends on the "affinity" between the biosorbent and the sorbate. Ultraviolet (UV) irradiation is a widely used method for disinfecting water in developed countries. An adequate exposure to ultraviolet light effectively inactivates the majority of microorganisms. This UV light is generated by creating an electric arc in mercury or, more recently, in xenon vapor, similar to standard fluorescent lamps. The UV spectrum ranges from 100 to 400 nano meters (nm), with the most effective wavelengths for bacterial disinfection falling between 200 and 280 nm (Singh & Thakur *et al.,* 2016).

3. NANOBIOTECHNOLOGY IN WATER PURIFICATION

Electrospinning presents a straightforward, economical, and effective technique for the fabrication of nanofibers. Such nanofibers are very porous, possess a large surface area, and have very complex pore structures. The physical and chemical deposition properties of electro spun nanofibers can be designed according to the need for various applications. As indicated in the current literature, including the studies of Crane and Scott, 2012. These nanofiber membranes are highly effective at removing micron-sized particles from water, exhibiting low levels of fouling. Although their possible applications are extensive, nanofibers play a particularly essential function in the development of materials for ultrafiltration and reverse osmosis. While the benefits of electrospinning in air purification are well-established, its use in wastewater treatment has not been thoroughly investigated (Gehrke et al., 2015). A continual challenge in membrane technology lies in achieving an optimal balance between selectivity and permeability. Due to its reliance on pressure, this method can require significant energy. Electro spun nanomembranes have the ability to eliminate viruses or bacteria through size exclusion. However, the small pore sizes of these membranes can complicate the effective filtration of viral agents, potentially restricting water flow.

Nevertheless, advancements are being made. A prominent example is a new composite cellulose-based membrane engineered to filter bacteria while preserving water flow, demonstrating an exceptional efficacy of 99.9999% in the removal of E. coli (Ali & Ahmad, 2020). This system utilizes positively charged fibres to draw in and capture negatively charged viruses, adhering to the rigorous 2 CFU/mL standards established by the American National Sanitation Foundation. Notably, poly (acrylonitrile-co-glycidyl methacrylate) nanofibers having a diameter of 100. nm were subjected to protein functionalization, increasing their diameter to 126 nm. When these biofunctionalized nanofibers come into contact with moisture during the filtration process, they undergo a transformation in shape. This change elevates the pH beyond the protein's isoelectric point, thereby exposing previously hidden functional groups and chemicals. Additionally, the swelling of the protein enhances the filtration process by creating a larger steric barrier that effectively captures nanosolids, including metal ions.

3.1. Introduction to Nanobiotechnology

Biotechnology and nanotechnology are among the brighter worlds of promise for the 21st century. Nanotechnology, for short-nanotech-is the art and science of developing and applying materials and devices at the nano scale. Nanotechnology deals with the making, manipulating, or assembling of materials, devices, or structures that have one or more dimensions in the range of 1-100 nanometers. In contrast, biotechnology examines the metabolic and physiological processes of biological entities, including microorganisms. The integration of these two fields, known as nanobiotechnology, has the potential to significantly advance the development and application of various tools in biological research. Nanotechnology is indeed versatile. It stretches from modifications of traditional device physics to completely new methods established using molecular self-assembly. It creates new materials at the nanoscale and senses the possibility of manipulating matter at the atomic level. This concept evokes virtually everything one can imagine about sciences- ranging from surface science, organic chemistry, molecular biology, and semiconductor physics to microfabrication.

3.2. Properties of Nanomaterials (Nanoparticles, Nanotubes, Nanofibers)

Nanomaterials are defined commonly as materials which possess dimensions that are at least 100 nanometers or less in one dimension. Such dimensions confer special and distinct physical and chemical characteristics, which are specific to the nanoscale. This includes examples such as nanofilaments, nanotubes, nanowires, nanocables, thin films, dendrimers, quantum dots, multi-phase materials, and all variants of nanomaterials in the powdered form. Due to their nanoscale dimensions (ranging from 1 to 100 nm), these materials exhibit significant differences in their physical, chemical, electrical, and magnetic properties compared to larger atomic and molecular structures, as well as macroscopic objects (Zhang 2008). The transition to nanoscale dimensions results in a dramatic alteration of the properties of macroscopic materials, leading to the emergence of exceptional characteristics. For instance, research conducted by Japanese scientists in the 1970s revealed that metals like copper and silver, known for their electrical and thermal conductivity, lose these properties and become nonconductive and nonthermal when reduced to the nanoscale (Shi and Guo, 2014). As the size of materials approaches the nanoscale, the surface-to-volume ratio increases significantly, resulting in a higher proportion of atoms residing on the surface of the material (Guisbiers et al., 2012; Koçak & Karasu, 2018).

3.3 Enzyme-Mediated Degradation and Microbial Fuel Cells

Microbial fuel cells (MFCs) are devices that utilize bacteria as catalysts to oxidize both organic and inorganic materials, thereby generating electrical current (Berk and Canfield, 1964). Electrons produced by bacteria from these substrates shunt the anode (the negative terminal) and subsequently flow to the cathode (the positive terminal) via a conductive medium, possibly including a resistor or under load. Even though current by convention current travels from the positive terminal to the negative terminal, the opposite is true for the electrons. In order for the system to perform properly, the substrate at the anode should be fed continuously or at intervals; not doing so would classify it as a biobattery. Electrons can reach the anode through various means, including electron mediators or shuttles (Rabaey et al., 2004), direct electron transfer associated with membranes (Bond & Lovley, 2003), or potentially through nanowires produced by the bacteria, along with other yetto-be-discovered methods. Chemical mediators, such as neutral red or anthraquinone-2,6disulfonate (AQDS), can be introduced to facilitate electricity generation by bacteria that cannot directly utilize the electrode (Bond et al., 2002). If no external mediators are incorporated, the MFC is termed a "mediator-less" MFC, even if the exact electron transfer mechanism remains unclear. Most of the time in an MFC, the electrons reaching the cathode will react with protons that diffused from the anode via a separator, together with oxygen from the air to form water. Additionally, chemical oxidizers like ferricyanide or Mn (IV) can be employed, although these substances require replacement or regeneration (6, 1820). In instances involving metal ions, such as the reduction of Mn (IV) to Mn (II), bacteria can assist in the reoxidation of the metal using dissolved oxygen (Shantaram et al., 2005).

4. KEY NANOBIOTECHNOLOGY-BASED PURIFICATION METHODS

4.1. Nanofiltration

Membrane technologies, including nanofiltration (NF), are becoming essential elements in the field of water purification (US Bureau of Reclamation and Sandia National Laboratories, 2003). Nanofiltration membranes (NF membranes) are extensively utilized in the treatment of both consuming water and wastewater. This process functions well under low pressure, segregating anything in the size range of 0.001 to 0.1 micrometeors. NF membranes are defined pressure-driven membranes which have features that fall between reverse osmosis and ultrafiltration membranes with a pore dimensions ranging from 0.2 to 4 nm. Research has It has been demonstrated that NF membranes can effectively eliminate turbidity, microorganisms, and inorganic ions such as calcium and sodium. Their applications can be described as softening groundwater to remove hardness, removing soluble organic substances and minute impurities from surface water, treating wastewaters for removal of both organic and inorganic pollutants, including organic carbon, and as a pretreatment stage in seawater desalination processes. Bruggen and Vandercasteele, 2003) examined the effectiveness of nanofiltration in removing cations, natural organic matter, biological contaminants, organic pollutants, nitrates, and arsenic from groundwater and surface water (Van der Bruggen and Vandercasteele, 2003). Carbon nanotube filters are becoming more known in water treatment processes, with recent publications indicating their successful development (Srivastava *et al.,*2004). These new filtration membranes are structured as empty tubes with walls made of radially aligned carbon nanotubes. Studies proved that these filters can efficiently remove bacteria, namely Escherichia coli and *Staphylococcus aureus*, from polluted water sources. These carbon nanotube filters are simple to clean through ultrasonication and autoclaving techniques. As compared to that, nanoceramic filters are made of a mixture of nano alumina fibbers and micro glass, having a high positive charge, through which they are able to catch negatively charged particles.

Nanoceramic filters have high effectiveness in the elimination of both viruses and bacteria, with a large capacity for particulates, lower clogging, and chemisorption capability for dissolved heavy metals (Shah & Ahmed, 2011).

4.2. Nano Adsorption

Iron nanoparticles play a important function in the domain of nano remediation. These particles are produced from Fe (II) and Fe (III) through the use of borohydride as a reducing agent (Peydayesh & Mezzenga, 2021). Nanoscale zero-valent iron particles generally range in sizes from 10 to 100 nm and possess core-shell structures. The core is primarily made of zero-valent or metallic iron, while the shell, formed by mixed-valent iron oxides (Fe(II) and Fe(III)) due to oxidation of the metallic iron, protects the core (Mahmoud et al., 2021). Nanoscale zero-valent iron is preferred in nano remediation due to its extensive surface area and enhanced reactivity compared to larger particles, along with its ability to perform both adsorption and reduction. The applications of Nanoscale Zero-valent Iron (nZVI) in groundwater remediation have emerged as the most researched method within the area of environmental nanotechnology (Gul et al., 2022).

4.3 Nano-Catalytic Degradation

Research has shown that nanoscale metallic iron is highly effective in removing a wide variety of common pollutants, such as chlorinated methane's, brominated methane's, trihalomethanes, chlorinated ethenes, chlorinated benzenes, various polychlorinated hydrocarbons, as well as pesticides and dyes (Zhang, 2003). Nanoscale zerovalent iron (nZVI) is really good in reducing both organic pollutions. In addition to inorganic anions such as nitrate, perchlorate, selenate, arsenate, and chromate. Kinetics of reactions are at least magnitudes faster for nZVI, and its sorption capacities are much greater compared to traditional granular iron. Furthermore, nZVI can effectively remove dissolved metals from solutions, such as lead (Pb) and nickel (Ni), by reducing them to zerovalent forms or lower oxidation states (USEPA, 2013). In recent years, nanoscale zerovalent iron has gained recognition as a promising technology in this field. These reactive materials are strategically positioned in underground trenches located downstream of the contaminated area, compelling the contaminated water to pass through them. This process allows for the treatment of contaminants without the need for water excavation. Overall, this economical removal technique tends to inflict less environmental damage compared to alternative methods.

4.4. Removal of Heavy Metals, Pathogens, and Pesticides

The incorporation of nanoparticles are utilized in environmental applications. inevitably result in their release into the ecosystem. An understanding of their mobility, bioavailability, toxicity, and persistence is essential for the evaluation of potential risks they may pose. Little is known regarding the revelation of both aquatic and terrestrial environments organisms to nanoparticles present in water and soil. The increasing utilization of engineered nanoparticles in many industrial applications and possibly in wastewater treatment and portable water purification raises urgent questions regarding their removal in the urban water cycle. Traditional techniques for the elimination of particulate matter in the process of wastewater treatment, such as sedimentation together with filtration, are generally practiced. However, due to the diminutive size of nanoparticles, their sedimentation velocities are relatively low, preventing significant sedimentation unless larger aggregates are formed (Frimmel & Niessner, 2010.). The majority of nanoparticles utilized in contemporary technical applications are functionalized, which suggests that research conducted with unmodified nanoparticles may not accurately reflect the behaviour of the particles currently in use. Functionalization is commonly employed to reduce agglomeration, thereby enhancing the mobility of the particles. Regrettably, there is still limited understanding regarding the impact of functionalization on the environmental behaviour of nanoparticles.

5. TYPES OF NANOMATERIALS

Nanotechnology is a contemporary and interdisciplinary field that encompasses a variety of emerging applications across multiple scientific areas, including materials science, physics, chemistry, biology, engineering, and medicine. The initial classification of nanomaterials is based on their origin, which can be either natural or synthetic. Additionally, they can be categorized based on whether they are the intended final product or by-products, and they may generally be classified as organic or inorganic. In terms of their origin, nanoparticles can result from advanced technological processes, but they are also produced through natural phenomena such as volcanic eruptions, wildfires, saline aerosols, or ultrafine mineral sand grains (e.g., oxides and carbonates). Furthermore, some ultrafine particles are unintentionally formed through the combustion of fossil fuels or during activities like barbecuing, which are a result of human actions (Jeevanandam et al., 2018). (i) carbon-based nanomaterials, (ii) inorganic-based nanomaterials, (iii) organic-based nanomaterials excluding the carbon nanomaterials, and (iv) composite-based nanomaterials. Additionally, nanomaterials can be categorized based on their dimensionality (size), morphology (shape), surface characteristics, composition, uniformity, functionalization, and agglomeration (Bowman, 2017)). In relation to In regard to the latter point, products derived from nanotechnology can be categorized based on their physical form: they may exist as powders, dispersions (which are powders suspended in liquids), coatings (thin film surface coverings), or macro solids (for instance, solid particles interconnected by nanoparticles, commonly referred to as nano whiskers). Within the context of nanotechnology, the term "nano" primarily pertains to the size of particles. As defined by the EU Commission SCENHIR2, the nanoworld encompasses "objects whose dimensions in all three axes range from 1 to 100 nm (Leitgeb et al., 2016). Nanotubes and particulate objects refer to entities that extend solely in two dimensions at the nanometre scale, while those with a single dimension measuring less than 100 nanometres are referred to as nano pellets. The subsequent discussion will highlight certain properties that are significant for classification, beginning with the most defining characteristic: dimensionality.

6. ROLE OF NANOMATERIALS IN WATER TREATMENT

Nanomaterials are categorized according to their size, composition, shape, and origin. There is significant interest in the research community to investigate novel types of nanomaterials featuring diverse structural designs suitable for new applications. Moreover, the unique properties of these materials facilitate the development of substances with customizable characteristics, providing improved performance that competes with established products currently on the market. The dimensions of nanomaterials can be influenced by various factors, including the synthesis method, temperature, pressure, duration, pH, and concentration. Depending on their specific applications, nanomaterials have been engineered in diverse shapes and sizes, including spheres, fibers, tubes, sheets, and interconnected forms. Zero-dimensional (0D) structures are characterized by their spherical shape, while one-dimensional (1D) structures are generally found in the form of fibres and tubes. Two-dimensional (2D) structures are represented as flat, sheet-like configurations, whereas three-dimensional (3D) structures are typically identified by their intricate interconnections (Singh *et al.,* 2018). The evolution of nanostructured materials across different dimensions results in intriguing morphologies, properties, and functionalities. Key features such as surface area, adhesion, adsorption, reflectance, and transport capabilities have been effectively utilized in various applications, including wastewater treatment. The most prevalent technologies for purifying water and wastewater involve the use of adsorbents or membrane-based separation methods. The process of regenerating adsorbents significantly enhances the cost-effectiveness of traditional adsorbents. Numerous nanosized adsorbents, including metal oxides, nanosized metals, graphene, nanocomposites, and carbon nanotubes (CNTs), exhibit remarkable selectivity and high reactivity, outperforming conventional adsorbents in adsorption efficiency by several orders of magnitude (Mohammed, 2011).

7. ADVANTAGES OF NANOTECHNOLOGY

Most nanoparticles used in modern technological applications are functionalized, indicating that studies involving unmodified nanoparticles may not truly represent the behavior of the particles in current use. Functionalization is often applied to minimize agglomeration, which in turn improves the mobility of the particles. Regrettably, there is still limited understanding regarding the impact of functionalization on the environmental behaviour of nanoparticles (Das *et al.,* 2014). Nanotechnology-based processes exhibit remarkable efficiency, modularity, and multifunctionality, offering effective and cost-efficient solutions for water and wastewater management. Materials composed of nano-objects are characterized by their durability and have a significant specific surface area. This means that the substantial surfaceto-volume ratio significantly influences their interaction with pollutants and bacteria (Qu et al., 2013). Processes for water treatment that use nanotechnology pose great challenges to conventional methods of water treatment. In addition, purifying and economically utilizing unconventional sources of water can make use of nanotechnology. The treatment of industrial wastewater using nanomaterials is both crucial and prevalent. Although current remediation technologies are effective, they tend to be costly and time-intensive. The application of nanotechnology in wastewater treatment offers benefits by removing contaminants and facilitating the recycling of water, which ultimately reduces labour, time, and costs for industries while addressing various environmental concerns (Kanchi, 2014). It is crucial to highlight that nanomaterials designed for the purification of drinking water should be environmentally friendly and nontoxic. Harmful particles can inflict significant harm to critical organs upon exposure to the human body. Given their size, nanomaterials have the potential to migrate to different organs, thereby increasing the risk of biological harm. Consequently, prior to their industrial application, toxicity assessments must be rigorously incorporated into safety data sheets, standard operating procedures, and other relevant regulatory documents. This chapter presents through a overview of the applications of nanotechnology in the treatment of water and wastewater.

8. LIMITATION

Nanomaterials present significant opportunities; however, they encounter challenges related to the selection of raw materials, reaction conditions, quality control of the final products, and their practical applications. These issues pose barriers to the advancement of nanomaterials for manufacturing and widespread utilization (Ying et al., 2022). The dimensions and morphology of nanoparticles generated from various extracts exhibit considerable diversity, and the evaluation of their quality is inadequate. Current literature indicates that the diameter of these particles fluctuates significantly, rendering green technology impractical for large-scale production. Additionally, managing particle size presents a substantial challenge during the manufacturing process (Chahardoli et al., 2018). Particulate matter limitations have been classified into two main categories: fibro genic and non-fibro genic. The measurement units for particulates have evolved over time, with certain fibro genic aerosols, including crystalline silica and asbestos, originally quantified in millions of particles per cubic foot (mppcf) of air (Higgins *et al.,* 1917). The influence of plant extracts in the process of synthesis was only recently evidenced in a study, yet the precise chemical routes involved remain unclear. The dimensions and morphology of nanoparticles produced from different extracts show significant variation, and the characteristics assessed are insufficient. Current literature indicates substantial discrepancies in particle size, rendering green technology impractical for large-scale production and complicating the regulation of particle size during the manufacturing process (Turakhia et al., 2018).

Nanotechnology has become a pivotal method for the synthesis of nanomaterials in recent years. In contrast to traditional techniques, the methods for synthesizing nanomaterials offer a safe, non-toxic, and environmentally sustainable approach. However, challenges such as low yield, inconsistent particle sizes, complex separation processes, periodic availability of local raw materials, and various other issues must be addressed to achieve sustainable nanomaterial synthesis and its applications. Currently, there exists a wide array of green nanomaterial synthesis methods and protocols, with additional innovations anticipated in the future. In contemporary society, advanced solutions to these challenges, which encompass both automotive and stationary systems, are increasingly vital. Consequently, the initial aspect of green nanotechnology focuses on adopting more environmentally friendly or energy-efficient manufacturing processes.

The subsequent phase involves the development of eco-friendly products. Indeed, environmental pollution and health crises, such as the COVID-19 pandemic, pose significant challenges globally, which can be partially mitigated through the application of nanotechnology. The production of nanomaterials holds considerable importance due to its cost-effectiveness and environmental benefits. This paves the way for a multitude of applications in areas such as water purification, environmental remediation, medical uses, sensing technologies, energy storage, and catalysis (Shafey, 2020). Furthermore, it has been advised that any new restrictions on nanomaterials must adhere to the following specified conditions (IFA, 2009).

1. In the absence of adequate information regarding the hazards and exposures associated with nanomaterials, precautionary principles should be implemented.
2. The established "general" dust exposure limits, as well as the exposure thresholds for the same substances in their macroscopic form, must not be surpassed.
3. The suggested exposure limits must not exceed those attained through cost-effective engineering controls.
4. The suggested exposure limits ought to be quantifiable using cost-effective monitoring methods.

OECD sponsorship program for the evaluation of produced nanomaterials: this initiative is the most significant for providing information on dose-response and other hazard-related matters. Through this program, OECD member nations, along with certain non-member economies and various stakeholders, are collaborating to assess the hazard potential of 14 manufactured nanomaterials that are either currently in use or nearing commercial application (Organization for Economic Co-operation and Development, 2008).

9. TOXICITY OF NANOMATERIALS

There are a number of ways that nanoparticles can be harmful to the body, but the majority of the intracellular and in real time toxicities result from the creation of excess oxygen species that are reactive. One way that nanoparticles can cause oxidative stress is when iron-based nanoparticles dissolve, which catalyzes the generation of ROS and the creation of OOH and radicals of OH from H2O2 via the Fenton process. Additionally, a few inert nanomaterials do not produce ROS on their own, but can induce ROS production in biological settings based on the NPs' capacity to target water purification (Li *et al.,* 2011). ROS have the potential to be harmful in addition to being physiologically necessary. Certain cellular processes, such as signal transmission, proliferative response, gene expression, and protein redox control, are modulated by moderate amounts of ROS.30,31 Elevated ROS levels are a sign of oxidative stress and can harm cells through lipid peroxidation, protein modification, DNA disruption, interference with signalling, and transcriptional modulation, ultimately leading to carcinoma, kidney damage, neurodegeneration, cardiovascular disease, or pulmonary disease. Cell death and a reduction in physiological function can arise from ROS stealing electrons from lipids in the cell membrane (Schulz *et al.,* 2014).

10. CONCLUSION

Nanomaterial-based technologies for water purification present considerable benefits in enhancing water quality, attributed to their remarkable efficiency, extensive surface area, and capability to eliminate a diverse array of contaminants. These materials have shown outstanding effectiveness in extracting toxic metals, organic contaminants, bacteria, and emerging contaminants from water sources. Nonetheless, obstacles persist regarding costeffectiveness, widespread implementation, and possible environmental and health hazards. The toxicity of certain nanomaterials, their long-term stability, and their behaviour post-disposal necessitate further research to guarantee their safe and sustainable utilization. To enhance advantages while reducing risks, future studies should prioritize the development of eco-friendly, biodegradable, and non-toxic nanomaterials. Furthermore, it is essential to establish regulatory frameworks to ensure the safe use of nanotechnology in water purification. With ongoing progress in materials science and engineering, Nanotechnology possesses considerable potential to revolutionize water treatment processes and guarantee access to clean, safe water for all individuals.

REFERENCES

Ahmadi, M. H., Ghazvini, M., Alhuyi Nazari, M., Ahmadi, M. A., Pourfayaz, F., Lorenzini, G., & Ming, T. (2019). Renewable energy harvesting with the application of nanotechnology: A review. *International Journal of Energy Research, 43*(4), 1387–1410. DOI: 10.1002/er.4282

Ali, Z., & Ahmad, R. (2020). Nanotechnology for water treatment. In *Environmental Nanotechnology* (Vol. 3, pp. 143–163). Springer., DOI: 10.1007/978-3-030-40363-4_7

Alon, G., & Adin, A. (1994). Mathematical modeling of particle size distribution in secondary effluent filtration. *Water Environment Research, 66*(7), 836–841. DOI: 10.2175/WER.66.6.11

Alshameri, A., ... —Understanding the role of natural clay minerals as effective adsorbents and alternative source of rare earth elements: Adsorption operative parameters,|| Hydrometallurgy, vol. 185, no. September 2018, pp. 149–161, 2019, DOI: 10.1016/j.hydromet.2019.02.016

Babatunde, D. E., Denwigwe, I. H., Babatunde, O. M., Gbadamosi, S. L., Babalola, I. P., & Agboola, O. (2020). Environmental and societal impact of nanotechnology. *IEEE Access : Practical Innovations, Open Solutions, 8*, 4640–4667. DOI: 10.1109/ACCESS.2019.2961513

Bain, R., Wright, J., Yang, H., Gundry, S., Pedley, S., & Bartram, J. (2014). Improved but not necessarily safe: Water access and the Millennium Development Goals. In *Global Water: Issues and Insights* (p. 89). Global Water Forum.

Bal, G., & Thakur, A. (2022).. . *Proceedings, 50*(5), 1575–1579.

Bayda, N. I., & Sutliff, G. (2020). Comparing extracted and stipulated definitions in Algebra 1 textbooks and *Khan Academy. International Electronic Journal of Mathematics Education, 15*(2). Advance online publication. DOI: 10.29333/iejme/7601

Bellona, C., & Drewes, J. (2005). The role of membrane surface charge and solute physicochemical properties in the rejection of organic acids by NF membranes. *Journal of Membrane Science, 249*(1–2), 227–234. DOI: 10.1016/j.memsci.2004.09.041

Berk, R. S., & Canfield, J. H. (1964). Bioelectrochemical energy conversion. *Applied Microbiology, 12*(1), 10–12. DOI: 10.1128/am.12.1.10-12.1964 PMID: 14106931

Bond, D. R., Holmes, D. E., Tender, L. M., & Lovley, D. R. (2002). Electrode-reducing microorganisms that harvest energy from marine sediments. *Science, 295*(5554), 483–485. DOI: 10.1126/science.1066771 PMID: 11799240

Bond, D. R., & Lovley, D. R. (2003). Electricity production by *Geobacter sulfurreducens* attached to electrodes. *Applied and Environmental Microbiology, 69*(3), 1548–1555. DOI: 10.1128/AEM.69.3.1548-1555.2003 PMID: 12620842

Bowman, D. M. (2017). More than a decade on: Mapping today's regulatory and policy landscapes following the publication of *Nanoscience and nanotechnologies: Opportunities and uncertainties. NanoEthics, 11*(2), 169–186. DOI: 10.1007/s11569-017-0281-x

Brinker, C. J., & Scherer, G. W. (1990). *Sol-gel science: The physics and chemistry of sol-gel processing*. Academic Press.

Chahardoli, A., Karimi, N., & Fattahi, A. (2018). *Nigella arvensis* leaf extract mediated green synthesis of silver nanoparticles: Their characteristic properties and biological efficacy. *Advanced Powder Technology, 29*(1), 202–210. DOI: 10.1016/j.apt.2017.11.003

Cory, D. C., & Rahman, T. (2009). Environmental justice and enforcement of the Safe Drinking Water Act: The Arizona arsenic experience. *Ecological Economics, 68*(6), 1825–1837. DOI: 10.1016/j.ecolecon.2008.12.010

Cotruvo, J. A. (1987). Risk assessment and control decisions for protecting drinking water quality. In I. H. Suffet & M. Malaiyandi (Eds.), *Organic pollutants in water* (Advances in Chemistry Series, No. 214, pp. [insert page range if known]). American Chemical Society. https://doi.org/DOI: 10.1021/ba1987-0214.ch034

Das, R., Ali, M. E., Hamid, S. B. A., Ramakrishna, S., & Chowdhury, Z. Z. (2014). Carbon nanotube membranes for water purification: A bright future in water desalination. *Desalination, 336*, 97–109. DOI: 10.1016/j.desal.2013.12.026

DiMascio, G., & Ganzi, J. (1999). *Electrodeionization apparatus & method* (U.S. Patent No. 5,858,191). United States Patent and Trademark Office.

Edelstein, A. S., & Cammarata, R. C. (Eds.). (1996). *Nanomaterials: Synthesis, properties and applications*. Institute of Physics Publishing. DOI: 10.1887/0750305789

Espey, M. J., & Shaw, W. D. (1997). Price elasticity of residential demand for water: A meta-analysis. *Water Resources Research, 33*(6), 1369–1374. DOI: 10.1029/97WR00571

Frimmel, F. H., & Niessner, R. (2010). *Nanoparticles in the water cycle: Properties, analysis and environmental relevance*. Springer., DOI: 10.1007/978-3-642-10318-6

Gabelich, K., Ishida, K. P., & Bold, R. (2005). Testing of water treatment copolymers for compatibility with polyamide reverse osmosis membranes. *Environmental Progress, 24*(4), 410–416. DOI: 10.1002/ep.10111

Gehrke, I., Geiser, A., & Somborn-Schulz, A. (2015). Innovations in nanotechnology for water treatment. *Nanotechnology, Science and Applications, 8*, 1–17. DOI: 10.2147/NSA.S43773 PMID: 25609931

Gleick, P. H. (2003). Global freshwater resources: Soft-path solutions for the 21st century. *Science, 302*(5650), 1524–1528. DOI: 10.1126/science.1089967 PMID: 14645837

Graham, N. J. D. (1988). *Slow sand filtration – Recent developments in water treatment technology*. Ellis Horwood Ltd.

Guisbiers, G., Mejía-Rosales, D. F., & Leonard, C. (2012). Nanomaterial properties: Size and shape dependencies. *Journal of Nanomaterials, 2012. Journal of Nanomaterials, 180976*(1), 1–2. DOI: 10.1155/2012/180976

Gul, S., Ahmad, Z., Asma, M., Ahmad, M., Rehan, K., Munir, M., Bazmi, A. A., Ali, H. M., Mazroua, Y., Salem, M. A., Akhtar, M. S., Khan, M. S., Chuah, L. F., & Asif, S. (2022). Effective adsorption of cadmium and lead using SO3H-functionalized Zr-MOFs in aqueous medium. *Chemosphere, 307*, 135633. Advance online publication. DOI: 10.1016/j.chemosphere.2022.135633 PMID: 35810866

Higgins, E., ... (1917). *Siliceous dust in relation to pulmonary diseases among miners in the Joplin District, Missouri* (Bulletin No. 132). U.S. Department of the Interior, Bureau of Mines.

Hinrichsen, D., Robey, B., & Upadhyay, U. (1997). *Solutions for a water short world* (Population Reports, Series M). Johns Hopkins Bloomberg School of Public Health.

Imani, M. M., & Safaei, M. (2019). Optimized synthesis of magnesium oxide nanoparticles as bactericidal agents. *Journal of Nanotechnology, 2019*, 6063832. Advance online publication. DOI: 10.1155/2019/6063832

Institute for Occupational Safety and Health of the German Social Accident Insurance (IFA). (2009). *Criteria for assessment of the effectiveness of protective measures*. https://www.dguv.de/ifa/en/fac/nanopartikel/beurteilungsmassstaebe/index.jsp

Jeevanandam, J., Barhoum, A., Chan, Y. S., Dufresne, A., & Danquah, M. K. (2018). Review on nanoparticles and nanostructured materials: History, sources, toxicity and regulations. *Beilstein Journal of Nanotechnology*, *9*(1), 1050–1074. DOI: 10.3762/bjnano.9.98 PMID: 29719757

Kanchi, S. (2014). Nanotechnology for water treatment. *Journal of Environmental & Analytical Chemistry*, *1*(2), e102. DOI: 10.4172/2380-2391.1000e102

Kansara, N.. (2016). Wastewater treatment by ion exchange method: A review of past and recent researches. *Environmental Science Journal*, *14*(4), 5706.

Kaye, B. H. (1981). *Direct characterisation of fine particles*. Wiley.

Khalid, N., Ahmad, A., Khalid, S., Ahmed, A., & Irfan, M. (2014). Mineral composition and health functionality of Zamzam water: A review. *International Journal of Food Properties*, *17*(3), 661–677. DOI: 10.1080/10942912.2012.660721

Koçak, A., & Karasu, B. (2018). Nanotaneciklerin Genel Değerlendirilmesi. *El-Cezeri*, *5*(1), 191–236.

Kuhn, S. P., & Pfister, R. M. (1990). Accumulation of cadmium by immobilized *Zoogloea ramigera* 115. *Journal of Industrial Microbiology*, *6*(2), 123–138. DOI: 10.1007/BF01576431

Leitgeb, N., Auvinen, A., Danker-Hopfe, H., & Mild, K. (2016). SCENIHR (Scientific Committee on Emerging and Newly Identified Health Risks): Potential health effects of exposure to electromagnetic fields (EMF). *Bioelectromagnetics*, *37*(8), 556–559. DOI: 10.1002/bem.21930

Li, X., Xu, H., Chen, Z.-S., & Chen, G. (2011). Biosynthesis of nanoparticles by microorganisms and their applications. *Journal of Nanomaterials*, *2011*, 270974. Advance online publication. DOI: 10.1155/2011/270974

Lum, L.-S. H., Malghan, S. G., & Schiller, S. B. (1996). Standard reference materials for particle-size analysis of ceramic powders by gravity sedimentation. *Powder Technology*, *87*(3), 233–238. DOI: 10.1016/0032-5910(96)03086-0

Mahmoud, A. S., Mostafa, M. K., & Peters, R. W. (2021). A prototype of textile wastewater treatment using coagulation and adsorption by Fe/Cu nanoparticles: Technoeconomic and scaling-up studies. *Nanomaterials and Nanotechnology*, *11*. Advance online publication. DOI: 10.1177/18479804211041181

Mehmood, S., Ahmad, A., Ahmed, A., Khalid, N., & Javed, T. (2013). Drinking water quality in capital city of Pakistan. *Open Access Scientific Reports*, *2*, 637–641. DOI: 10.4172/scientificreports.637

Michalak, I., Chojnacka, K., & Witek-Krowiak, A. (2013). State of the art for the biosorption process—A review. *Applied Biochemistry and Biotechnology*, *170*(6), 1389–1416. DOI: 10.1007/s12010-013-0269-0 PMID: 23666641

Mir, U. S. (2001). *[Patent title, if available]* (U.S. Patent No. 6,241,867). United States Patent and Trademark Office.

Mishel, L., & Shierholz, H. (2013, August). *A decade of flat wages: The key barrier to shared prosperity and a rising middle class* (Briefing Paper No. 365). Economic Policy Institute. https://www.epi.org/publication/a-decade-of-flat-wages-the-key-barrier-to-shared-prosperity-and-a-rising-middle-class/

Mohammed, M. B. (2011). *Low cost nanomaterials for water desalination and purification: Final technical report.* United Nations–UNESCO. https://www.scirp.org/(S(lz5mqp453edsnp55rrgjct55.))/reference/referencespapers.aspx?referenceid=2529208

Mondal, C., Hsiao, S., & Wickramasinghe, S. R. (2008). Nanofiltration/reverse osmosis for treatment of coproduced waters. *Environmental Progress*, *27*(2), 173–179. DOI: 10.1002/ep.10271

Organisation for Economic Co-operation and Development (OECD). (2008). *Manufactured nanomaterials: List of manufactured nanomaterials and list of endpoints for phase one of the OECD testing programme* (ENV/JM/MONO(2008)13/REV). https://www.olis.oecd.org/olis/2008doc.nsf/LinkTo/NT000034C6/$FILE/JT03248749.PDF

Organisation for Economic Co-operation and Development (OECD). (2008).

Peydayesh, M., & Mezzenga, R. (2021). Protein nanofibrils for next generation sustainable water purification. *Nature Communications*, *12*(1), 3248. DOI: 10.1038/s41467-021-23388-2 PMID: 34059677

Qu, X., Alvarez, P. J. J., & Li, Q. (2013). Applications of nanotechnology in water and wastewater treatment. *Water Research*, *47*(12), 3931–3946. DOI: 10.1016/j.watres.2012.09.058 PMID: 23571110

Rabaey, K., Boon, N., Siciliano, S. D., Verhaege, M., & Verstraete, W. (2004). Biofuel cells select for microbial consortia that self-mediate electron transfer. *Applied and Environmental Microbiology*, *70*(9), 5373–5382. DOI: 10.1128/AEM.70.9.5373-5382.2004 PMID: 15345423

Radjenović, M., Petrović, M., Ventura, F., & Barceló, D. (2008). Rejection of pharmaceuticals in nanofiltration and reverse osmosis membrane drinking water treatment. *Water Research*, *42*(13), 3601–3610. DOI: 10.1016/j.watres.2008.05.020 PMID: 18656225

Safaei, M., Taran, M., & Imani, M. M. (2019). Preparation, structural characterization, thermal properties and antifungal activity of alginate-CuO bionanocomposite. *Materials Science and Engineering C, 101*, 323–329. DOI: 10.1016/j.msec.2019.03.108 PMID: 31029325

Safe Drinking Water Act, 42 U.S.C. §§ 300f–300j (1986).

Savage, N., & Diallo, M. S. (2005). Nanomaterials and water purification: Opportunities and challenges. *Journal of Nanoparticle Research*, *7*(3), 331–342. DOI: 10.1007/s11051-005-7523-5

Schulz, M. J., Ruff, B., Johnson, A., Vemaganti, K., Li, W., Sundaram, M. M., Hou, G., Krishnaswamy, A., Li, G., & Fialkova, S. (2014). New applications and techniques for nanotube superfiber development. In *Nanotube superfiber materials* (pp. 33–59). Elsevier. DOI: 10.1016/B978-1-4557-7863-8.00002-5

Shafey, A. M. (2020). Green synthesis of metal and metal oxide nanoparticles from plant leaf extracts and their applications: A review. *Green Processing and Synthesis*, *9*(1), 304–339. DOI: 10.1515/gps-2020-0031

Shah, M. A., & Ahmed, T. (2011). *Principles of nanoscience and nanotechnology*. Narosa Publishing House.

Shantaram, A., Beyenal, H., Raajan, R., Veluchamy, A., & Lewandowski, Z. (2005). Wireless sensors powered by microbial fuel cells. *Environmental Science & Technology*, *39*(13), 5037–5042. DOI: 10.1021/es0480668 PMID: 16053108

Sharma, S., Yadav, A., & Ahmad, W. (2018). The classification, characterization, and application of ion exchange resins: A general survey. SSRN *Electronic Journal*. https://doi.org/DOI: 10.2139/ssrn.3299226

Shi, Z., & Guo, N. (2014). *Nanomaterials and Devices*., DOI: 10.1016/C2012-0-03204-8

Singh, B., Thakur, V., Bhatia, G., Verma, D., & Singh, K. (2016). Bioremediation and biodegradation. *Journal of Bioremediation & Biodegradation*, *7*, 374. Advance online publication. DOI: 10.4172/2155-6199.1000374 PMID: 29142804

Singh, N. B., Nagpal, G., Agrawal, S., & Rachna, . (2018). Water purification by using adsorbents: A review. *Environmental Technology & Innovation*, *11*, 187–240. DOI: 10.1016/j.eti.2018.05.006

Srivastava, A., Srivastava, O. N., Talapatra, S., Vajtai, R., & Ajayan, P. M. (2004). Carbon nanotube filters. *Nature Materials*, *3*(9), 610–614. DOI: 10.1038/nmat1192 PMID: 15286755

Trevors, J. T. (1989). The role of microbial metal resistance and detoxification mechanisms in environmental bioassay research. *Hydrobiologia, 188*(1), 143–147. DOI: 10.1007/BF00027779

Turakhia, B., Turakhia, P., & Shah, S. (2018). Green synthesis of zero valent iron nanoparticles from *Spinacia oleracea* (spinach) and its application in wastewater treatment. *Journal of Advanced Research in Applied Sciences*, *5*(1), 46–51.

United States Environmental Protection Agency (USEPA). (2013). *Nanotechnology and radiation cleanup*. Retrieved from https://www.epa.gov/radiation/cleanup/nanotechnology.html

U.S. Bureau of Reclamation & Sandia National Laboratories. (2003). *Desalination and water purification technology roadmap: A report of the executive committee*. Water Purification.

U.S. Environmental Protection Agency. (2011). *Drinking water and ground water statistics* (EPA 816-R-13-003). Office of Water. Retrieved from http://water.epa.gov/scitech/datait/databases/drink/sdwisfed/upload.epa816r13003.pdf

Van der Bruggen, B., & Vandercasteele, C. (2003). Removal of pollutants from surface water and groundwater by nanofiltration: Overview of the possible applications in the drinking water industry. *Environmental Pollution*, *122*(3), 435–445. DOI: 10.1016/S0269-7491(02)00308-1 PMID: 12547533

Werber, J. R., Osuji, C. O., & Elimelech, M. (2016). Materials for next-generation desalination and water purification membranes. *Nature Reviews. Materials*, *1*(5), 1–15. DOI: 10.1038/natrevmats.2016.18

Yeoman, S., Lester, J. N., & Perry, R. (1992). Phosphorus removal and its influence on particle size and heavy metal distribution during wastewater treatment. *Environmental Technology*, *13*(9), 901–924. DOI: 10.1080/09593339209385227

Ying, S., Guan, Z., Ofoegbu, P. C., Clubb, P., Rico, C., He, F., & Hong, J. (2022). Green synthesis of nanoparticles: Current developments and limitations. *Environmental Technology & Innovation*, *26*, 102336. DOI: 10.1016/j.eti.2022.102336

Zhang, W. X. (2003). Nano-scale iron particles for environmental remediation: An overview. *Journal of Nanoparticle Research*, *5*(3–4), 323–332. DOI: 10.1023/A:1025520116015

Zhang. (2008). *Physical Fundamentals of Nanomaterials*. https://doi.org/DOI: 10.1016/C2012-0-02139-4

Chapter 9
Policy and Regulatory Landscape:
A Legal Framework for the Regulation of Advanced Nano Biomaterials: Responsibility, Risk, Reflexivity

Aftab Haider
https://orcid.org/0009-0009-3339-5057
Southwest University of Political Science and Law, China

Farouq Saber Al-Shibli
https://orcid.org/0000-0002-2935-0405
University of Kalba, UAE

Ibrar Ahmad
https://orcid.org/0009-0001-0846-0218
Southwest University of Political Science and Law, China

ABSTRACT

The governance of advanced nano biomaterials is failing due to obsolete regulations (like REACH and TSCA), which transforms technological risk into legal hazard and impedes the adoption of materials that could save \$2.5 trillion annually in corrosion costs and cut waste by 60-70%. These frameworks suffer from definitional ambiguity and lifecycle myopia. The proposed solution is the 4-R Governance Framework (Responsibility, Risk, Reflexivity, Regeneration). Crucially, this involves the Doctrine of Perpetual Regulatory Design (PRD), featuring the Obligatory Reflexivity Clause and Trans-Generational Liability Ledger. This design mandates continuous legal ad-

DOI: 10.4018/979-8-3373-2617-7.ch009

aptation and perpetual manufacturer accountability, providing a Model International Protocol to solve the trilemma of environmental safety, innovation, and global trade by aligning governance with circular economy principles.

1. INTRODUCTION

Corrosion, the degradation of metals because of chemical or electrochemical reactions with their environment, is an emerging global economic problem (Bender et al., 2022). The global economic cost of corrosion is estimated to be 3.5% of total global GDP, which is around $2.5 trillion annually (Masroor, 2023). Economic cost estimates that leave out costs from corrosion, lost productivity, and environmental damage are probably half of the true total cost. In the past, corrosion was usually controlled using chemical inhibitors made from toxic metals and other harmful materials. Although they are partially effective in protecting metals from degradation, they are environmentally dangerous. The permanent and bioaccumulative characteristics of these inhibitors could result in permanent adverse ecological consequences (Ramasamy & Palanisamy, 2021).

Recently, an eco-friendlier way to control corrosion has appeared in nanobiotechnology, which uses advances in both nanotechnology and biotechnology. The creation of new materials with enzyme-based nanocoatings and inhibitors is an example of these modern techniques. The solutions are better than traditional methods in many ways, as they act more efficiently, are more specific to nipping corrosive sites in the bud, and are more biodegradable. This approach is also aligned with the circular economy, which aims to promote resource efficiency, minimize waste, and reduce environmental impacts. According to data from the Organization for Economic Co-operation and Development (OECD) from 2021 to 2023, there were more than 3,200 nano-enabled biotechnological innovations commercialized (OECD, 2023). The fast development of biotechnology shows the flexibility and potential of this area to be used commercially. For example, enzyme-encapsulated nanocontainers are advanced delivery systems designed to stay inactive and safe under normal conditions. They open and discharge corrosion inhibitors when something changes, such as a change in pH levels during corrosion. This controlled release is quite unlike the traditional inhibitors, which are always active and release everywhere, whether they are required or not. But the very same "intelligent" feature of nano-bio materials to respond or to be triggered in some circumstances is a significant challenge to the current legislation. The existing rules were not formulated to handle materials that alter their characteristics or behavior once they are used. This presents a fundamental classification issue: *are these materials single chemicals, complex manufactured products, or something completely new that only forms when activated?* This is not merely a technical point–it shows that the modern laws are obsolete. The EU and the U.S. have significant regulatory frameworks for controlling the production, use, and disposal of stable chemicals. The EU, especially, has strong systems for evaluating the risks of chemicals to health and the environment. However, the special properties, lack of knowledge of toxicity, and complicated life cycle of new nanobiotech materials are too broad and outdated to be dealt with by these rules (Lee & Choi, 2019).

The consequences of this regulatory mismatch are severe. Research indicates that nano-enhanced corrosion inhibitors have the potential to cut greenhouse gas emissions and material waste by 60 to 70% when properly designed (OECD, 2023). However, unclear and conflicting regulations are slowing the adoption of these technologies. The gap between science and regulation is now holding back sustainability and safety. It is further complicated by the fact of what experts refer to as the trilemma in nanobiotechnology regulation. Policymakers must strike a balance between three objectives: safeguarding human and environmental health, promoting technological advancement, and maintaining fair and consistent global trade. It is almost impossible to achieve all three simultaneously. When regulations are either too strict or too weak, they may be harmful. Excessive regulation may restrict research, prevent new products, and establish trade barriers. However, excessive laxity may cause health and environmental hazards. To make matters worse, various nations have varying regulations. As an example, the EU, the U.S., and the UAE have their own testing and certification systems, it implies that companies will have to undergo the same procedures in each of the regions.

This has already led to trade tensions. The Committee on Technical Barriers to Trade (TBT) at the World Trade Organization has documented the disagreements regarding nano-labeling requirements, which demonstrates the disintegration of the market (WTO, 2023; Foreign Trade Barriers, 2025). According to the United Nations Environment Program (UNEP, 2023), the growing gap between innovation and oversight risks creating systemic risks for long-term liability and public trust (Klinke & Renn, 2021). The failure of good governance hampers the adoption of sustainable technologies and undermines considerable gains in resource efficiency. This inertia is economically irrational. According to a study by the International Monetary Fund, clarity in regulations about green technologies leads to 15-20% higher private R&D investment (Basterra & López, 2024). The cost of not acting will be more expensive than the cost of developing new, adaptive governance systems, as it will hinder innovation, damage the environment further, and create more trade resistance. The trilemma does not just create regulatory inconvenience but also causes material damage to global economic resilience and the switch to the circular economy. The primary aim of modern environmental law must be to design regulation policies that will develop science and technology while ensuring the safety of humans and the environment.

Two primary yet restricted methods of regulatory governance are connected in this chapter. First, it criticizes the procedural inflexibility of the conventional command-and-control regulations, including Registration, Evaluation, Authorization and Restriction of Chemicals (REACH) and the fixed chemical lists in the Toxic Substances Control Act (TSCA), and the purely reflexive law approach, which tends to focus on process rather than results. It also doubts the efficacy of gradual adaptive governance, which, though beneficial, does not have the powerful guidance required to steer innovation in the direction of sustainability and circularity.

We believe that the 4-R Framework integrates the advantages of the current regulatory models and adds a new normative aspect to them, such as a deliberate orientation of technological development towards a regenerative economy. The main issue, which is called the Trilemma, is resolved with the help of the 4-pillars of the framework. These pillars combined make the system both ethically progressive and legally viable. The 4-R Framework encourages nanobiotechnology to actively recover the environment, improve economic sustainability, and promote social equity. We contend that this framework is proactive, unlike the traditional systems that respond to problems once they have occurred–it tries to prevent problems before they can happen, yet it is long-term effective and ethically governed.

2. METHODOLOGY

The chapter takes a prescriptive approach. Instead of only providing a critical analysis, it also presents a practical legal framework. The study makes three main contributions. First, the chapter provides an updated evaluation of the global regulatory landscape. It identifies systemic flaws and governance gaps. The 4-R Governance Framework is introduced through the Responsibility, Risk, Reflexivity, and Regeneration solutions (Figure 1). The framework combines four established but siloed principles, risk governance, adaptive management, ethical standards, and circular design into one. This model is entirely new as it does not analyze the four main R's individually, unlike the other scholars. It shows how these elements work together in a reinforcing cycle to solve the trilemma by combining safety, innovation, and trade. The approach moves from reacting to problems to designing institutions that close the gap between rapid technological advances and slow legal adaptation. The chapter provides a more powerful rationale for reform than earlier descriptive studies by presenting the problem in the form of a trilemma.

The chapter is split into four main sections to find a problem and provide a solution. The second part is our literature review, focusing on "Lex Lata: A Post-Mortem on Regulatory Failure." It looks at how regulatory systems in the EU, US, and UAE respond when problems occur. We found common failures in managing product life cycles and integrating circularity. Section three discusses our main idea, "Lex Ferenda: The 4-Pillar Nano-Bio Governance Framework," explaining each pillar and how they connect. Part four, "Doctrine of Perpetual Regulatory Design (PRD)," answers the important final question: how will it endure for the next 50 years? Our two original and logical concepts are introduced here.

Figure 1. The 4-R Governance Framework

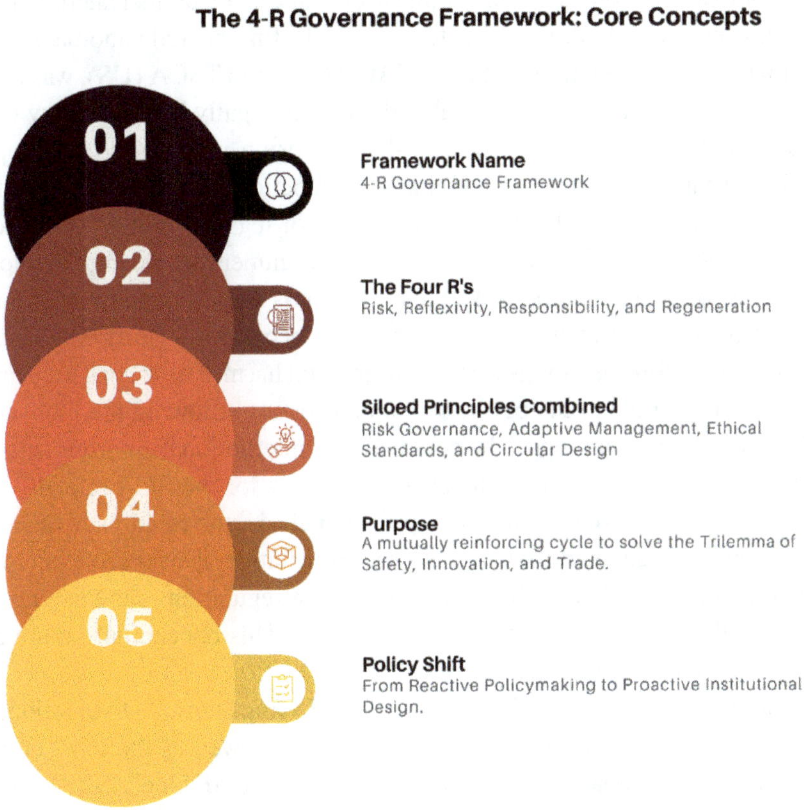

(Source: Author's own Design)

The 4-R Governance Framework has four pillars—Risk, Reflexivity, Responsibility, and Regeneration—forming an ongoing cycle for managing nanobiomaterials. Each pillar supports the others, creating a loop of data collection, legal updates, accountability, and circular innovation.

3. LITERATURE REVIEW

This section provides an analysis of the current regulatory framework (lex lata—the law as it is). It identifies the reasons behind failures in governing nano-bio corrosion inhibitors. Global regulation of nanobiotechnology for corrosion inhibitors currently reacts to problems after they arise and lacks coordination (Shahcheraghi et al., 2022). The main problem is a category error: applying legal frameworks which is for discrete, conventional chemical substances to sophisticated, dynamic, and in many instances, biologically active materials. Engineered nanomaterials did not exist when core legislation, such as REACH (EU) and TSCA (US), was created (Widmer & Meili, 2010). Internationally, there is no legally binding treaty to deal specifically with the life cycle of a nano-bio corrosion inhibitor. For example, a protein-based nano capsule. The claim that it can function as a 'chemical substance', 'biocidal product', or 'novel entity' creates a gap of legal certainty. The outcome is not just an administrative hassle, but also misalignment between technological progression and legal adjustment. The liability gap is caused by the model failure. When a product's legal identity is unclear, it can become very difficult to assign liability for any unforeseen long-term environmental harm. These gaps transfer risk from private innovators to the public sector and society, allowing the innovator to profit from the new technology. The core belief of international environmental law is "polluter pays". The failure of the generator to pay for the full cycle also gives rise to a moral hazard. According to empirical analysis, the rapid pace of innovation is widening the gap and causing policy fragmentation. This disconnect has led to commercial releases happening much faster than regulations can keep up, with regulators still relying on outdated safety concepts. This conceptual failure helps explain the jurisdictional failure discussed below.

The EU regulatory system based on REACH (Registration, Evaluation, Authorization and Restriction of Chemicals) is widely viewed as the most complex and the most precautionary regulatory system in the world. One of the important strengths is that the legislation binds nanoforms. Thus, data must be supplied on particle size, surface area, and surface chemistry. However, that strong point creates a big weakness: the compliance vortex. The cost and technical difficulty of generating nano-specific data under REACH is a high barrier to entry for small and medium-sized enterprises (SMEs) developing innovative, sustainable inhibitors (Subramanian, 2017). Research suggests that these costs can give an advantage to large companies with strong regulatory departments, making it harder for new, more sustainable technologies to compete. Also, the European Green Deal supports the circular economy, REACH contains no obligation for producers to make nano-bio inhibitors recyclable, safely degradable, or able to re-enter biological cycles (Saxena, 2023). This is a design flaw. It means the regulation focuses mainly on the first risk assessment and not on the entire product lifecycle. The EU's innovative framework is, therefore, undermined through its own rigidity and its failure to impose circular-by-design principles starting from R&D. A system meant to provide safety through compliance is blocking innovation needed for sustainability. For example, there was a delay in launching nano-bio inhibitors made from farm waste. These nanoform products help the circular economy; however, small companies producing them often cannot fulfill all the REACH nanoform requirements. The new green tech that converts waste into protective coatings is unable to penetrate the market. Even though eco-friendly products are popular now, big chemical companies still sell regular inhibitors. It imposes an unintentional "green penalty" on the most environmentally friendly solutions when approving such solutions.

Most of these substances are regulated in the United States through the Toxic Substances Control Act (TSCA), which adopts a risk-benefit approach as opposed to the precautionary principle applied in the EU (Wagner et al., 2024). The regulatory approach of TSCA is lenient and is designed to promote innovation, and in the past, was based on voluntary data submission programs whereby companies volunteered to provide information. Although this method leads to innovation, it also leads to a lack of information regarding the environmental effects of nano-bio inhibitors in the long run. Research is always keen to point out that TSCA still has many gaps (Rayasam et al., 2022), especially since there is no compulsory post-market surveillance, and recurring data gaps remain unresolved. The use of case-by-case Significant New Use Rules (SNURs) is a slow process and does not build the full dataset needed for proactive governance. Because the system is designed to be non-intrusive, it cannot force data to be created to handle new or ongoing risks. This method allows products to enter the market without complete safety records, instigating risk. The different access to information of the innovators and regulators creates a scenario in which the environmental damage is likely discovered too late to intervene effectively. This basic philosophical difference with the EU not only hinders innovation at home but also creates real problems for exports. As a result, chemical companies must deal with two incompatible compliance systems. To capitalize on the experience of regulation, the UAE has established a strong governance framework to use international standards in its domestic policymaking. One of the foundations of this framework is the *Cabinet Resolution No. 10 of 2022* on the Technical Regulations of the Control of the Quantity of Product in Pre-packages, which is aimed at protecting consumer rights and improving the quality and competitiveness of industrial products by ensuring adherence to the approved regulations (UAE Legislation: Cabinet Resolution No. 10 of 2022 on Pre-packages) (*United Arab Emirates Legislations | Cabinet Resolution On the Technical Regulations for the Control of the Quantity of Product in Pre-Packages*, 2025). Specialized national programs, including the UAE Nano Registry and standards like ESMA UAE.S 5025 (a name given to technical standards), are used to support this governance to control the safety and management of advanced materials (United Arab Emirates - Standards for Trade, 2025). They are included in the larger scope of the Ministry of Industry and Advanced Technology (MoIAT), which absorbed the former Emirates Authority for Standardization and Metrology (ESMA) to create and implement UAE Standards and Technical Regulations (MoIAT/ESMA Standards). The United Arab Emirates is establishing itself as a center of sustainable technology by deliberately developing its structure of control on the principles of OECD (Organization for Economic Co-operation and Development) and ISO (International Organization for Standardization) standards. This alignment facilitates international best practices in tax transparency and product quality and safety (OECD and ISO alignment). However, the fact that the new system is based on data and best practices of other regulatory settings can be a significant challenge, since frameworks such as Environmental Impact Assessments (EIAs) do not always include sufficient engagement of regional data that is specific to the local ecological and socio-economic situation. Depending on outside information shows a weakness and an area for development in the capacity to work with. The rules are designed to protect local ecosystems, which are often fragile. This approach, grounded in established standards, represents a third alternative beyond the adaptive strategies of both the EU and US systems. This model provides actors with generous rewards and is highly risky as it relies on the continuing strength of international standard setters (Figure 2). This dependence can be easily taken advantage of when countries have different policies. Figure 2 is an analytical comparison of the three dominant regulatory philosophies—precautionary (EU), risk–benefit (USA), and standards-based leapfrogging (UAE)—and their implications for nanobiomaterial innovation, safety oversight, and trade consistency.

Figure 2. Comparative Regulatory Approaches in the EU, USA, and UAE

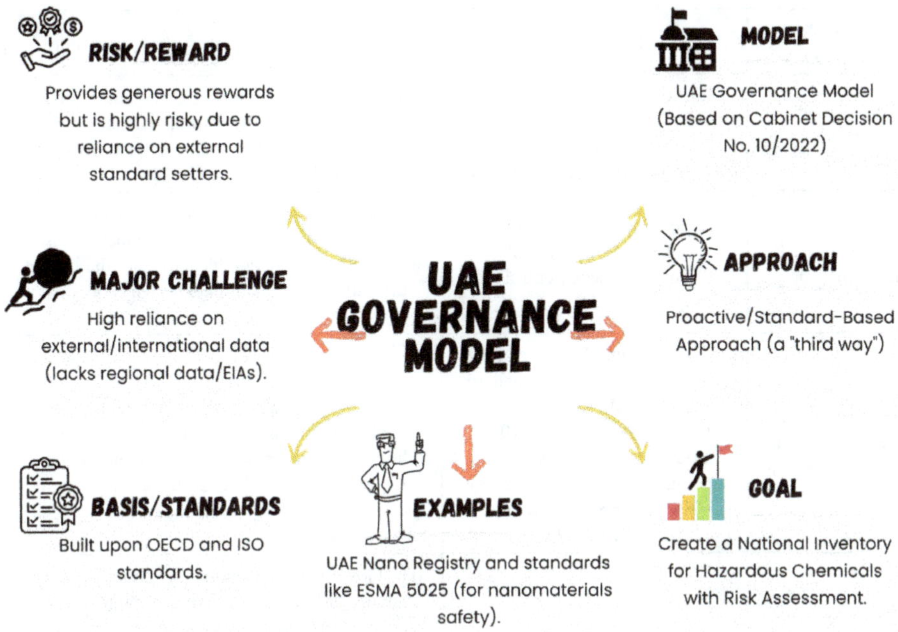

(Source: Author's own Creation)

It is not a matter of one case of regulatory failure, but a systemic and interconnected issue. Table 1 brings out trends that transcend individual problems, demonstrating that the Governance Trilemma is a reality and a working concept. As an illustration, the EU focuses on high safety standards, which may slow innovation. Conversely, the U.S. is more innovative, but it does not always have good safety data. The UAE model can ignore local ecological considerations. Regardless of these differences, none of the three approaches is a clear advocate of global trade opportunities. This part will discuss the key areas and demonstrate how the existing legislation that limits some technologies can complicate and become expensive to integrate technologies. This is a very important point that will be discussed in the following section. These systemic gaps have far-reaching and persistent consequences, which impact the resilience of global supply chains to all.

Table 1. Comparative Diagnostic of Regulatory Frameworks for Nano-Bio Corrosion Inhibitors

Jurisdiction & Core Philosophy	Key Legislative Instruments	Consequence for Nano-Bio Innovation
European Union Precautionary Principle	REACH Regulation (ECHA, 2022) Biocidal Products Regulation (BPR) (Regulation (EC) No 1907/2006)	Sacrificed Innovation: The compliance vortex and procedural rigidity disproportionately affect SMEs, which unintentionally helps to preserve more incumbent, less sustainable technologies.
United States Risk-Benefit Analysis	Toxic Substances Control Act (TSCA) (EPA, 2022) (Toxic Substances Control Act, 15)	Traded in Long-Term Safety: The data-scarce reactive environment permits products whose safety dossier is not complete to enter the market, offloading the risk of unexpected ecological damage.
United Arab Emirates Standards-Based Leapfrogging	UAE Cabinet Decision No. 10/2022 ESMA 5025 Standard (ESMA, 2021). (Cabinet Decision No. 10 of 2022; Emirates Authority for Standardization and Metrology (ESMA), ESMA Standard 5025:2021)	Lost Local Specificity: The high reliance on global standards poses a risk of regulations failing to provide protection to particular regional ecosystems (e.g. arid, marine).
Common Systemic Failure		The Governance Trilemma in Action: All three systems, although with divergent philosophies, impose an unsustainable trade-off between safety, innovation and trade, which makes the existing paradigm obsolete.

4. ANALYSIS AND DISCUSSION

After critically assessing the fundamental weaknesses of the current regulatory framework, this section presents the foundational framework for a more effective model. The 4-R Governance Framework is a comprehensive framework that is interconnected and inclusive with respect to risk, reflexivity, responsibility, and regeneration. This framework aims to function as a self-reinforcing cycle rather than serve as a checklist. The goal is to turn general governance rules into clear laws to address the main issues mentioned earlier. The first pillar of the approach is the Risk, which marks a substantial shift from the ad hoc one-time evaluation of hazards in existing regulations. The model of Dynamic Risk Governance predicts that risk assessment will become a continuous and iterative process during the life cycle of a nano-bio corrosion inhibitor (Figure 3). This assessment is carried out with the help of two key innovations, such as, Advanced Material Passports (AMPs) and Adaptive Tiered Testing (ATT). An AMP serves as an electronic file for each nano-bio product with standardized information on their physicochemical properties, use, and environmental exposure. An AMP might specify the catalyst's functionality, the behavior of the protein coating at different pH levels, and the nature of the eventual degradation products. For example, an enzyme-based nanocapsule would carry an AMP, which gives details about its options, the catalytic core, the stability of its protein shell, the pH range it can handle, and what kinds of degradation products it produces. This directly addresses definitional ambiguity by creating a targeted and data-rich identity for the substance.

Figure. 3 is a schematic representation of the proposed risk governance system combining Advanced Material Passports (AMPs) and Adaptive Tiered Testing (ATT). The figure shows how real-time data collection, iterative evaluation, and progressive testing thresholds create a continuous, evidence-based risk-management cycle.

Figure 3. Dynamic Risk Governance Mechanism

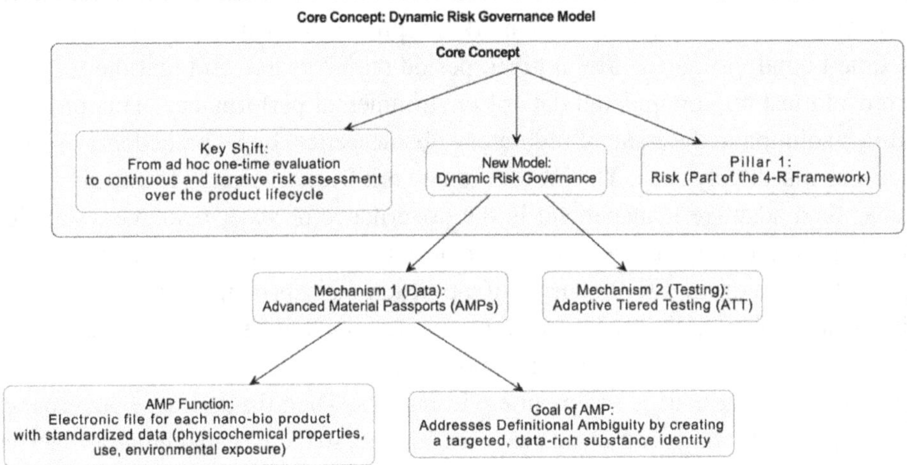

(Source: Author's own creation)

At the same time, ATT uses the recommendation of OECD and takes an active approach to testing. According to OECD (2023), it permits entry into the market with a limited data package, notably appropriate for small-scale or low-risk applications, but requires more extensive post-market assessments as production quantities or environmental release information. The approach reduces the compliance burden on small and medium-sized enterprises (SMEs), which is a significant drawback of the EU model. Furthermore, it would help fix the persistent shortage of environmental data (Data Desert) that afflicts the regulatory landscape in the United States. For example, in tier 1, a newly formulated nano-bio corrosion inhibitor may only need to provide basic ecotoxicity assessments and predictive models of its environmental behavior for low-volume approvals. If the production exceeds a certain level, for instance, 10 tons a year in the EU, an automatic upper-tier assessment will take place. The Tier 2 tests are basically simulated tests that do look at environmental impacts, but they are not quite enough to evaluate anything complex. Signs of persistence or bioaccumulation mark the start of tier 3. This can require full pilot-scale monitoring. The tiered approach makes sure the level of scrutiny from regulations matches the possible level of environmental damage. The system is "dynamic" because it regularly adds more science-based data, helping to create a responsive regulatory framework.

The second pillar, Reflexivity, institutionalizes the capacity for legal and regulatory adaptation. Current systems lack reflexivity; the science-based data is not adopted in the new legal requirements. Through legal mandate, this pillar enables adaptive management through Triennial Scientific Reviews and Sunset Clauses for approvals. A Triennial Review would require ECHA and the EPA to assess and modify their nano-specific testing requirements based on the current peer-reviewed literature. Sunset Clauses would mean that the market approvals for nano-bio inhibitors would be time-bound, primarily after a given period (say 7 years), and that the renewed approval must rely on updated data of environmental performance. This provides a direct solution to the issue of regulatory obsolescence. Legal precedents for such mechanisms already exist. The Water Framework Directive in the European Union has applied adaptive management in the governance of water resources, showing that legally required reassessments are feasible and meaningful. Furthermore, Sunset Clauses exemplify the virtue of a precautionary principle where temporary dispositions taken in the face of uncertainty are not allowed to become permanent dispositions. If new evidence indicates possible ecological harm, such as a nano-bio inhibitor affecting marine sediment environments. Then the reflexive governance system put in place would make it canceled on time, so there is no regulatory inertia, which is currently restraining possible harm.

The third pillar, which is Responsibility, seeks to address the Liability Gap and Enforcement Fragmentation by strengthening and codifying the 'polluter pays' principle. The document proposes a strong Extended Producer Responsibility (EPR) framework for nanomaterials, requiring manufacturers and importers to be responsible for their products from production to disposal. This involves collection, environmental monitoring, and safe disposal or recycling. Unlike traditional EPR, which only covers packaging waste, this EPR model also requires continuous post-market environmental monitoring and the establishment of a Nano-Remediation Fund. A tax would be applied to all the nano-bio products and would be utilized for creating the fund. Thus, the financial burden of environmental damage will not lie with the taxpayers but with the ones who create and profit from it. An example is a supplier of an anti-corrosion coating made from graphene, who will have to fund the long-term research on what happens to the coating as infrastructure ages, including its degradation and impacts on nature. This strategy encourages making safer, longer-lasting, and recyclable products, fixing the short-term thinking in current rules and business practices. The regulations under this pillar will require products to be Circular-by-Design, meaning they will be created for disassembly, recovery, and reuse in industrial or biological metabolisms. The metrics to measure are the ability to recover and become usable again at end-of-life, the ability to break down into safe nutrients that support nature, or the capability to increase the durability and lifespan of the asset they protect. For example, a nano-inhibitor that prevents corrosion and seals micro-cracks will continually upgrade the structural health of materials. This makes the regulation's role shift from a barrier to the driver of radical sustainability. The regeneration pillar requires standardized, quantifiable thresholds for a net-positive claim for going from principle to practice. This pillar significantly shifts the function of regulation from limiting innovation to enabling sustainable innovation. To turn these aspirations into action, the Regeneration pillar will require the definition of the measurable indicators of what a "net-positive" impact is. For a corrosion inhibitor, the cost would be to show that the energy and resources saved by extending the life of protected infrastructure equals or exceeds the environmental costs of making and applying this product, which would pass an LCA aligned with ISO. For instance, the product will need to show an improvement of 20% in the life of the protected asset compared to the standard. It changes sustainability from marketing rhetoric to a verifiable and regulatory condition to gain access to the market.

The 4-R Framework (Table 2) is only effective if all four components are linked together. They create an integrated self-reinforcing governance cycle, which can break the core regulatory deadlock. The three-way interface between the pillars is multiplicative rather than merely additive, and each one gives strength and enhanced ability to the other one. The net effect is a practical-oriented governance system that has high practical potential, which can be used to convert high-level regulatory principles into effective implementation.

Pillar 1: Dynamic Risk Governance is the core of this cycle as it makes sure there is a continuous stream of trusted and standardized data, expressed in Advanced Materials Properties (AMPs) and Adaptive Trigger Thresholds (ATTs). The information created in Pillar 1 is the foundation of *Pillar 2:* Reflexive Regulation, which allows evidence-based renewal of policies with Triennial Reviews and Sunset Clauses. These adaptive tools provide a feedback loop that reinforces risk assessment practices and gradually improves testing and monitoring standards as time goes by.

Pillar 3: Responsibility improves the whole system of data-driven and adaptive governance by introducing legal responsibility and financial instruments like Extended Producer Responsibility (EPR) and Nano-Remediation Fund. These are the tools that facilitate compliance and responsible conduct in the industry (Sahu et al., 2025). Collectively, these pillars result in the objectives of *Pillar 4:* Regeneration, which aims at promoting innovation and creating technologies that not only avoid harm but also provide a restoration of environmental and social value. By doing so, the 4 pillars create a system of governance that is interdependent and integrates risk management, flexibility, and accountability to achieve sustainable development.

This combined strategy also contributes to the solution of the trilemma of governance mentioned above. The framework improves environmental protection by ensuring that the risks are continuously monitored (Pillars 1 and 3) and, at the same time, encourages innovation by providing a transparent, predictable, and flexible regulatory process (Pillar 2). It also facilitates international trade through a single science-based framework rather than the current fragmented approaches. The framework's closed loop, where information, adaptation, and enforcement inform each other, overcomes the usual design trade-offs for regulation – the best outcome of the 4-R Framework (Table 2). This shows that achieving strong environmental standards, promoting ongoing innovation, and facilitating international trade are possible if the governance architecture has been designed and developed inclusively.

Table 2. The 4-R Framework – Operational Mechanisms and Their Target Flaws

Pillar	Core Objective	Key Operational Mechanisms	Fatal Flaws Addressed
1. Risk (Dynamic Governance)	To manage uncertainty through continuous, data-driven learning.	Advanced Material Passports (AMPs) - Adaptive Tiered Testing (ATT)	Definitional Ambiguity - Lifecycle Myopia - Data Silos
2. Reflexivity (Adaptive Learning)	To ensure regulations evolve with scientific and technical progress.	Mandatory Triennial Scientific Reviews - Time-limited Sunset Clauses	Reflexivity Deficit - Enforcement Fragmentation
3. Responsibility (Liability & Stewardship)	To internalize costs and assign clear accountability for the entire lifecycle.	Cradle-to-Cradle EPR Scheme - Nano-Remediation Fund	Liability Vacuum - Lifecycle Myopia - Enforcement Fragmentation
4. Regeneration (Circular & Positive Impact)	To transition from harm reduction to positive environmental contribution.	Mandatory "Circular-by-Design" Criteria - Net-Positive LCA Requirements	Circularity Blindness - Lifecycle Myopia

To implement the 4-R Framework in reality, a specific and enforceable legal approach is necessary. The best way forward is to develop a Model International Protocol on Advanced Materials, which we recommend. This protocol would be legally binding, unlike a non-legal declaration. Also, a newly designed, dedicated protocol will have the advantage of being purpose-built. This is different from changing existing international treaties—such as the Stockholm or Basel Conventions—which were made for stable pollutants and hazardous wastes and cannot easily adapt to new, changing 'smart' materials (Casoli & Ramkumar, 2020). Doing so will ensure that it is not impeded by the narrow definitions and limited scopes conferred by older agreements. It would not work to retrofit the 4-R Framework into an existing regulatory framework, such as REACH in the EU (Registration, Evaluation, Authorization and Restriction of Chemicals) or TSCA in the USA (Toxic Substances Control Act), because it may undermine its key principles (Botos et al., 2019). The aim of the Protocol is not to replace the existing framework. However, it will enhance the existing framework by providing a unified, harmonized system that does not exist.

The Protocol's governance structure would be designed to clearly reflect the 4-R Framework's four key pillars (Figure 4). The Advanced Materials Properties (AMP) data schema would be included in the Annex to the framework, and Adaptive Trigger Thresholds (ATT) would be defined under Pillar 1. It would also involve three-year review cycles to make sure that the regulations are in tandem with the existing scientific knowledge (Pillar 2). It would also create an administrative body that would oversee the Nano-Remediation Fund (Pillar 3) and encourage the use of international standards to conduct Regenerative Life Cycle Assessments (LCA) (Pillar 4). This approach is similar to the Montreal Protocol, which aims to preserve the ozone layer and solve the global environmental hazards of technological advancement (Gonzalez et al., 2015). The two frameworks are similar in the sense that they are proactive and science-based in the identification and control of hazardous substances. The success of the Montreal Protocol shows that even the most complicated, technology-associated environmental issues can be successfully solved with the help of global governance. It is thus a good example of the governance model that is suggested here. Figure 4 illustrates the institutional architecture that implements the 4-R Framework within an international protocol, highlighting the coordination of review cycles, liability, standards, and oversight for global governance.

Figure 4. Governance Structure of the Model International Protocol on Advanced Materials

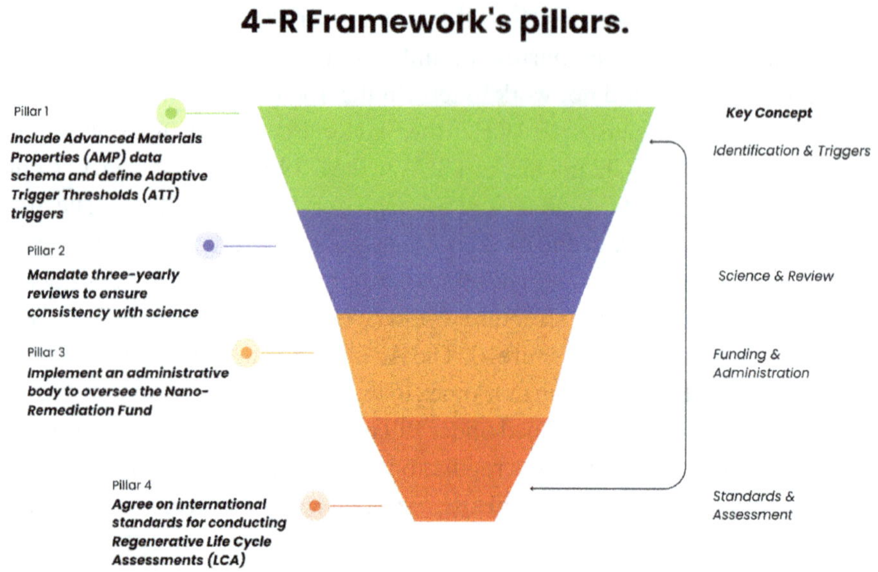

(Source: Author's own Creation)

As it has been mentioned above, the 4-R Framework is conceptually robust. Nevertheless, its influence is not as much as it could be as long as it is a mere theoretical model that lacks a specific roadmap on how it can be integrated into international law and policy. To solve this, the next section presents a staged implementation roadmap, which is well planned to capture the geopolitical and economic realities that tend to drag significant regulatory changes. An organized plan is not only needed to demonstrate the practical viability of the framework but also to instill confidence in policymakers and industry leaders about the advantages and feasibility of the plan. The roadmap is separated into three interrelated stages, each having its own goal, major players, and quantifiable results. This is a gradual process that guarantees a gradual progress and a continuity towards the final goal of having a global governance system on advanced materials. This approach is a conscious and logical step between what is only desirable (lex ferenda) and what is legally binding (lex lata). It is also a reasonable reaction to the increasing costs and inefficiencies brought about by the existing fragmented regulatory systems.

The first step is to build strong knowledge and expertise in the community. This group can lower political risks later by presenting a well-developed, evidence-based proposal instead of just an idea. The consensus-building process must happen very deliberately from the beginning to avoid the potential for geopolitical fragmentation where different international interests turn the framework into another fault line in the bigger global tech rivalry. The OECD and traditional institutions will not suffice for this. It will have to involve countries like India, Brazil, and Indonesia early in the co-development of standards, rather than as a consulted party later on. Therefore, key stakeholders should include the OECD, ISO, and G77+China group in UNEP to ensure their developmental and environmental priorities are integrated into the AMP and ATT (Farias, 2023). Making digital assets accessible to underserved communities around the world is a first step toward global inclusivity. A regulatory standard accepted by a few Western nations can create an exclusive "green" regulatory zone. This will worsen the market segmentation that the 4-R Trilemma seeks to resolve. Therefore, during this phase, the goal is not to create a technical program specification but to develop a Model Protocol that is politically credible and reflects a multipolar international system.

The second stage in establishing technical standards is to prove that the system is practically manageable and cost-effective using pilot projects in other countries. The key objective of this phase is to demonstrate that the 4-R system can be efficient in minimizing health and environmental risks without being impractical, costly, and not aligned with technological advancements. To do this, voluntary pilot projects may be initiated, with high-priority applications being nano-bio-based corrosion inhibitors to critical maritime infrastructure. Areas such as the European Union, Canada, and South Korea that already have good regulatory systems are the best places to pilot them first. Such areas would provide involved companies with an easier regulatory approval process in exchange of complete adherence to the 4-R Protocol. After the pilot phase is done, a third-party report should be made by an independent third party. In this report, the administrative effort, compliance efficiency, environmental outcomes, and innovation indicators will be evaluated. As an illustration, an effective pilot might demonstrate how the Sunset Clause enabled the timely re-evaluation and ongoing product enhancement, or how the Regenerative Life Cycle Assessment (LCA) prompted the development of a new recyclable corrosion inhibitor (Vlasopoulos, 2024).

An effective pilot would give solid arguments for the advantages and viability of the 4-R Framework over the existing disjointed and obsolete regulatory framework. The choice of pilot locations is a strategic decision. Piloting in places such as the EU and South Korea will enable testing in areas where there is a strong regulatory framework. The pilots should also collect information on the economic effect, which concerns the industry issues on competitiveness, especially in nations such as the

U.S. As an example, performance metrics may be used to monitor whether the 4-R authorization process allows quicker product approvals and reduced research and development (R&D) expenses than the conventional regulations. Such evidence is necessary to address the arguments that Extended Producer Responsibility (EPR) and circular design regulations decrease the competitiveness of industries. The pilot must also look at the possibility of offsetting the initial compliance costs or even surpassing them by the advantages of regulatory clarity and easier market access under the 4-R Framework. This would demonstrate that the framework is not a hindrance to competitiveness, as it would be proven. Those countries that embrace the 4-R Framework early enough can provide their home industries with a lead in sustainable nanotechnology.

The last phase of this roadmap is to officially incorporate the framework into national legislations in key economic regions to facilitate the harmonization of regulations across borders. This may involve the development of new, tailored laws, such as a possible EU Regulation on Advanced Materials, or the revision of existing laws, such as the U.S. Toxic Substances Control Act (TSCA), to incorporate the 4-R Framework. One of the factors would be the establishment of a global Nano-Remediation Fund to handle environmental restoration. Other helpful actions are compulsory audits, a three-year review, and documentation of the customs rules. One of the difficulties during this stage will be the handling of older products that are already in the market, and which might require special regulations and strict deadlines to meet the new requirements. The success of this step will be based on the strong data of the second stage, which will indicate the practical value of the 4-R Framework. When done properly, the process will result in a stable, efficient, and safe nano-bio inhibitor market globally. Goods that are accepted in the 4-R Framework in one nation would automatically be accepted as meeting the minimum standards in other nations, which would ease trade barriers and provide uniform regulation across the globe. This system would eliminate the danger of a splinternet of different standards by a common, science-based approach. The 4-R Framework is therefore a collaborative, evidence-based, and non-confrontational approach to global regulatory issues (Table 3). Although the threat of incompatible systems is still present, the suggested implementation roadmap is expected to encourage cooperation as the most rational, economical, and environmentally friendly way to go in all the key economies.

Table 3. Phased Implementation Roadmap for the 4-R Governance Framework

Phase	Timeline	Primary Objective	Key Actors	Outputs & Deliverables
I. Foundation and Consensus-Building	Years 0-3	To translate the 4-R Framework into a detailed technical and legal instrument.	OECD, ISO, UNEP, Academic Consortia.	Publication of the ISO 24000 Series on Advanced Material Passports. - Formal OECD adoption of Adaptive Tiered Testing Guidelines. - Drafting and public consultation of a Model International Protocol on Advanced Materials.
II. Piloting and Evidence-Generation	Years 2-5	To validate the framework's operational feasibility, administrative workability, and economic benefits.	Pioneer Regulators (e.g., EU, CA), Volunteer Industry Partners, Research Institutions.	Pioneer Jurisdiction Report on administrative feasibility. - Participant Industry Case Studies quantifying the innovation and circularity benefits of the 4-R model.
III. Full Integration and Global Harmonization	Years 4-8	To codify the validated framework into binding law and achieve regulatory convergence.	EU Commission, U.S. Congress/ EPA, National Governments, International Trade Bodies.	Enactment of bespoke "Advanced Materials Regulations" in key jurisdictions. - Establishment of the operational International Nano-Remediation Fund. - International recognition of Advanced Material Passports for customs and regulatory purposes.

For a global governance initiative to succeed, it must have a straightforward assessment of the political and economic impediments to its realization.

Geopolitical Fragmentation

The biggest threat is that nations are fighting to establish international regulations regarding high-tech materials due to political strains. The same competitions have already occurred in such fields as 5G and artificial intelligence (Li, 2025). The initial move towards this solution is to establish actual international collaboration. The 4-R Framework must not become another sphere of rivalry among the strong states. When global standards are being developed, developing nations like India, Brazil and Indonesia should be involved at the very outset. This is not only a fair approach but also a good approach to global business. When the governance standards are established by the Western countries only, they are not likely to reach the actual global cooperation. The case of the United Arab Emirates demonstrates how a nation can implement the framework in a manner that helps it to develop without seeming to exert external authority.

Competitiveness Concerns

Other nations, such as the United States, might be concerned that stringent regulations, such as Extended Producer Responsibility (EPR) and production that is circular-by-design, will cause their local businesses to be less competitive than those in other areas with less strict regulations (Dawson, 2024). This issue can be addressed by introducing the 4-R Framework as a means of exercising regulatory leadership. The quicker the countries implement the framework, the quicker they will establish the global standard and draw investment in sustainable nanotechnology.

A fragmented or "splinternet" approach towards advanced materials governance can pose a real danger. However, the 4-R Framework design is modular and principle-based, which is the solution. It creates a neutral, evidence-backed platform for international communication. Organizing these discussions will help organizations like UNEP and OECD build a common understanding and avoid regulatory fragmentation. By offering a clear, effective, and science-based framework, it will reduce a country's incentives to create competing, proprietary standards. Following a single framework efficiently performs much better than adhering to several national regulations that are inconsistent and conflicting. Additionally, the 4-R model builds regulatory resilience in a nation to deal with the future and fosters long-term economic competitiveness rather than being a heavy regulatory burden.

4. THE PRINCIPLE OF EVER-EVOLVING REGULATORY DESIGN (ERED)

Even though the 4-R Framework has a solid foundation for the current governance of nano biomaterials, its ultimate face will not emerge soon, but soon in the future. Over time, even the best regulatory systems tend to become old, inefficient, and finally detrimental, becoming the issue they were meant to solve. The 4-R framework must be anchored by one absolute principle: the Principle of Ever-Evolving Regulatory Design (ERED), to not become irrelevant or ineffectual. The principle ensures that the policy framework will be regularly updated in accordance with the latest scientific evidence.

Political complacency is the biggest hurdle for adaptive regulation (Amutha et al., 2024). When a law is passed, it is usually ignored instead of being updated. The fundamental element of countering this is the Obligatory Reflexivity Clause. This is a mechanism that compels the legislation to demonstrate that it is updated. We recommend that any protocol based on the 4-R Framework must include a mandatory sunset clause. The law should be reviewed every ten years and carefully examined by an independent panel of experts. The main question is whether the current regulations are still the best based on the latest science, or if they are unnecessarily preventing safe and beneficial practices. If the regulation is found lacking, the law panel will automatically void it, and Parliament will have to make changes. This is a safety net that is required to remove the fragmentation and conservative framework. By removing the political ways to delay reform, these measures will help keep the framework open to change.

The ethical issue of nanomaterials has a special challenge because the impacts of these substances are usually unpredictable, or the negative impacts of nanomaterials might require decades to manifest (Inbathamizh et al., 2023). The materials which we perceive as harmless now might be ecologically harmful in the future. The manufacturers of such nano biomaterials must ensure that they bear the responsibility for all consequences that their product goes through in its life cycle (Nizam et al., 2021). As such, we suggest creating a Trans-Generational Liability Ledger (TGLL). This can be proposed as a permanent, tamper-proof record (similar to blockchain) for every regulated batch of nano biomaterials. The TGLL records the location of each product, keeping the manufacturer perpetually accountable. We consider this the most reasonable approach to managing long-term risk. The system will be funded through the imposition of a Perpetuity Tax, which is a small fee built upon the registration of a product batch on the TGLL. All money goes to a global Remediation Fund that will pay for future clean-up or health costs that may arise in 30, 50, or even 100-year time frames. This helps to shift the costs of potential future harm to the environment from taxpayers to the firms that profited the most from the original innovation, ensuring lasting corporate accountability (R-1).

The last component of ERED is the introduction of the Regulatory Sentinel System (RSS). Current human-led oversight methods can't keep up with the pace of scientific discovery. To address this, RSS uses AI to constantly collect data from worldwide sources like scientific articles, environmental monitors, and patent records to detect early signs of possible nanotoxicity. The RSS alerts framework systems about any emerging risks. It is somewhat like an intelligent alarm system that catches signals and patterns signaling new risks. If three credible reports show a material could pose an unknown danger, the system would not wait until the next review cycle. The ARM requires the independent expert panel to undertake an urgent reassessment of the product under review. The incorporation of new machine learning into regulations will turn static regulations into dynamic, anticipatory ones that can preserve public health and the environment at the same pace as technology.

5. CONCLUSION

This chapter provides a direct criticism of the existing legal framework. The inability of the world to control high-tech materials is not a coincidence–it is the predictable and harmful consequence of the outdated legislation. The effects, which cost trillions of dollars and caused serious environmental damage, require urgent intervention. The inability of political systems to react efficiently should come to an end. The 4-R Governance Framework, which is based on the 4-R of Responsibility, Risk, Reflexivity, and Regeneration, is proposed as the way forward. This is not just a policy idea, but a plan for making innovation more ethical. It suggests that instead of proving something is harmful, the burden should be on proving it is safe. Also, we believe that this approach should be protected by a system—called the Doctrine of Perpetual Regulatory Design—that keeps rules flexible and able to fix themselves over time. We contend that the Obligatory Reflexivity Clause requires every nano-biomaterial law to be reviewed on a 10-year basis to ensure that it remains relevant and effective. The Trans-Generational Liability Ledger would have the producers shoulder the entire long-term expenses of their products so that future taxpayers do not have to shoulder this financial and environmental burden. Bureaucratic inertia should be replaced by scientific progress. The solution suggested is thorough, logical, and practical. We thus encourage international institutions to embrace the 4-R Framework and PRD immediately. Any failure to take these measures is no longer negligence, but a deliberate increase in systemic risk.

REFERENCES

Amutha, C., Gopan, A., Pushbalatatha, I., Ragavi, M., & Reneese, J. A. (2024). Nanotechnology and Governance: Regulatory Framework for Responsible Innovation. In George, S. C., & Tawiah, B. (Eds.), *Nanotechnology in Societal Development* (pp. 481–503). Springer Nature Singapore., DOI: 10.1007/978-981-97-6184-5_14

Basterra, M. L., & López, S. M. (2024). *INCENTIVES FOR INVESTMENT IN CLEAN TECHNOLOGIES.* https://www.orkestra.deusto.es/images/investigacion/publicaciones/informes/cuadernos-orkestra/240091-Incentives-for-_investment-in_clean-technologies.pdf

Bender, R., Féron, D., Mills, D., Ritter, S., Bäßler, R., Bettge, D., De Graeve, I., Dugstad, A., Grassini, S., Hack, T., Halama, M., Han, E., Harder, T., Hinds, G., Kittel, J., Krieg, R., Leygraf, C., Martinelli, L., Mol, A., & Zheludkevich, M. (2022). Corrosion challenges towards a sustainable society. *Materials and Corrosion*, *73*(11), 1730–1751. DOI: 10.1002/maco.202213140

Botos, Á., Graham, J. D., & Illés, Z. (2019). Industrial chemical regulation in the European Union and the United States: A comparison of REACH and the amended TSCA. *Journal of Risk Research*, *22*(10), 1187–1204. DOI: 10.1080/13669877.2018.1454495

Casoli, G., & Ramkumar, S. (2020). Plastic Pollution in the Oceans: A Systemic Analysis—Status Quo and Possible Sustainable Solutions. In M. Streit-Bianchi, M. Cimadevila, & W. Trettnak (Eds.), *Mare Plasticum—The Plastic Sea* (pp. 221–243). Springer International Publishing. DOI: 10.1007/978-3-030-38945-1_11

Dawson, L. (2024). *Extended producer responsibility and its role in facilitating a circular economy in the European Union: An analysis of four directives* [PhD Thesis, University of Birmingham]. https://etheses.bham.ac.uk/id/eprint/14893/

Farias, D. B. L. (2023). Country differentiation in the global environmental context: Who is 'developing' and according to what? *International Environmental Agreement: Politics, Law and Economics*, *23*(3), 253–269. DOI: 10.1007/s10784-023-09596-9

Foreign Trade Barriers. (n.d.). Retrieved October 27, 2025, from https://ustr.gov/sites/default/files/files/Press/Reports/2025NTE.pdf

Gonzalez, M., Taddonio, K. N., & Sherman, N. J. (2015). The Montreal Protocol: How today's successes offer a pathway to the future. *Journal of Environmental Studies and Sciences*, *5*(2), 122–129. DOI: 10.1007/s13412-014-0208-6

Inbathamizh, L., Harsha Varthan, M. K., Kumar, R. S. R., Rohinth, M., & Tawfeeq Ahmed, Z. H. (2023). Nanotechnology: Ethical Impacts, Health Issues, and Safety Issues. In Malik, J. A., & Sadiq Mohamed, M. J. (Eds.), *Modern Nanotechnology* (pp. 455–477). Springer Nature Switzerland., DOI: 10.1007/978-3-031-31104-8_20

Klinke, A., & Renn, O. (2021). The coming of age of risk governance. *Risk Analysis*, *41*(3), 544–557. DOI: 10.1111/risa.13383 PMID: 31379003

Lee, D., & Choi, K. (2019). Comparison of regulatory frameworks of environmental risk assessments for human pharmaceuticals in EU, USA, and Canada. *The Science of the Total Environment*, *671*, 1026–1035. https://www.sciencedirect.com/science/article/pii/S0048969719313713. DOI: 10.1016/j.scitotenv.2019.03.372

Li, J. (2025). Governing High-Risk Technologies in a Fragmented World: Geopolitical Tensions, Regulatory Gaps, and Institutional Barriers to Global Cooperation. *Fudan Journal of the Humanities and Social Sciences*. Advance online publication. DOI: 10.1007/s40647-025-00445-4

Masroor, S. (2023). Corrosion: Fundamental and Economic Impact. In *Carbon Allotropes* (pp. 3–14). CRC Press. https://www.taylorfrancis.com/chapters/edit/10.1201/9781003323976-2/corrosion-sheerin-masroor

Nizam, N. U. M., Hanafiah, M. M., & Woon, K. S. (2021). A content review of life cycle assessment of nanomaterials: Current practices, challenges, and future prospects. *Nanomaterials (Basel, Switzerland)*, *11*(12), 3324. https://www.mdpi.com/2079-4991/11/12/3324. DOI: 10.3390/nano11123324 PMID: 34947673

OECD. U. S. M. to the. (2023, October 31). About the OECD. *U.S. Mission to the Organization For Economic Cooperation & Development*. https://usoecd.usmission.gov/about-the-oecd/

Ramasamy, B. S. S., & Palanisamy, S. (2021). A review on occurrence, characteristics, toxicology and treatment of nanoplastic waste in the environment. *Environmental Science and Pollution Research International*, *28*(32), 43258–43273. DOI: 10.1007/s11356-021-14883-6 PMID: 34173146

Rayasam, S. D. G., Koman, P. D., Axelrad, D. A., Woodruff, T. J., & Chartres, N. (2022). Toxic Substances Control Act (TSCA) Implementation: How the Amended Law Has Failed to Protect Vulnerable Populations from Toxic Chemicals in the United States. *Environmental Science & Technology*, *56*(17), 11969–11982. DOI: 10.1021/acs.est.2c02079 PMID: 35980084

Regulation (EC) No. 1907/2006 of the European Parliament and of the Council concerning the Registration, Evaluation, Authorisation and Restriction of Chemicals (REACH) [2006] OJ L396/1 (ECHA, 2022). Toxic Substances Control Act, 15 USC §2601 et seq (EPA, 2022). Cabinet Decision No. 10 of 2022 on the Control of Substances Depleting the Ozone Layer and their Alternatives (United Arab Emirates). Emirates Authority for Standardization and Metrology (ESMA), ESMA Standard 5025:2021 – Safety Requirements for Chemical Products (ESMA, 2021).

Sahu, A., Swain, L. P., Behera, I. D., & Basak, G. (2025). Regulatory Framework and Policies to Achieve Zero Hydrocarbon Waste Management Through Circular Economy. In Behera, I. D., & Das, A. P. (Eds.), *Environmental Hydrocarbon Pollution and Zero Waste Approach Towards a Sustainable Waste Management* (pp. 385–408). Springer Nature Switzerland., DOI: 10.1007/978-3-031-81615-4_16

Saxena, N. (2023). Bio-nanotechnology in waste to energy conversion in a circular economy approach for better sustainability. In *Bionanotechnology towards Green Energy* (pp. 253–274). CRC Press., https://api.taylorfrancis.com/content/chapters/edit/download?identifierName=doi&identifierValue=10.1201/9781003316374-13&type=chapterpdf DOI: 10.1201/9781003316374-13

Shahcheraghi, N., Golchin, H., Sadri, Z., Tabari, Y., Borhanifar, F., & Makani, S. (2022). Nano-biotechnology, an applicable approach for sustainable future. *3 Biotech, 12*(3), 65. DOI: 10.1007/s13205-021-03108-9

Subramanian, V. (2017). *Development and testing of a framework for sustainable nanotechnology*. https://iris.unive.it/handle/10579/10350

United Arab Emirates Legislations | Cabinet Resolution On the Technical Regulations for the Control of the Quantity of Product in Pre-packages. (2025). https://uaelegislation.gov.ae/en/legislations/2594

United Arab Emirates—Standards for Trade. (2025, August 26). https://www.trade.gov/country-commercial-guides/united-arab-emirates-standards-trade

Vlasopoulos, A. (2024). *Assessment of life cycle impacts of sustainable technologies in industrial environments* [PhD Thesis, Brunel University London]. https://bura.brunel.ac.uk/handle/2438/31458

Wagner, W. E., Gold, S. C., & McGarity, T. O. (2024). Breaking the Corporate Stranglehold over Toxics Regulation: A Possible Path Forward. *DePaul L. Rev., 74*, 723. https://heinonline.org/hol-cgi-bin/get_pdf.cgi?handle=hein.journals/deplr74&section=25

Widmer, M., & Meili, C. (2010). 12 Approaching the nanoregulation problem in chemicals legislation in the EU and US. *International Handbook on Regulating Nanotechnologies*, 238. http://innovationsgesellschaft.ch/wp-content/uploads/2013/11/Hodge-et-al-2010-Chapter-12-Widmer-and-Meili.pdf

Compilation of References

A, P., & Xavier, J. R. (2025). A Review on Functional nanocomposites for enhanced anticorrosion and flame retardant properties. *Surfaces and Interfaces, 70*, 106850. https://doi.org/DOI: 10.1016/j.surfin.2025.106850

Ahmadi, M. H., Ghazvini, M., Alhuyi Nazari, M., Ahmadi, M. A., Pourfayaz, F., Lorenzini, G., & Ming, T. (2019). Renewable energy harvesting with the application of nanotechnology: A review. *International Journal of Energy Research*, *43*(4), 1387–1410. DOI: 10.1002/er.4282

Ahmad, T., Ganguly, A., Ahmed, J., Ganguli, A. K., & Alhartomy, O. A. (2010). Nanorods of transition metal oxalates: A versatile route to the oxide nanoparticles. *Arabian Journal of Chemistry*, *4*(2), 125–134. DOI: 10.1016/j.arabjc.2010.06.041

Ahmed, S. F., Mehejabin, F., Momtahin, A., Tasannum, N., Faria, N. T., Mofijur, M., Hoang, A. T., Vo, D.-V. N., & Mahlia, T. M. I. (2022). Strategies to improve membrane performance in wastewater treatment. *Chemosphere*, *306*, 135527. DOI: 10.1016/j.chemosphere.2022.135527 PMID: 35780994

Ahmed, S., Ahmad, M., Swami, B. L., & Ikram, S. (2016). A review on plants extract mediated synthesis of silver nanoparticles for antimicrobial applications: A green expertise. *Journal of Advanced Research*, *7*(1), 17–28. DOI: 10.1016/j.jare.2015.02.007 PMID: 26843966

Al-Amiery, A. A., Yousif, E., Isahak, W. N. R. W., & Al-Azzawi, W. K. (2023). A review of Inorganic corrosion Inhibitors: Types, Mechanisms, and applications. *Tribology in Industry*, *45*(1), 313–339. DOI: 10.24874/ti.1456.03.23.06

Al-Gethami, W., Qamar, M. A., Shariq, M., Alaghaz, A.-N. M. A., Farhan, A., Areshi, A. A., & Alnasir, M. H. (2024). Emerging environmentally friendly bio-based nanocomposites for the efficient removal of dyes and micropollutants from wastewater by adsorption: A comprehensive review. *RSC Advances*, *14*(4), 2804–2834. DOI: 10.1039/D3RA06501D PMID: 38234871

Ali, K. A., Roy, P. K., Hossain, C. M., Dutta, D., Vichare, R., & Biswal, M. R. (2021). Starch-based nanomaterials in drug delivery applications. In Biopolymer-Based Nanomaterials in Drug Delivery and Biomedical Applications. https://doi .org/DOI: 10.1016/B978-0-12-820874-8.00023-3

Ali, R. (2019). Deep learning- and infrared thermography-based subsurface damage detection in a steel bridge.

Ali, S., Naseer, S., Rehman, M., & Wei, Z. (2024). Recent trends and sources of lead toxicity: A review of state-of-the-art nano-remediation strategies. *Journal of Nanoparticle Research*, *26*(7), 168. DOI: 10.1007/s11051-024-06081-5

Ali, Z., & Ahmad, R. (2020). Nanotechnology for water treatment. In *Environmental Nanotechnology* (Vol. 3, pp. 143–163). Springer., DOI: 10.1007/978-3-030-40363-4_7

Alkhadra, M. A., Su, X., Suss, M. E., Tian, H., Guyes, E. N., Shocron, A. N., Conforti, K. M., de Souza, J. P., Kim, N., Tedesco, M., Khoiruddin, K., Wenten, I. G., Santiago, J. G., Hatton, T. A., & Bazant, M. Z. (2022). Electrochemical Methods for Water Purification, Ion Separations, and Energy Conversion. *Chemical Reviews*, *122*(16), 13547–13635. DOI: 10.1021/acs.chemrev.1c00396 PMID: 35904408

Alon, G., & Adin, A. (1994). Mathematical modeling of particle size distribution in secondary effluent filtration. *Water Environment Research*, *66*(7), 836–841. DOI: 10.2175/WER.66.6.11

Alshameri, A., ... —Understanding the role of natural clay minerals as effective adsorbents and alternative source of rare earth elements: Adsorption operative parameters,|| Hydrometallurgy, vol. 185, no. September 2018, pp. 149–161, 2019, DOI: 10.1016/j.hydromet.2019.02.016

Alver, A., Baştürk, E., Altaş, L., & Işık, M. (2022). A solution of taste and odor problem with activated carbon adsorption in drinking water: Detailed kinetics and isotherms. *Desalination and Water Treatment*, *252*, 300–318. DOI: 10.5004/ dwt.2022.28269

Amann, R. I., Ludwig, W., & Schleifer, K. H. (1995). Phylogenetic identification and in situ detection of individual microbial cells without cultivation. *Microbiological Reviews*, *59*(1), 143–169. DOI: 10.1128/mr.59.1.143-169.1995 PMID: 7535888

Amutha, C., Gopan, A., Pushbalatatha, I., Ragavi, M., & Reneese, J. A. (2024). Nanotechnology and Governance: Regulatory Framework for Responsible Innovation. In George, S. C., & Tawiah, B. (Eds.), *Nanotechnology in Societal Development* (pp. 481–503). Springer Nature Singapore., DOI: 10.1007/978-981-97-6184-5_14

Angon, P. B., Islam, M. S., Kc, S., Das, A., Anjum, N., Poudel, A., & Suchi, S. A. (2024). Sources, effects and present perspectives of heavy metals contamination: Soil, plants and human food chain. *Heliyon, 10*(7), e28357. DOI: 10.1016/j.heliyon.2024.e28357 PMID: 38590838

Anwer, A. H., Ahtesham, A., Shoeb, M., Mashkoor, F., Ansari, M. Z., Zhu, S., & Jeong, C. (2023). State-of-the-art advances in nanocomposite and bio-nanocomposite polymeric materials: A comprehensive review. *Advances in Colloid and Interface Science, 318*, 102955. DOI: 10.1016/j.cis.2023.102955 PMID: 37467558

Armelin, E., Alemán, C., & Iribarren, J. I. (2009). Anticorrosion performances of epoxy coatings modified with polyaniline: A comparison between the emeraldine base and salt forms. *Progress in Organic Coatings, 65*(1), 88–93. DOI: 10.1016/j.porgcoat.2008.10.001

Arun, S., Sinharoy, A., Pakshirajan, K., & Lens, P. N. (2020). Algae based microbial fuel cells for wastewater treatment and recovery of value-added products. *Renewable & Sustainable Energy Reviews, 132*, 110041. DOI: 10.1016/j.rser.2020.110041

Aslam, R., Zhao, J., Sun, X., Zhou, X., Wang, Q., Aslam, J., & Yan, Z. (2024). Bio-based ionic liquid as a corrosion inhibitor for mild steel in 5% HCl solution: Experimental and theoretical investigation. *Sustainable Chemistry and Pharmacy, 39*, 101614. DOI: 10.1016/j.scp.2024.101614

Astruc, D. (2020). Introduction: Nanoparticles in catalysis. *Chemical Reviews, 120*(2), 461–463. DOI: 10.1021/acs.chemrev.8b00696 PMID: 31964144

Atalah, J., Blamey, L., Amenábar, M. J., Kelley-Loughnane, N., & Blamey, J. M. (2022). Biochemical and microbiological characterization of a thermotolerant bacterial consortium involved in the corrosion of Aluminum Alloy 7075. *Research Square*. DOI: 10.21203/rs.3.rs-2371670/v1

Avossa, J., Pota, G., Vitiello, G., Macagnano, A., Zanfardino, A., di Napoli, M., Pezzella, A., D'Errico, G., Varcamonti, M., & Luciani, G. (2021). Multifunctional mats by antimicrobial nanoparticles decoration for bioinspired smart wound dressing solutions. *Materials Science and Engineering C, 123*, 111954. Advance online publication. DOI: 10.1016/j.msec.2021.111954 PMID: 33812582

Aziz, S., Mazhar, A. R., Ubaid, A., Shah, S. M. H., Riaz, Y., Talha, T., & Jung, D.-W. (2024). A comprehensive review of membrane-based water filtration techniques. *Applied Water Science, 14*(8), 169. DOI: 10.1007/s13201-024-02226-y

Azmana, M., Mahmood, S., Hilles, A. R., Rahman, A., Arifin, M. A. B., & Ahmed, S. (2021). A review on chitosan and chitosan-based bionanocomposites: Promising material for combatting global issues and its applications. *International Journal of Biological Macromolecules, 185*, 832–848. DOI: 10.1016/j.ijbiomac.2021.07.023 PMID: 34237361

Babatunde, D. E., Denwigwe, I. H., Babatunde, O. M., Gbadamosi, S. L., Babalola, I. P., & Agboola, O. (2020). Environmental and societal impact of nanotechnology. *IEEE Access : Practical Innovations, Open Solutions*, *8*, 4640–4667. DOI: 10.1109/ACCESS.2019.2961513

Bagwell, C. E., Adams, S., & Barnhart, E. P. (2016). Radiation-resistant bacteria and their implications for nuclear waste disposal. *Reviews in Environmental Science and Biotechnology*, *15*(3), 353–364. DOI: 10.1007/s11157-016-9401-2

Bain, R., Wright, J., Yang, H., Gundry, S., Pedley, S., & Bartram, J. (2014). Improved but not necessarily safe: Water access and the Millennium Development Goals. In *Global Water: Issues and Insights* (p. 89). Global Water Forum.

Bairagi, H., Vashishth, P., Ji, G., Shukla, S. K., Ebenso, E. E., & Mangla, B. (2024). Polymers and their composites for corrosion inhibition application: Development, advancement, and future scope–A critical review. *Corrosion Communications*, *15*, 79–94. DOI: 10.1016/j.corcom.2023.10.006

Bakhtiari, S., Salari, M., Shahrashoub, M., Zeidabadinejad, A., Sharma, G., & Sillanpää, M. (2024). A Comprehensive Review on Green and Eco-Friendly Nano-Adsorbents for the Removal of Heavy Metal Ions: Synthesis, Adsorption Mechanisms, and Applications. *Current Pollution Reports*, *10*(1), 1–39. DOI: 10.1007/s40726-023-00290-7

Bal, G., & Thakur, A. (2022).. . *Proceedings*, *50*(5), 1575–1579.

Banerjee, S., Mittal, A., & Kumar, S. (2017). Microbiologically influenced corrosion: Perspectives and approaches. *Journal of Materials Engineering and Performance*, *26*(7), 3113–3125. DOI: 10.1007/s11665-017-2744-1

Bao, W., Deng, Z., Zhang, S., Ji, Z., & Zhang, H. (2019). Next-Generation composite coating system: Nanocoating. *Frontiers in Materials*, *6*, 72. Advance online publication. DOI: 10.3389/fmats.2019.00072

Barman, J., Tirkey, A., Batra, S., Paul, A. A., Panda, K., Deka, R., & Babu, P. J. (2022). The role of nanotechnology based wearable electronic textiles in biomedical and healthcare applications. *Materials Today. Communications*, *32*, 104055. DOI: 10.1016/j.mtcomm.2022.104055

Barrera-Díaz, C. E., Lugo-Lugo, V., & Bilyeu, B. (2012). A review of chemical, electrochemical and biological methods for aqueous Cr (VI) reduction. *Journal of Hazardous Materials, 223*, 1–12. DOI: 10.1016/j.jhazmat.2012.04.054 PMID: 22608208

Basera, P., Lavania, M., & Lal, B. (2019). Potential of dynamic bacterial communities in the bio-corrosion process: A proof study with surface morphology of metal coupons. *RSC Advances, 9*(30), 17040–17050. DOI: 10.1039/C9RA01959F PMID: 35519851

Basterra, M. L., & López, S. M. (2024). *INCENTIVES FOR INVESTMENT IN CLEAN TECHNOLOGIES.* https://www.orkestra.deusto.es/images/investigacion/publicaciones/informes/cuadernos-orkestra/240091-Incentives-for-_investment-in_clean-technologies.pdf

Bayat, M., Nasernejad, B., & Falamaki, C. (2021). Preparation and characterization of nano-galvanic bimetallic Fe/Sn nanoparticles deposited on talc and its enhanced performance in Cr(VI) removal. *Scientific Reports, 11*(1), 7715. Advance online publication. DOI: 10.1038/s41598-021-87106-0 PMID: 33833296

Bayda, N. I., & Sutliff, G. (2020). Comparing extracted and stipulated definitions in Algebra 1 textbooks and *Khan Academy. International Electronic Journal of Mathematics Education, 15*(2). Advance online publication. DOI: 10.29333/iejme/7601

Beech, I. B., & Sunner, J. (2004). Biocorrosion: Towards understanding interactions between biofilms and metals. *Current Opinion in Biotechnology, 15*(3), 181–186. DOI: 10.1016/j.copbio.2004.05.001 PMID: 15193324

Bellona, C., & Drewes, J. (2005). The role of membrane surface charge and solute physicochemical properties in the rejection of organic acids by NF membranes. *Journal of Membrane Science, 249*(1–2), 227–234. DOI: 10.1016/j.memsci.2004.09.041

Bender, R., Féron, D., Mills, D., Ritter, S., Bäßler, R., Bettge, D., De Graeve, I., Dugstad, A., Grassini, S., Hack, T., Halama, M., Han, E., Harder, T., Hinds, G., Kittel, J., Krieg, R., Leygraf, C., Martinelli, L., Mol, A., & Zheludkevich, M. (2022). Corrosion challenges towards a sustainable society. *Materials and Corrosion, 73*(11), 1730–1751. DOI: 10.1002/maco.202213140

Berk, R. S., & Canfield, J. H. (1964). Bioelectrochemical energy conversion. *Applied Microbiology, 12*(1), 10–12. DOI: 10.1128/am.12.1.10-12.1964 PMID: 14106931

Bermont-Bouis, D., Janvier, M., Grimont, P. A. D., Dupont, I., & Vallaeys, T. (2007). Both sulfate-reducing bacteria and Enterobacteriaceae take part in marine biocorrosion of carbon steel. *The Journal of Applied Bacteriology*, *102*(1), 161–168. DOI: 10.1111/j.1365-2672.2006.03053.x PMID: 17184331

Bernardes, A. F., Meng, Z., Campos, L. C., & Coppens, M.-O. (2025). Bio-inspired anti-fouling strategies for membrane-based separations. *Chemical Communications*, *61*(27), 5064–5071. DOI: 10.1039/D4CC05149A PMID: 40084840

Besegatto, S. V., Costa, F. N., Damas, M. S. P., Colombi, B. L., De Rossi, A. C., de Aguiar, C. R. L., & Immich, A. P. S. (2018). Enzyme treatment at different stages of textile processing: A review. *Industrial Biotechnology (New Rochelle, N.Y.)*, *14*(6), 298–307. DOI: 10.1089/ind.2018.0018

Beura, V. K., Karanth, Y., Darling, K., & Solanki, K. (2022). Role of gradient nanograined surface layer on corrosion behavior of aluminum 7075 alloy. *NPJ Materials Degradation*, *6*(1), 62. Advance online publication. DOI: 10.1038/s41529-022-00271-z

Bhagawati, M., Datta, B. K., Newar, R., Sonowal, S., Buragohain, K., Tamuly, D., & Nath, R. (2025). Nanoparticle-Assisted Microbial Removal of Arsenic (As) from Drinking Water Sources. In *Nano-Bioremediation for Wastewater Treatment* (pp. 249–289). Wiley. DOI: 10.1002/9781394271641.ch10

Bhatt, P., Pandey, S. C., Joshi, S., Chaudhary, P., Pathak, V. M., Huang, Y., Wu, X., Zhou, Z., & Chen, S. (2022). Nanobioremediation: A sustainable approach for the removal of toxic pollutants from the environment. *Journal of Hazardous Materials*, *427*, 128033. DOI: 10.1016/j.jhazmat.2021.128033 PMID: 34999406

Bide, M. (2014). Sustainable dyeing with synthetic dyes. Roadmap to sustainable textiles and clothing: eco-friendly raw Materials, technologies, and processing methods, 81-107.

Bigham, A., Zarepour, A., Safarkhani, M., Huh, Y., Khosravi, A., Rabiee, N., Iravani, S., & Zarrabi, A. (2025). Inspired by nature: Bioinspired and biomimetic photocatalysts for biomedical applications. *Nano Materials Science*, *7*(1), 1–23. DOI: 10.1016/j.nanoms.2024.02.006

Binnig, G., Quate, C. F., & Gerber, C. (1986). Atomic force microscope. *Physical Review Letters*, *56*(9), 930–933. DOI: 10.1103/PhysRevLett.56.930 PMID: 10033323

Biranje, S. S., Kaushik, S., Marewad, D., Yadav, A., Vankundre, V., Panse, M., Joshi, I., Goli, A., Shahid, M., Kulkarni, K., Liu, J., & Adivarekar, R. (2024). Applications of nanocellulose and its derivatives in developing sustainable textiles. *Cellulose (London, England)*, *31*(9), 5343–5379. DOI: 10.1007/s10570-024-05935-4

Bond, D. R., Holmes, D. E., Tender, L. M., & Lovley, D. R. (2002). Electrode-reducing microorganisms that harvest energy from marine sediments. *Science*, *295*(5554), 483–485. DOI: 10.1126/science.1066771 PMID: 11799240

Bond, D. R., & Lovley, D. R. (2003). Electricity production by *Geobacter sulfurreducens* attached to electrodes. *Applied and Environmental Microbiology*, *69*(3), 1548–1555. DOI: 10.1128/AEM.69.3.1548-1555.2003 PMID: 12620842

Boruah, J. S., Devi, C., Hazarika, U., Reddy, P. V. B., Chowdhury, D., Barthakur, M., & Kalita, P. (2021). Green synthesis of gold nanoparticles using an antiepileptic plant extract: In vitro biological and photo-catalytic activities. *RSC Advances*, *11*(45), 28029–28041. DOI: 10.1039/D1RA02669K PMID: 35480751

Botos, Á., Graham, J. D., & Illés, Z. (2019). Industrial chemical regulation in the European Union and the United States: A comparison of REACH and the amended TSCA. *Journal of Risk Research*, *22*(10), 1187–1204. DOI: 10.1080/13669877.2018.1454495

Bowman, D. M. (2017). More than a decade on: Mapping today's regulatory and policy landscapes following the publication of *Nanoscience and nanotechnologies: Opportunities and uncertainties*. *NanoEthics*, *11*(2), 169–186. DOI: 10.1007/s11569-017-0281-x

Briggs, D., & Grant, J. T. (2003). *Surface analysis by auger and X-ray photoelectron spectroscopy*. IM Publications.

Brinker, C. J., & Scherer, G. W. (1990). *Sol-gel science: The physics and chemistry of sol-gel processing*. Academic Press.

Brune, K. D., & Bayer, T. S. (2012). Engineering microbial consortia to enhance biomining and bioremediation. *Frontiers in Microbiology*, *3*, 1–6. DOI: 10.3389/fmicb.2012.00203 PMID: 22679443

Cairone, S., Hasan, S. W., Choo, K.-H., Li, C.-W., Zarra, T., Belgiorno, V., & Naddeo, V. (2024). Integrating artificial intelligence modeling and membrane technologies for advanced wastewater treatment: Research progress and future perspectives. *The Science of the Total Environment*, *944*, 173999. DOI: 10.1016/j.scitotenv.2024.173999 PMID: 38879019

Cai, Y., Zhao, Y., Ma, X., Zhou, K., & Chen, Y. (2018). Influence of environmental factors on atmospheric corrosion in dynamic environment. *Corrosion Science*, *137*, 163–175. DOI: 10.1016/j.corsci.2018.03.042

Cao, S., Wang, J., & Chen, H. (2015). Progress in biopolymer-based anticorrosive coatings. *Progress in Organic Coatings*, *87*, 1–8. DOI: 10.1016/j.porgcoat.2015.04.019

Carofiglio, M., Laurenti, M., & Cauda, V. (2022). *The World Big Challenges Faced by Nanoscience: Examples of How Integrated Science Can Save the World* (pp. 153–181). DOI: 10.1007/978-3-030-94651-7_8

Casoli, G., & Ramkumar, S. (2020). Plastic Pollution in the Oceans: A Systemic Analysis—Status Quo and Possible Sustainable Solutions. In M. Streit-Bianchi, M. Cimadevila, & W. Trettnak (Eds.), *Mare Plasticum—The Plastic Sea* (pp. 221–243). Springer International Publishing. DOI: 10.1007/978-3-030-38945-1_11

Cattant, F., Crusset, D., & Féron, D. (2008). Corrosion issues in nuclear industry today. *Materials Today*, *11*(10), 32–37. DOI: 10.1016/S1369-7021(08)70205-0

Chahardoli, A., Karimi, N., & Fattahi, A. (2018). *Nigella arvensis* leaf extract mediated green synthesis of silver nanoparticles: Their characteristic properties and biological efficacy. *Advanced Powder Technology*, *29*(1), 202–210. DOI: 10.1016/j.apt.2017.11.003

Chakrabarty, S., & Jasuja, K. (2022). Applications of nanomaterials in the textile industry. In *Nanoscale Engineering of Biomaterials: Properties and Applications* (pp. 567–587). Springer Nature Singapore. DOI: 10.1007/978-981-16-3667-7_20

Chakraborty, J., Das, S., & Sarkar, P. K. (2020). Enzyme-based green approaches for the mitigation of microbiologically influenced corrosion. *Journal of Materials Engineering and Performance*, *29*(5), 2795–2806. DOI: 10.1007/s11665-020-04799-6

Chami, A., Benabbou, R., Taleb, M., Rais, Z., & Haji, M. E. (2021). Analysis of survey data on corrosion in the automotive industry. *Materials Today: Proceedings*, *45*, 7636–7642. DOI: 10.1016/j.matpr.2021.03.113

Chatterjee, M., Fan, Y., Cao, F., Jones, A. A., Pilloni, G., & Zhang, X. (2021). Proteomic study of *Desulfovibrio ferrophilus* IS5 reveals overexpressed extracellular multi-heme cytochrome associated with severe microbiologically influenced corrosion. *Scientific Reports*, *11*(1), 15458. Advance online publication. DOI: 10.1038/s41598-021-95060-0 PMID: 34326431

Chaudhary, S., Tanvi, R. D., & Goyal, S. (2019). Different applications of sulphur oxidizing bacteria: A review. *International Journal of Current Microbiology and Applied Sciences*, *8*(11), 770–778. DOI: 10.20546/ijcmas.2019.811.091

Chauhan, M. S., Rahul, A. K., Shekhar, S., & Kumar, S. (2023). Removal of heavy metal from wastewater using ion exchange with membrane filtration from Swarnamukhi river in Tirupati. *Materials Today: Proceedings*, *78*, 1–6. DOI: 10.1016/j.matpr.2022.08.280

Chavez, M. L., Kato, C., & Zhang, D. (2019). Application of scanning electron microscopy in microbiologically influenced corrosion investigations. *Materials and Corrosion*, *70*(10), 1725–1733. DOI: 10.1002/maco.201910758

Chen, H., Kimyon, Ö., Gunawan, C., Ramandi, H. L., Craig, P., Chen, R., et al. (2024). An effective barrier coating technology against premature bolt failures in underground mines. Rock Mechanics and Rock Engineering. https://doi.org/10.1007/s00603-024-03909-3

Chen, S., Wang, P., & Zhang, D. (2014). Corrosion behavior of copper under biofilm of sulfate-reducing bacteria. *Corrosion Science*, *87*, 407–415. DOI: 10.1016/j.corsci.2014.07.001

Chen, X., Liu, J., Yuan, T., Zhang, Z., Song, C., Yang, S., Gao, X., Wang, N., & Cui, L. (2022). Recent advances in earth-abundant first-row transition metal (Fe, Co and Ni)-based electrocatalysts for the oxygen evolution reaction. *Energy Materials*, *2*(4), 28. DOI: 10.20517/energymater.2022.30

Chen, X., Schippers, A., & Sand, W. (2011). Progress in bioleaching: Fundamentals and mechanisms of microbial metal sulfide oxidation—part A. *Applied Microbiology and Biotechnology*, *97*(17), 7529–7541. DOI: 10.1007/s00253-011-3556-0

Chithra, N., & Bhambri, P. (2024). Ethics in Sustainable Technology. In *Handbook of Technological Sustainability* (pp. 245–256). CRC Press., DOI: 10.1201/9781003475989-21

Chugh, B., Sheetal, , Singh, M., Thakur, S., Pani, B., Singh, A. K., & Saji, V. S. (2022). Extracellular electron transfer by Pseudomonas aeruginosa in biocorrosion: A review. *ACS Biomaterials Science & Engineering*, *8*(3), 1049–1059. DOI: 10.1021/acsbiomaterials.1c01645 PMID: 35199512

Chung, H. C., Lee, O. O., Huang, Y. L., Mok, S. Y., Kolter, R., & Qian, P. Y. (2010). Bacterial community succession and chemical profiles of subtidal biofilms in relation to larval settlement of the polychaete *Hydroides elegans*. *The ISME Journal*, *4*(6), 817–828. DOI: 10.1038/ismej.2009.157 PMID: 20090788

Coenye, T., Ahonen, M., Anderson, S., Cámara, M., Chundi, P., Fields, M. W., Foidl, I., Gnimpieba, E. Z., Griffin, K., Hinks, J., Loka, A. R., Lushbough, C., MacPhee, C., Nater, N., Raval, R., Slater-Jefferies, J., Teo, P., Wilks, S., Yung, M., & Webb, J. S. (2024). Global challenges and microbial biofilms: Identification of priority questions in biofilm research, innovation and policy. *Biofilm*, *8*, 100210. DOI: 10.1016/j.bioflm.2024.100210 PMID: 39221168

Cory, D. C., & Rahman, T. (2009). Environmental justice and enforcement of the Safe Drinking Water Act: The Arizona arsenic experience. *Ecological Economics*, *68*(6), 1825–1837. DOI: 10.1016/j.ecolecon.2008.12.010

Costa, C., Viana, A., Silva, C., Marques, E. F., & Azoia, N. G. (2022). Recycling of textile wastes, by acid hydrolysis, into new cellulosic raw materials. *Waste Management (New York, N.Y.)*, *153*, 99–109. DOI: 10.1016/j.wasman.2022.08.019 PMID: 36067549

Costa, F. C. R., Moreira, V. R., Guimarães, R. N., Moser, P. B., Santos, L. V. S., de Paula, E. C., & Amaral, M. C. S. (2024). Pre-oxidation and coagulation-flocculation as a pretreatment to UF-RO applied for surface water treatment and arsenic removal. *Desalination*, *586*, 117855. DOI: 10.1016/j.desal.2024.117855

Cotruvo, J. A. (1987). Risk assessment and control decisions for protecting drinking water quality. In I. H. Suffet & M. Malaiyandi (Eds.), *Organic pollutants in water* (Advances in Chemistry Series, No. 214, pp. [insert page range if known]). American Chemical Society. https://doi.org/DOI: 10.1021/ba1987-0214.ch034

Cui, X.-J., Lin, X.-Z., Liu, C.-H., Yang, R.-S., Zheng, X.-W., & Gong, M. (2015). Fabrication and corrosion resistance of a hydrophobic micro-arc oxidation coating on AZ31 Mg alloy. *Corrosion Science*, *90*, 402–412. DOI: 10.1016/j.corsci.2014.10.041

Custódio, V., Gonin, M., Stabl, G., Bakhoum, N., Oliveira, M. M., Gutjahr, C., & Castrillo, G. (2021). Sculpting the soil microbiota. *The Plant Journal*, *109*(3), 508–522. DOI: 10.1111/tpj.15568 PMID: 34743401

da Silva, R. G. L., & Blasimme, A. (2024). From lab to society: Fostering clinical translation of molecular systems engineering. *Bioengineering & Translational Medicine*, *9*(1), e10564. Advance online publication. DOI: 10.1002/btm2.10564 PMID: 38193130

Dagbert, C., Meylheuc, T., & Bellon-Fontaine, M.-N. (2006). Corrosion behaviour of AISI 304 stainless steel in presence of a biosurfactant produced by *Pseudomonas fluorescens*. *Electrochimica Acta*, *51*(24), 5221–5227. DOI: 10.1016/j.electacta.2006.03.063

Dang, H., & Lovell, C. R. (2016). Microbial surface colonization and biofilm development in marine environments. *Microbiology and Molecular Biology Reviews : MMBR*, *80*(1), 91–138. DOI: 10.1128/MMBR.00037-15 PMID: 26700108

Daoudi, W., Dagdag, O., Nik, W. B. W., Oussaid, A., & Aatiaoui, A. E. (2024). Introduction to nano-hybrid smart coatings. In *ACS symposium series* (pp. 1–16). https://doi.org/DOI: 10.1021/bk-2024-1469.ch001

Das, R., Ali, M. E., Hamid, S. B. A., Ramakrishna, S., & Chowdhury, Z. Z. (2014). Carbon nanotube membranes for water purification: A bright future in water desalination. *Desalination*, *336*, 97–109. DOI: 10.1016/j.desal.2013.12.026

Dawson, L. (2024). *Extended producer responsibility and its role in facilitating a circular economy in the European Union: An analysis of four directives* [PhD Thesis, University of Birmingham]. https://etheses.bham.ac.uk/id/eprint/14893/

de Jesús-González, A., Rangel-Vázquez, N.-A., & de Velasco-Maldonado, P.-S. (2022). Thermodynamic and structural analysis of graphene/polymer nanocomposites for the adsorption of azo dyes present in binary mixed. In *Novel Materials for Environmental Remediation Applications*. Adsorption and Beyond., DOI: 10.1016/B978-0-323-91894-7.00015-3

Dec, W., Mosiałek, M., Socha, R. P., Jaworska-Kik, M., Simka, W., & Michalska, J. (2017). Characterization of *Desulfovibrio desulfuricans* biofilm on high-alloyed stainless steel: XPS and electrochemical studies. *Materials Chemistry and Physics*, *195*, 28–39. DOI: 10.1016/j.matchemphys.2017.04.011

Deland, E., An, B. A., Yao, J., Sobol, O., Skovhus, T. L., & Koerdt, A. (2021). *Environmental conditions impact the corrosion layer composition of methanogen induced microbiologically influenced corrosion (MI-MIC)*. Research Portal Denmark.

Deng, X., Dohmae, N., Kaksonen, A. H., & Okamoto, A. (2020). Biogenic iron sulfide nanoparticles to enable extracellular electron uptake in sulfate-reducing bacteria. *Angewandte Chemie*, *132*(15), 6051–6055. DOI: 10.1002/ange.201915196 PMID: 31875491

DiMascio, G., & Ganzi, J. (1999). *Electrodeionization apparatus & method* (U.S. Patent No. 5,858,191). United States Patent and Trademark Office.

Ding, W., Zhang, W., Alikunhi, N. M., Batang, Z., Pei, B., Wang, R., Chen, L., Al-Suwailem, A., & Qian, P. Y. (2019). Metagenomic analysis of zinc surface–associated marine biofilms. *Microbial Ecology*, *77*(2), 406–416. DOI: 10.1007/s00248-018-01313-3 PMID: 30612183

Dinh, H. T., Kuever, J., Mussmann, M., Hassel, A. W., Stratmann, M., & Widdel, F. (2004). Iron corrosion by novel anaerobic microorganisms. *Nature*, *427*(6977), 829–832. DOI: 10.1038/nature02321 PMID: 14985759

Dong, Y., He, Z., & Guo, N. (2015). The formation process of biofilm of marine microorganisms and the influence on the corrosion of ship structural steel. *Journal of Marine Science*, *33*(1), 39–44. DOI: 10.3969/j.issn.1001-909X.2015.01.006

dos Santos, A. J., Kronka, M. S., Fortunato, G. V., & Lanza, M. R. V. (2021). Recent advances in electrochemical water technologies for the treatment of antibiotics: A short review. *Current Opinion in Electrochemistry*, *26*, 100674. DOI: 10.1016/j.coelec.2020.100674

Doshvarpassand, S., Wu, C., & Wang, X. (2018). An overview of corrosion defect characterization using active infrared thermography. *Infrared Physics & Technology*, *96*, 366–375. DOI: 10.1016/j.infrared.2018.12.006

Doyle, L. E., Marsili, E., & Bond, D. R. (2021). Metagenomic analysis of electroactive microbial communities in marine sediments. *Frontiers in Microbiology*, *12*, 1–15. DOI: 10.3389/fmicb.2021.705004

Duan, J., Wu, S., Zhang, X., Huang, G., Du, M., & Hou, B. (2008). Corrosion of carbon steel influenced by anaerobic biofilm in natural seawater. *Electrochimica Acta*, *54*(1), 22–28. DOI: 10.1016/j.electacta.2008.04.085

Duggireddy, R. P., & Pisharody, L. (2024). Traditional methods of water purification in rural areas. In *Water Resources Management for Rural Development* (pp. 55–64). Elsevier., DOI: 10.1016/B978-0-443-18778-0.00003-9

Du, J. S., He, K., Xu, Y., Wahl, C. B., Xu, D. D., Dravid, V. P., & Mirkin, C. A. (2021). Galvanic transformation dynamics in heterostructured nanoparticles. *Advanced Functional Materials*, *31*(46), 2105866. Advance online publication. DOI: 10.1002/adfm.202105866

Dwivedi, D., Lepková, K., & Becker, T. (2017). Carbon steel corrosion: A review of key surface properties and characterization methods. *RSC Advances*, *7*(8), 4580–4610. DOI: 10.1039/C6RA25094G

Edelstein, A. S., & Cammarata, R. C. (Eds.). (1996). *Nanomaterials: Synthesis, properties and applications*. Institute of Physics Publishing. DOI: 10.1887/0750305789

Edyvean, R. G., & Videla, H. A. (1991). Biological corrosion. *Interdisciplinary Science Reviews*, *16*(3), 267–282. DOI: 10.1179/isr.1991.16.3.267

Eid, M. M., Duncan, K. E., & Tanner, R. S. (2018). A semi-continuous system for monitoring microbially influenced corrosion. *Journal of Microbiological Methods*, *150*, 55–62. DOI: 10.1016/j.mimet.2018.05.018 PMID: 29803719

Eliaz, N. (2019). Corrosion of Metallic Biomaterials: A Review. *Materials (Basel)*, *12*(3), 407. DOI: 10.3390/ma12030407 PMID: 30696087

El-Mahdy, O. M., Mohamed, H. I., & Mogazy, A. M. (2021). Biosorption effect of Aspergillus niger and Penicillium chrysosporium for Cd- and Pb-contaminated soil and their physiological effects on Vicia faba L. *Environmental Science and Pollution Research International*, *28*(47), 67608–67631. DOI: 10.1007/s11356-021-15382-4 PMID: 34258698

Enning, D., & Garrelfs, J. (2014). Corrosion of iron by sulfate-reducing bacteria: New views of an old problem. *Applied and Environmental Microbiology*, *80*(4), 1226–1236. DOI: 10.1128/AEM.02848-13 PMID: 24317078

Enning, D., Venzlaff, H., Garrelfs, J., Dinh, H. T., Meyer, V., Mayrhofer, K., Hassel, A. W., Stratmann, M., & Widdel, F. (2012). Marine sulfate-reducing bacteria cause serious corrosion of iron under electroconductive biogenic mineral crust. *Environmental Microbiology*, *14*(7), 1772–1787. DOI: 10.1111/j.1462-2920.2012.02778.x PMID: 22616633

Enning, D., Venzlaff, H., Garrelfs, J., Dinh, H. T., Meyer, V., Mayrhofer, K., & Widdel, F. (2012). Marine sulfate-reducing bacteria cause serious corrosion of iron under an electroconductive biogenic mineral crust. *Proceedings of the National Academy of Sciences of the United States of America*, *109*(30), 11950–11955. DOI: 10.1073/pnas.1200072109 PMID: 22616633

Espey, M. J., & Shaw, W. D. (1997). Price elasticity of residential demand for water: A meta-analysis. *Water Resources Research*, *33*(6), 1369–1374. DOI: 10.1029/97WR00571

Farias, D. B. L. (2023). Country differentiation in the global environmental context: Who is 'developing' and according to what? *International Environmental Agreement: Politics, Law and Economics*, *23*(3), 253–269. DOI: 10.1007/s10784-023-09596-9

Farooq, S. A., Raina, A., Mohan, S., Singh, R. A., Jayalakshmi, S., & Haq, M. I. U. (2022). Nanostructured Coatings: Review on processing techniques, corrosion behaviour and tribological performance. *Nanomaterials (Basel, Switzerland)*, *12*(8), 1323. DOI: 10.3390/nano12081323 PMID: 35458032

Feder-Kubis, J., Wirwis, A., Policht, M., Singh, J., & Kim, K.-H. (2024). Principles and practice of greener ionic liquid–nanoparticles biosystem. *Green Chemistry, 26*(6), 3072–3124. DOI: 10.1039/D3GC04387H

Fernandes, M., Souto, A. P., Dourado, F., & Gama, M. (2021). Application of bacterial cellulose in the textile and shoe industry: Development of biocomposites. *Polysaccharides, 2*(3), 566–581. DOI: 10.3390/polysaccharides2030034

Figueira, R. B. (2020). Hybrid sol–gel coatings for corrosion mitigation: A critical review. *Polymers, 12*(3), 689. DOI: 10.3390/polym12030689 PMID: 32204462

Fischer, M., Friedrichs, G., & Lachnit, T. (2014). Fluorescence-based quasicontinuous and in situ monitoring of biofilm formation dynamics in natural marine environments. *Applied and Environmental Microbiology, 80*(12), 3721–3728. DOI: 10.1128/AEM.00298-14 PMID: 24727266

Flemming, H. C., & Wuertz, S. (2019). Bacteria and archaea on Earth and their abundance in biofilms. *Nature Reviews. Microbiology, 17*(4), 247–260. DOI: 10.1038/s41579-019-0158-9 PMID: 30760902

Foreign Trade Barriers. (n.d.). Retrieved October 27, 2025, from https://ustr.gov/sites/default/files/files/Press/Reports/2025NTE.pdf

Frimmel, F. H., & Niessner, R. (2010). *Nanoparticles in the water cycle: Properties, analysis and environmental relevance*. Springer., DOI: 10.1007/978-3-642-10318-6

Fuentes, K. M., Gómez, M., Rebolledo, H., Figueroa, J. M., Zamora, P., & Naranjo-Briceño, L. (2022). Nanomaterials in the future biotextile industry: A new cosmovision to obtain smart biotextiles. *Frontiers in Nanotechnology, 4*, 1056498. Advance online publication. DOI: 10.3389/fnano.2022.1056498

Gabelich, K., Ishida, K. P., & Bold, R. (2005). Testing of water treatment copolymers for compatibility with polyamide reverse osmosis membranes. *Environmental Progress, 24*(4), 410–416. DOI: 10.1002/ep.10111

Galedari, S. A., Mahdavi, A., Azarmi, F., Huang, Y., & McDonald, A. (2019). A comprehensive review of corrosion resistance of thermally-sprayed and thermally-diffused protective coatings on steel structures. *Journal ofThermal Spray Technology, 28*(4), 645–677. DOI: 10.1007/s11666-019-00855-3

Garg, S. (2022). *Industrial Wastewater: Characteristics*. Treatment Techniques and Reclamation of Water., DOI: 10.1007/978-3-030-83811-9_1

Garrido, I., Lagüela, S., Cфappa, C., Saavedra, F. J. M., & Arias, P. (2019). Automatic detection of moistures in different construction materials from thermographic images. *Journal of Thermal Analysis and Calorimetry*, *138*(2), 1649–1660. DOI: 10.1007/s10973-019-08264-y

Gaur, V. K., Gupta, S., Sharma, P., Gupta, P., Varjani, S., Srivastava, J. K., Chang, J.-S., & Bui, X.-T. (2022). Metabolic cascade for remediation of plastic waste: A case study on microplastic degradation. *Current Pollution Reports*, *8*(1), 30–50. DOI: 10.1007/s40726-021-00210-7

Geesey, G., & White, D. C. (1990). Determination of bacterial growth and activity at solid-liquid interfaces. *Annual Review of Microbiology*, *44*(1), 579–597. DOI: 10.1146/annurev.mi.44.100190.003051 PMID: 2252395

Geetha Varma, V., Jha, S., Himesh Karthik Raju, L., Lalith Kishore, R., & Ranjith, V. (2022). A review on decentralized wastewater treatment systems in India. *Chemosphere*, *300*, 134462. DOI: 10.1016/j.chemosphere.2022.134462 PMID: 35390407

Gehrke, I., Geiser, A., & Somborn-Schulz, A. (2015). Innovations in nanotechnology for water treatment. *Nanotechnology, Science and Applications*, *8*, 1–17. DOI: 10.2147/NSA.S43773 PMID: 25609931

Ghafari, M. D., Bahrami, A., Rasooli, I., Arabian, D., & Ghafari, F. (2013). Bacterial exopolymeric inhibition of carbon steel corrosion. *International Biodeterioration & Biodegradation*, *80*, 29–33. DOI: 10.1016/j.ibiod.2013.02.007

Ghosh, S., Debsarkar, A., Dutta, A., Bhandary, S., & Chopra, D. (2021). Delivering Arsenic-free Drinking Water-Made Practically Possible: Continuous Scale Electrochemical Arsenic Remediation Process Furnished, based on Experimental Studies and ANN Simulation. *Environment, Development and Sustainability*, *23*(9), 13087–13112. DOI: 10.1007/s10668-020-01200-3

Giangeri, G. (2024). Novel approaches for steering microbial interactions in anaerobic digestion: Focusing on the role of direct interspecies electron transfer. *Research Portal Denmark*, 165.

Gleick, P. H. (2003). Global freshwater resources: Soft-path solutions for the 21st century. *Science*, *302*(5650), 1524–1528. DOI: 10.1126/science.1089967 PMID: 14645837

Golubović-Bugarski, V., Petković, S., & Globočki-Lakić, G. (2021). The effect of corrosion on a structural integrity and vehicle safety. *Mobility and Vehicle Mechanics*, *47*(3), 49–62. DOI: 10.24874/mvm.2021.47.03.05

Gonzalez, M., Taddonio, K. N., & Sherman, N. J. (2015). The Montreal Protocol: How today's successes offer a pathway to the future. *Journal of Environmental Studies and Sciences*, *5*(2), 122–129. DOI: 10.1007/s13412-014-0208-6

Gonzalez, R., Curtis, K., & Bixler, G. (2020). Metagenomics and metabolomics approaches to microbiologically influenced corrosion. *Current Opinion in Biotechnology*, *62*, 1–7. DOI: 10.1016/j.copbio.2019.09.005

Googan, C. (2022). *Marine Corrosion and Cathodic Protection* (1st ed.). CRC Press., DOI: 10.1201/9781003216070

Goswami, M., Pal, D., Ghosh, M. K., Dwivedi, S., & Chandraker, S. K. (2025). Bimetallic Nanoparticles: A Comprehensive review of synthesis methods and their Potential application. *ChemistrySelect*, *10*(29), e01493. Advance online publication. DOI: 10.1002/slct.202501493

Graham, N. J. D. (1988). *Slow sand filtration – Recent developments in water treatment technology*. Ellis Horwood Ltd.

Guisbiers, G., Mejía-Rosales, D. F., & Leonard, C. (2012). Nanomaterial properties: Size and shape dependencies. *Journal of Nanomaterials, 2012. Journal of Nanomaterials*, *180976*(1), 1–2. DOI: 10.1155/2012/180976

Gul, S., Ahmad, Z., Asma, M., Ahmad, M., Rehan, K., Munir, M., Bazmi, A. A., Ali, H. M., Mazroua, Y., Salem, M. A., Akhtar, M. S., Khan, M. S., Chuah, L. F., & Asif, S. (2022). Effective adsorption of cadmium and lead using SO3H-functionalized Zr-MOFs in aqueous medium. *Chemosphere*, *307*, 135633. Advance online publication. DOI: 10.1016/j.chemosphere.2022.135633 PMID: 35810866

Guo, K., Wu, Z., Chen, C., & Fang, J. (2022). UV/Chlorine Process: An Efficient Advanced Oxidation Process with Multiple Radicals and Functions in Water Treatment. *Accounts of Chemical Research*, *55*(3), 286–297. DOI: 10.1021/acs.accounts.1c00269 PMID: 35025201

Guo, X., Liu, J., & Wang, J. (2018). XPS analysis of the chemical composition of the corrosion product layers of carbon steel in the presence of sulfate-reducing bacteria. *Journal of Materials Engineering and Performance*, *27*(11), 5595–5602. DOI: 10.1007/s11665-018-3636-8

Guo, Z., Liu, T., Cheng, Y. F., Guo, N., & Yin, Y. (2017). Adhesion of *Bacillus subtilis* and *Pseudoalteromonas lipolytica* to steel in a seawater environment and their effects on corrosion. *Colloids and Surfaces. B, Biointerfaces*, *157*, 157–165. DOI: 10.1016/j.colsurfb.2017.05.045 PMID: 28586728

Gupta, P., Singh, J., Verma, S., Chandel, A. S., & Bhatla, R. (2021). Impact of climate change and water quality degradation on food security and agriculture. In *Water Conservation in the Era of Global Climate Change* (pp. 1–22). Elsevier., DOI: 10.1016/B978-0-12-820200-5.00006-3

Gu, T., Jia, R., Unsal, T., & Xu, D. (2019). Toward a better understanding of microbiologically influenced corrosion caused by sulfate reducing bacteria. *Journal of Materials Science and Technology*, *35*(4), 631–636. DOI: 10.1016/j.jmst.2018.10.026

Gu, T., Wang, D., Lekbach, Y., & Xu, D. (2021). Extracellular electron transfer in microbial biocorrosion. *Current Opinion in Electrochemistry*, *29*, 100763. DOI: 10.1016/j.coelec.2021.100763

Hagarová, M., Brezinová, J., Baranová, G., Viňáš, J., & Maruschak, P. (2021). Degradation of components in cars due to bimetallic corrosion. *Materials (Basel)*, *14*(12), 3323. DOI: 10.3390/ma14123323 PMID: 34208438

Haidar, Z., Fatema, K., Shoily, S. S., & Sajib, A. A. (2023). Disease-associated metabolic pathways affected by heavy metals and metalloid. *Toxicology Reports*, *10*, 554–570. DOI: 10.1016/j.toxrep.2023.04.010 PMID: 37396849

Hamilton, W. A. (2003). Microbially influenced corrosion as a model system for the study of metal microbe interactions: A unifying electron transfer hypothesis. *Biofouling*, *19*(1), 65–76. DOI: 10.1080/0892701021000041078 PMID: 14618690

Handelsman, J. (2004). Metagenomics: Application of genomics to uncultured microorganisms. *Microbiology and Molecular Biology Reviews : MMBR*, *68*(4), 669–685. DOI: 10.1128/MMBR.68.4.669-685.2004 PMID: 15590779

He, B., Liu, X., Qi, S., Zheng, R., Chang, M., Lin, Q., & Ren, J. (2021). A review of water-resistant cellulose-based materials in pharmaceutical and biomedical application. *Current Medicinal Chemistry*, *28*(40), 8296–8318. DOI: 10.2174/0929867328666210208113354 PMID: 33557729

He, G., Zhang, J., Zhang, Y., Chen, H., & Wang, X. (2013). Fast and efficient removal of cationic dye using graphite oxide, adsorption, and kinetics studies. *Journal of Dispersion Science and Technology*, *34*(9), 1223–1229. DOI: 10.1080/01932691.2012.739946

He, J., Tan, Y., Liu, H., Jin, Z., Zhang, Y., He, F., Yan, Z., Liu, H., Meng, G., & Liu, H. (2022). Extracellular polymeric substances secreted by marine fungus *Aspergillus terreus*: Full characterization and detailed effects on aluminum alloy corrosion. *Corrosion Science*, *209*, 110703. DOI: 10.1016/j.corsci.2022.110703

He, Z., Fan, X., Jin, W., Gao, S., Yan, B., Chen, C., Ding, W., Yin, S., Zhou, X., Liu, H., Li, X., & Wang, Q. (2023). Chlorine-resistant bacteria in drinking water: Generation, identification and inactivation using ozone-based technologies. *Journal of Water Process Engineering*, *53*, 103772. DOI: 10.1016/j.jwpe.2023.103772

Higgins, E., ... (1917). *Siliceous dust in relation to pulmonary diseases among miners in the Joplin District, Missouri* (Bulletin No. 132). U.S. Department of the Interior, Bureau of Mines.

Hinrichsen, D., Robey, B., & Upadhyay, U. (1997). *Solutions for a water short world* (Population Reports, Series M). Johns Hopkins Bloomberg School of Public Health.

Hiromoto, S. (2010). Corrosion of metallic biomaterials. In *Elsevier eBooks* (pp. 99–121). https://doi.org/DOI: 10.1533/9781845699246.2.99

Hornibrook, E. R., Longstaffe, F. J., & Fyfe, W. S. (2000). Evolution of stable carbon isotope compositions for methane and carbon dioxide in freshwater wetlands and other anaerobic environments. *Geochimica et Cosmochimica Acta*, *64*(6), 1013–1027. DOI: 10.1016/S0016-7037(99)00321-X

Hou, B., Li, X., Ma, X., Du, C., Zhang, D., Zheng, M., & Ma, F. (2017). The cost of corrosion in China. *NPJ Materials Degradation*, *1*(1), 1–10. DOI: 10.1038/s41529-017-0005-2

Huang, Y., Xu, D., Huang, L., Lou, Y., Muhadesi, J.-B., Qian, H., et al. (2021). Responses of soil microbiome to steel corrosion. *npj Biofilms and Microbiomes, 7*, 6. **https://doi.org/**DOI: 10.1038/s41522-020-00175-3

Huang, Y., Liu, S., & Jing, C. (2017). Microbiologically influenced corrosion and mechanisms. *Wei Sheng Wu Xue Tong Bao*, *44*, 1699–1713.

Huang, Y., Zhou, E., Jiang, C., Jia, R., Liu, S., Xu, D., Gu, T., & Wang, F. (2018). Endogenous phenazine-1-carboxamide encoding gene *PhzH* regulated the extracellular electron transfer in biocorrosion of stainless steel by marine *Pseudomonas aeruginosa. Electrochemistry Communications*, *94*, 9–13. DOI: 10.1016/j.elecom.2018.07.019

Hua, W., Sun, R., Wang, X., Zhang, Y., Li, J., Qiu, R., & Gao, Y. (2023). Corrosion of Q235 carbon steel induced by sulfate-reducing bacteria in groundwater: Corrosion behavior, corrosion product, and microbial community structure. *Environmental Science and Pollution Research International*, *31*(3), 4269–4283. DOI: 10.1007/s11356-023-31422-7 PMID: 38097840

Hussain, M. K., Khatoon, S., Nizami, G., Fatma, U. K., Ali, M., Singh, B., Quraishi, A., Assiri, M. A., Ahamad, S., & Saquib, M. (2024). Unleashing the power of bio-adsorbents: Efficient heavy metal removal for sustainable water purification. *Journal of Water Process Engineering*, *64*, 105705. DOI: 10.1016/j.jwpe.2024.105705

Huynh, K., Pham, X., Kim, J., Lee, S. H., Chang, H., Rho, W., & Jun, B. (2020). Synthesis, properties, and biological applications of metallic alloy nanoparticles. *International Journal of Molecular Sciences*, *21*(14), 5174. DOI: 10.3390/ijms21145174 PMID: 32708351

Idumah, C. I., Obele, C. M., Emmanuel, E. O., & Hassan, A. (2020). Recently emerging nanotechnological advancements in polymer nanocomposite coatings for anti-corrosion, anti-fouling and self-healing. *Surfaces and Interfaces*, *21*, 100734. DOI: 10.1016/j.surfin.2020.100734 PMID: 34957345

Ielo, I., Giacobello, F., Sfameni, S., Rando, G., Galletta, M., Trovato, V., Rosace, G., & Plutino, M. R. (2021). Nanostructured Surface Finishing and Coatings: Functional properties and applications. *Materials (Basel)*, *14*(11), 2733. DOI: 10.3390/ma14112733 PMID: 34067241

Imani, M. M., & Safaei, M. (2019). Optimized synthesis of magnesium oxide nanoparticles as bactericidal agents. *Journal of Nanotechnology*, *2019*, 6063832. Advance online publication. DOI: 10.1155/2019/6063832

Inbathamizh, L., Harsha Varthan, M. K., Kumar, R. S. R., Rohinth, M., & Tawfeeq Ahmed, Z. H. (2023). Nanotechnology: Ethical Impacts, Health Issues, and Safety Issues. In Malik, J. A., & Sadiq Mohamed, M. J. (Eds.), *Modern Nanotechnology* (pp. 455–477). Springer Nature Switzerland., DOI: 10.1007/978-3-031-31104-8_20

Inobeme, A., Mathew, J. T., Adetunji, C. O., Ajai, A. I., Inobeme, J., Maliki, M., Okonkwo, S., Adekoya, M. A., Bamigboye, M. O., Jacob, J. O., & Eziukwu, C. A. (2023). Recent advances in nanotechnology for remediation of heavy metals. *Environmental Monitoring and Assessment*, *195*(1), 111. DOI: 10.1007/s10661-022-10614-7 PMID: 36378336

Institute for Occupational Safety and Health of the German Social Accident Insurance (IFA). (2009). *Criteria for assessment of the effectiveness of protective measures*. https://www.dguv.de/ifa/en/fac/nanopartikel/beurteilungsmassstaebe/index.jsp

Iravani, S. (2011). Green synthesis of metal nanoparticles using plants. *Green Chemistry*, *13*(10), 2638–2650. DOI: 10.1039/c1gc15386b

Işıtan, A., Pasquardini, L., Bersani, M., Gök, C., Fioravanti, S., Lunelli, L., Çağlarer, E., & Koluman, A. (2025). Sustainable Production of Microcrystalline and Nanocrystalline Cellulose from Textile Waste Using HCl and NaOH/Urea Treatment. *Polymers, 17*(1), 48. Advance online publication. DOI: 10.3390/polym17010048 PMID: 39795454

Islander, R. L., Devinny, J. S., Mansfeld, F., Postyn, A., & Shih, H. (1991). Microbial ecology of crown corrosion in sewers. *Journal of Environmental Engineering, 117*(6), 751–770. DOI: 10.1061/(ASCE)0733-9372(1991)117:6(751)

Jagaba, A. H., Lawal, I. M., Birniwa, A. H., Affam, A. C., Usman, A. K., Soja, U. B., Saleh, D., Hussaini, A., Noor, A., & Yaro, N. S. A. (2023). Sources of Water Contamination by Heavy Metals. In *Membrane Technologies for Heavy Metal Removal from Water* (pp. 3–27). CRC Press., DOI: 10.1201/9781003326281-2

Jamir, I., Ezung, L. Y., Merry, L., Tikendra, L., Devi, R. S., & Nongdam, P. (2024). Heavy Metals Clean Up: The Application of Fungi for Biosorption. *Geomicrobiology Journal, 41*(3), 201–212. DOI: 10.1080/01490451.2024.2307899

Javaherdashti, R. (2016). Microbiologically influenced corrosion (MIC). In *Engineering materials and processes* (p. 29). Springer Nature. DOI: 10.1007/978-3-319-44306-5_4

Javed, M. A., Neil, W. C., McAdam, G., & Wade, S. A. (2017). Effect of sulphate reducing bacteria on the microbiologically influenced corrosion of ten different metals using constant test conditions. *International Biodeterioration & Biodegradation, 125*, 73–85. DOI: 10.1016/j.ibiod.2017.08.011

Javed, M. A., Stoddart, P. R., & Wade, S. A. (2015). Corrosion of carbon steel by sulphate reducing bacteria: Initial attachment and the role of ferrous ions. *Corrosion Science, 93*, 48–57. DOI: 10.1016/j.corsci.2015.01.006

Jeevanandam, J., Barhoum, A., Chan, Y. S., Dufresne, A., & Danquah, M. K. (2018). Review on nanoparticles and nanostructured materials: History, sources, toxicity and regulations. *Beilstein Journal of Nanotechnology, 9*(1), 1050–1074. DOI: 10.3762/bjnano.9.98 PMID: 29719757

Jelil, R. A. (2015). A review of low-temperature plasma treatment of textile materials. *Journal of Materials Science, 50*(18), 5913–5943. DOI: 10.1007/s10853-015-9152-4

Jeong, H., Alarcón-Correa, M., Mark, A. G., Son, K., Lee, T., & Fischer, P. (2017). Corrosion-Protected hybrid nanoparticles. *Advancement of Science, 4*(12), 1700234. Advance online publication. DOI: 10.1002/advs.201700234 PMID: 29270338

Jiang, G., Keller, J., & Bond, P. L. (2016). Determining the long-term effects of H2S concentration, relative humidity, and air temperature on concrete sewer corrosion. *Water Research, 88*, 136–144. DOI: 10.1016/j.watres.2015.10.002 PMID: 25108169

Jiang, Y., Qiao, Y., Jin, R., Jia, M., Liu, J., He, Z., & Liu, Z. (2024). Application of chlorine dioxide and its disinfection mechanism. *Archives of Microbiology, 206*(10), 400. DOI: 10.1007/s00203-024-04137-7 PMID: 39256286

Jia, R., Ünsal, T., Xu, D., Lekbach, Y., & Gu, T. (2018). Microbiologically influenced corrosion and current mitigation strategies: A state of the art review. *International Biodeterioration & Biodegradation, 137*, 42–58. DOI: 10.1016/j.ibiod.2018.11.007

Jia, R., Yang, D., & Xu, D. (2019). Electrochemical impedance spectroscopy (EIS): A powerful tool for the study of microbiologically influenced corrosion (MIC). *Journal of Materials Science and Technology, 35*(10), 2113–2119. DOI: 10.1016/j.jmst.2019.05.007

Jomova, K., Alomar, S. Y., Nepovimova, E., Kuca, K., & Valko, M. (2025). Heavy metals: Toxicity and human health effects. *Archives of Toxicology, 99*(1), 153–209. DOI: 10.1007/s00204-024-03903-2 PMID: 39567405

Jones, L., Salta, M., Skovhus, T. L., Thomas, K., Illson, T., & Wharton, J. A.. (2024). Dual anaerobic reactor model to study biofilm and microbiologically influenced corrosion interactions on UNS G10180 carbon steel. *Research Square*. **https://doi.org/**DOI: 10.21203/rs.3.rs-5216435/v1

Joseph, T. M., Al-Hazmi, H. E., Śniatała, B., Esmaeili, A., & Habibzadeh, S. (2023). Nanoparticles and nanofiltration for wastewater treatment: From polluted to fresh water. *Environmental Research, 238*, 117114. DOI: 10.1016/j.envres.2023.117114 PMID: 37716387

Juikar, S. J., & Vigneshwaran, N. (2017). Extraction of nanolignin from coconut fibers by controlled microbial hydrolysis. *Industrial Crops and Products, 109*, 420–425. DOI: 10.1016/j.indcrop.2017.08.067

Kadyrzhanov, K. K., Kozlovskiy, A. L., Egizbek, K., Kenzhina, I. E., Abdinov, R. S., & Zdorovets, M. V. (2022). Study of corrosion mechanisms in corrosive media and their influence on the absorption capacity of Fe_2O_3/NDFeO_3 nanocomposites. *Nanomaterials (Basel, Switzerland), 12*(13), 2302. DOI: 10.3390/nano12132302 PMID: 35808138

Kamali, M., Capela, I., & Costa, M. E. (2018). Ultrasonic Synthesis of Zero Valent Iron Nanoparticles for the Efficient Discoloration of Aqueous Solutions Containing Methylene Blue Dye. In Nanomaterials in the Wet Processing of Textiles. DOI: 10.1002/9781119459804.ch8

Kanchi, S. (2014). Nanotechnology for water treatment. *Journal of Environmental & Analytical Chemistry*, *1*(2), e102. DOI: 10.4172/2380-2391.1000e102

Kansara, N.. (2016). Wastewater treatment by ion exchange method: A review of past and recent researches. *Environmental Science Journal*, *14*(4), 5706.

Karadoğan, A., Azgin, F., Yiğittekin, E. S., & Dinçer, S. (2025). Biofilms and their multidimensional impacts: From combating industrial and clinical risks to unlocking opportunities in sustainable agriculture. In *IntechOpen eBooks*. IntechOpen., DOI: 10.5772/intechopen.1010579

Kato, S. (2016). Microbial extracellular electron transfer and its environmental application. *Journal of Bioscience and Bioengineering*, *122*(1), 1–8. DOI: 10.1263/jbb.106.1 PMID: 26777236

Kato, S., Hashimoto, K., & Watanabe, K. (2012). Microbial interspecies electron transfer via electric currents through conductive minerals. *Proceedings of the National Academy of Sciences of the United States of America*, *109*(25), 10042–10046. DOI: 10.1073/pnas.1117592109 PMID: 22665802

Kaur, S., Gallei, M., & Ionescu, E. (2015). Polymer–ceramic nanohybrid materials. In *Organic–Inorganic Hybrid Nanomaterials* (pp. 143–185). doi:DOI: 10.1016/B978-0-444-63317-8.00006-3

Kaye, B. H. (1981). *Direct characterisation of fine particles*. Wiley.

Khalid, N., Ahmad, A., Khalid, S., Ahmed, A., & Irfan, M. (2014). Mineral composition and health functionality of Zamzam water: A review. *International Journal of Food Properties*, *17*(3), 661–677. DOI: 10.1080/10942912.2012.660721

Khan, M. S. H., Rahaman, M. T., Pranta, A. D., & Hasan, M. K. (2025). Eco-friendly organic nanomaterials for multifunctional textiles: Sources, applications, recent advances and future prospects towards sustainability. *International Journal of Environmental Science and Technology*, *22*(8), 7353–7410. Advance online publication. DOI: 10.1007/s13762-024-06299-9

Khan, M. S., Liang, T., Liu, Y., Shi, Y., Zhang, H., Li, H., Guo, S., Pan, H., Yang, K., & Zhao, Y. (2022). Microbiologically influenced corrosion mechanism of ferrous alloys in marine environment. *Metals*, *12*(9), 1458. DOI: 10.3390/met12091458

Khan, N. A., Singh, S., López-Maldonado, E. A., N, P., Méndez-Herrera, P. F., López-López, J. R., Baig, U., Ramamurthy, P. C., Mubarak, N. M., Karri, R. R., & Aljundi, I. H. (2023). Emerging membrane technology and hybrid treatment systems for the removal of micropollutants from wastewater. *Desalination*, *565*, 116873. DOI: 10.1016/j.desal.2023.116873

Kilthau, M., Henkel, V., Wagner, L. P., Gehlhoff, F., & Fay, A. (2025). A decentralized optimization approach for scalable agent-based energy dispatch and congestion management. *Applied Energy*, *377*, 124606. DOI: 10.1016/j.apenergy.2024.124606

Kim, S., Kim, C., Lee, S., & Park, S. (2024). Carbon materials and their metal composites for biomedical applications: A short review. *Nanoscale*, *16*(35), 16313–16328. DOI: 10.1039/D4NR02059F PMID: 39110002

Kip, N., & van Veen, J. A. (2014). The dual role of microbes in corrosion. *The ISME Journal*, *9*(3), 542–551. DOI: 10.1038/ismej.2014.169 PMID: 25259571

Kiran, M. G., Pakshirajan, K., & Das, G. (2020). A comprehensive review on biosurfactant assisted bioelectrochemical systems for microbially influenced corrosion control. *Journal of Hazardous Materials*, *402*, 123795. DOI: 10.1016/j.jhazmat.2020.123795

Kishore, S., Malik, S., Shah, M. P., Bora, J., Chaudhary, V., Kumar, L., Sayyed, R. Z., & Ranjan, A. (2024). A comprehensive review on removal of pollutants from wastewater through microbial nanobiotechnology -based solutions. *Biotechnology & Genetic Engineering Reviews*, *40*(4), 3087–3112. DOI: 10.1080/02648725.2022.2106014 PMID: 35923085

Klenam, D. E. P., Bodunrin, M. O., Akromah, S., Gikunoo, E., Andrews, A., & McBagonluri, F. (2021). Ferrous materials degradation: Characterisation of rust by colour – an overview. *Corrosion Reviews*, *39*(4), 297–311. DOI: 10.1515/corrrev-2021-0005

Klinke, A., & Renn, O. (2021). The coming of age of risk governance. *Risk Analysis*, *41*(3), 544–557. DOI: 10.1111/risa.13383 PMID: 31379003

Km, S., Praveen, B., & Devendra, B. K. (2024). A review on corrosion inhibitors: Types, mechanisms, electrochemical analysis, corrosion rate and efficiency of corrosion inhibitors on mild steel in an acidic environment. *Results in Surfaces and Interfaces*, *16*, 100258. DOI: 10.1016/j.rsurfi.2024.100258

Knisz, J., Eckert, R. B., Gieg, L. M., Koerdt, A., Lee, J., Silva, E. R., Skovhus, T. L., An Stepec, B. A., & Wade, S. A. (2023). Microbiologically influenced corrosion—More than just microorganisms. *FEMS Microbiology Reviews*, *47*(5), fuad041. Advance online publication. DOI: 10.1093/femsre/fuad041 PMID: 37437902

Koçak, A., & Karasu, B. (2018). Nanotaneciklerin Genel Değerlendirilmesi. *El-Cezeri*, *5*(1), 191–236.

Kodama, Y., & Watanabe, K. (2004). *Sulfuricurvum kujiense* gen. nov., sp. nov., a facultatively anaerobic, chemolithoautotrophic, sulfur-oxidizing bacterium isolated from an underground crude-oil storage cavity. *International Journal of Systematic and Evolutionary Microbiology*, *54*(6), 2297–2300. DOI: 10.1099/ijs.0.63243-0 PMID: 15545474

Kolossváry, E., Farkas, K., Karahan, O., Golledge, J., Schernthaner, G.-H., Karplus, T., Bernardo, J. J., Marschang, S., Abola, M. T., Heinzmann, M., Edmonds, M., & Catalano, M. (2023). The importance of socio-economic determinants of health in the care of patients with peripheral artery disease: A narrative review from VAS. *Vascular Medicine*, *28*(3), 241–253. DOI: 10.1177/1358863X231169316 PMID: 37154387

Kubista, M., Andrade, J. M., Bengtsson, M., Forootan, A., Jonák, J., Lind, K., Sindelka, R., Sjöback, R., Sjögreen, B., Strömbom, L., Ståhlberg, A., & Zoric, N. (2006). The real-time polymerase chain reaction. *Molecular Aspects of Medicine*, *27*(2-3), 95–125. DOI: 10.1016/j.mam.2005.12.007 PMID: 16460794

Kugel, A., Stafslien, S., & Chisholm, B. J. (2011). Antimicrobial coatings produced by "tethering" biocides to the coating matrix: A comprehensive review. *Progress in Organic Coatings*, *72*(3), 222–252. DOI: 10.1016/j.porgcoat.2011.07.004

Kuhn, S. P., & Pfister, R. M. (1990). Accumulation of cadmium by immobilized *Zoogloea ramigera* 115. *Journal of Industrial Microbiology*, *6*(2), 123–138. DOI: 10.1007/BF01576431

Kumar, N., & Sharma, A. (2022). Surface coatings and functionalization strategies for corrosion mitigation. In *ACS symposium series* (pp. 291–316). https://doi.org/ DOI: 10.1021/bk-2022-1418.ch014

Kumar, A. S., Sivakumar, L., Rajadesingu, S., Sathish, S., Malik, T., & Parthipan, P. (2025). Sustainable corrosion inhibition approaches for the mitigation of microbiologically influenced corrosion - a systematic review. *Frontiers in Materials*, *12*, 1545245. Advance online publication. DOI: 10.3389/fmats.2025.1545245

Kumar, A., Kumar, A., Seema, K., & Kumari, K. (2025a). *Detection and Remediation Techniques for Emerging Contaminants.*, DOI: 10.1007/978-3-031-82579-8_10

Kumari, P. P., & Mulky, L. (2024). Optimization strategies for corrosion management in industries with artificial neural network and response surface technology: A comprehensive review. *Journal of Bio- and Tribo-Corrosion*, *10*(3), 59. Advance online publication. DOI: 10.1007/s40735-024-00863-z

Kumar, R., Rai, B., Gahlyan, S., & Kumar, G. (2021). A comprehensive review on production, surface modification and characterization of nanocellulose derived from biomass and its commercial applications. *Express Polymer Letters*, *15*(2), 104–120. DOI: 10.3144/expresspolymlett.2021.11

Kumar, S., Arora, A., Mathur, D., Chaudhary, A., Pant, V., Guchhait, S., & Singh, B. K. (2025). A review on chitosan and chitosan-based bionanocomposites: Promising biological macromolecules for sustainable corrosion inhibition. *International Journal of Biological Macromolecules*, *140392*, 140392. Advance online publication. DOI: 10.1016/j.ijbiomac.2025.140392 PMID: 39880227

Kumar, V., Verma, P., de Freitas, F. A., Srivastava, P. K., Vashishth, A., & Américo-Pinheiro, J. H. P. (2025b). A critical review on biofuels generation from pulp-paper mill sludge with emphasis on pretreatment methods: Renewable energy for environmental sustainability. *BMC Environmental Science*, *2*(1), 2. DOI: 10.1186/s44329-024-00016-0

Kundu, D. K., Gupta, A., Mol, A. P. J., Rahman, M. M., & van Halem, D. (2018). Experimenting with a novel technology for provision of safe drinking water in rural Bangladesh: The case of sub-surface arsenic removal (SAR). *Technology in Society*, *53*, 161–172. DOI: 10.1016/j.techsoc.2018.01.010

Lago, A., Rocha, V., Barros, O., Silva, B., & Tavares, T. (2024). Bacterial biofilm attachment to sustainable carriers as a clean-up strategy for wastewater treatment: A review. *Journal of Water Process Engineering*, *63*, 105368. DOI: 10.1016/j.jwpe.2024.105368

Lanneluc, I., Langumier, M., Sabot, R., Jeannin, M., Refait, P., & Sablé, S. (2015). On the bacterial communities associated with the corrosion product layer during the early stages of marine corrosion of carbon steel. *International Biodeterioration & Biodegradation*, *99*, 55–65. DOI: 10.1016/j.ibiod.2015.01.003

Lasserus, M., Knez, D., Schnedlitz, M., Hauser, A. W., Hofer, F., & Ernst, W. E. (2019). On the passivation of iron particles at the nanoscale. *Nanoscale Advances*, *1*(6), 2276–2283. DOI: 10.1039/C9NA00161A PMID: 36131962

Lee, A. K., Newman, D. K., & Coates, J. D. (2019). Microbial corrosion of iron-based materials: A review. *Corrosion*, *75*(4), 339–359. DOI: 10.5006/3129

Lee, D., & Choi, K. (2019). Comparison of regulatory frameworks of environmental risk assessments for human pharmaceuticals in EU, USA, and Canada. *The Science of the Total Environment*, *671*, 1026–1035. https://www.sciencedirect.com/science/article/pii/S0048969719313713. DOI: 10.1016/j.scitotenv.2019.03.372

Leitgeb, N., Auvinen, A., Danker-Hopfe, H., & Mild, K. (2016). SCENIHR (Scientific Committee on Emerging and Newly Identified Health Risks): Potential health effects of exposure to electromagnetic fields (EMF). *Bioelectromagnetics*, *37*(8), 556–559. DOI: 10.1002/bem.21930

Lens, (2025). *Search results for "Corrosion AND Nano"*. The Lens. Retrieved September 19, 2025, from https://www.lens.org/lens/search/scholar/analysis?q=(corrosion%20nano)&p=0&n=10&s=_score&d=%2B&f=false&e=false&l=en&authorField=author&dateFilterField=publishedYear&orderBy=%2B_score&presentation=false&preview=true&stemmed=true&useAuthorId=false

Lens, (2025). *Search results for "Corrosion in Metal Nanoparticles (MNPs)"*. The Lens. Retrieved September 19, 2025, from https://www.lens.org/lens/search/scholar/list?q=Corrosion%20of%20Metal%20Nanoparticles&n=10

Li, X., Duan, J., Xiao, H., Li, Y., Liu, H., Guan, F. & Zhai, X. (2017).

Liamleam, W., & Annachhatre, A. P. (2007). Electron donors for biological sulfate reduction. *Biotechnology Advances*, *25*(5), 452–463. DOI: 10.1016/j.biotechadv.2007.05.002 PMID: 17572039

Li, G., Li, Y., Chen, G., He, J., Han, Y., Wang, X., & Kaplan, D. L. (2015). Silk-based biomaterials in biomedical textiles and fiber-based implants. *Advanced Healthcare Materials*, *4*(8), 1134–1151. DOI: 10.1002/adhm.201500002 PMID: 25772248

Li, J. (2025). Governing High-Risk Technologies in a Fragmented World: Geopolitical Tensions, Regulatory Gaps, and Institutional Barriers to Global Cooperation. *Fudan Journal of the Humanities and Social Sciences*. Advance online publication. DOI: 10.1007/s40647-025-00445-4

Li, J., Tian, X., Hua, T., Fu, J., Koo, M., Chan, W., & Poon, T. (2021). Chitosan natural polymer material for improving antibacterial properties of textiles. *ACS Applied Bio Materials*, *4*(5), 4014–4038. DOI: 10.1021/acsabm.1c00078 PMID: 35006820

Li, L., Ma, C., Yang, Y., Wang, B., Liu, X., Wang, Y., Bian, X., Zhang, G., & Zhang, N. (2024a). Exploring the potential of plant-derived metal ion binding peptides: Preparation, structure-activity relationship, and biological activities. *Trends in Food Science & Technology*, *152*, 104650. DOI: 10.1016/j.tifs.2024.104650

Li, M., Nan, L., Liang, C., Sun, Z., Yang, L., & Yang, K. (2021). Antibacterial behavior and related mechanisms of martensitic Cu-bearing stainless steel evaluated by a mixed infection model of *Escherichia coli* and *Staphylococcus aureus* in vitro. *Journal of Materials Science and Technology*, *62*, 139–147. DOI: 10.1016/j.jmst.2020.05.030

Lin, X., Gong, X., Ruan, Q., Xu, W., Zhang, C., & Zhao, K. (2023). Antimicrobial Application of Chitosan Derivatives and their Nanocom-posites. *Current Medicinal Chemistry*, *30*(15), 1736–1755. DOI: 10.2174/0929867329666220803114729 PMID: 35927801

Li, Q., Gong, L., Chen, X.-G., Gadd, G. M., & Liu, D. (2025). Dual role of microorganisms in metal corrosion: A review of mechanisms of corrosion promotion and inhibition. *Frontiers in Microbiology*, *16*, 1552103. Advance online publication. DOI: 10.3389/fmicb.2025.1552103 PMID: 40270819

Little, B. J., Blackwood, D. J., Hinks, J., Lauro, F. M., Marsili, E., Okamoto, A., Rice, S. A., Wade, S. A., & Flemming, H.-C. (2020). Microbially influenced corrosion—Any progress? *Corrosion Science*, *170*, 108641. DOI: 10.1016/j.corsci.2020.108641

Little, B., Wagner, P., & Mansfeld, F. (1991). Microbiologically influenced corrosion of metals and alloys. *International Materials Reviews*, *36*(1), 253–272. DOI: 10.1179/imr.1991.36.1.253

Liu, H., Gu, T., Asif, M., Zhang, G., & Liu, H. (2017). The corrosion behavior and mechanism of carbon steel induced by extracellular polymeric substances of iron-oxidizing bacteria. *Corrosion Science*, *114*, 102–111. DOI: 10.1016/j.corsci.2016.10.025

Liu, H., Zhang, W., Zhao, Z., Lai, H., & Liu, H. (2023). Broken passive film and subsequent pitting corrosion behavior of 2205 duplex stainless steel induced by marine fungus *Aspergillus terreus* in artificial seawater. *Corrosion Science*, *218*, 111147. DOI: 10.1016/j.corsci.2023.111147

Liu, L., Guo, X., Liu, W., & Lee, C. (2021). Recent Progress in the Energy Harvesting Technology—From Self-Powered Sensors to Self-Sustained IoT, and New Applications. *Nanomaterials (Basel, Switzerland)*, *11*(11), 2975. DOI: 10.3390/nano11112975 PMID: 34835739

Liu, L., Wu, X., Wang, Q., Yan, Z., Wen, X., Tang, J., & Li, X. (2022). An overview of microbiologically influenced corrosion: Mechanisms and its control by microbes. *Corrosion Reviews*, *40*(2), 103–117. DOI: 10.1515/corrrev-2021-0039

Liu, M., Guo, Y., Wang, J., & Yergin, M. (2018). Corrosion avoidance in lightweight materials for automotive applications. *NPJ Materials Degradation*, *2*(1), 24. Advance online publication. DOI: 10.1038/s41529-018-0045-2

Liu, M., & Zhao, Z. (2025). Corrosion and mechanical Micro-Interaction behavior of metal materials. *Materials (Basel)*, *18*(17), 4114. DOI: 10.3390/ma18174114 PMID: 40942539

Liu, T., Guo, Z., Zeng, Z., Guo, N., Lei, Y., Liu, T., & Wang, X. (2018). Marine bacteria provide lasting anti-corrosion activity for steel via biofilm-induced mineralization. *ACS Applied Materials & Interfaces*, *10*, 14991. DOI: 10.1021/acsami.8b03317

Liu, W., Herrmann, A., Bigall, N. C., Rodriguez, P., Wen, D., Oezaslan, M., Schmidt, T. J., Gaponik, N., & Eychmüller, A. (2015). Noble Metal Aerogels—Synthesis, Characterization, and Application as Electrocatalysts. *Accounts of Chemical Research*, *48*(2), 154–162. DOI: 10.1021/ar500237c PMID: 25611348

Li, X., Che, Y., Lyu, Y., Liu, F., Wang, Y., & Zhao, C. (2018). Antimicrobial Properties and Applications of Chitosan/Inorganic Nanocomposites: An Overview | / . Cailiao Daobao. *Cailiao Daobao*, *32*(11), 3823–3830. DOI: 10.11896/j.issn.1005-023X.2018.21.020

Li, X., Duan, J., Xiao, H., Li, Y., Liu, H., Guan, F., & Zhai, X. (2017). Analysis of bacterial community composition of corroded steel immersed in Sanya and Xiamen seawaters in China via method of Illumina MiSeq sequencing. *Frontiers in Microbiology*, *8*, 1737. Advance online publication. DOI: 10.3389/fmicb.2017.01737 PMID: 28955315

Li, X., Li, Z., Yang, D., Li, S., Yang, L., Yan, W., Xiang, W., Pei, L., Ye, Z., & Xu, H. (2024b). Electrochemical disinfection for human health protection: Disinfection mechanisms, system innovations, applications. *Journal of Environmental Chemical Engineering*, *12*(5), 114073. DOI: 10.1016/j.jece.2024.114073

Li, X., Xu, H., Chen, Z.-S., & Chen, G. (2011). Biosynthesis of nanoparticles by microorganisms and their applications. *Journal of Nanomaterials*, *2011*, 270974. Advance online publication. DOI: 10.1155/2011/270974

Li, X., Zhang, D., Liu, Z., Li, Z., Du, C., & Dong, C. (2015). Materials science: Share corrosion data. *Nature*, *527*(7579), 441–442. DOI: 10.1038/527441a PMID: 26607528

Li, Y., Xu, D., Chen, C., Li, X., Jia, R., Zhang, D. W., Wolfgang, S., Wang, F. H., & Gu, T. (2018). Anaerobic microbiologically influenced corrosion mechanisms interpreted using bioenergetics and bioelectrochemistry: A review. *Journal of Materials Science and Technology, 34*(10), 1713–1718. DOI: 10.1016/j.jmst.2018.02.023

Li, Z., Chang, W., Cui, T., Xu, D., Zhang, D., Lou, Y., Qian, H., Song, H., Mol, A., Cao, F., Gu, T., & Li, X. (2021). Adaptive bidirectional extracellular electron transfer during accelerated microbiologically influenced corrosion of stainless steel. *Communications Materials, 2*(1), 67. Advance online publication. DOI: 10.1038/s43246-021-00173-8

Lloyd, S. M., & Lave, L. B. (2003). Life cycle economic and environmental implications of using nanocomposites in automobiles. *Environmental Science & Technology, 37*(15), 3458–3466. DOI: 10.1021/es026023q PMID: 12966996

Logan, B. E., & Regan, J. M. (2006). Microbial fuel cells—Challenges and applications. *Environmental Science & Technology, 40*(17), 5172–5180. DOI: 10.1021/es0627592 PMID: 16999086

López-Maldonado, E. A., Khan, N. A., Singh, S., Ramamurthy, P. C., Kabak, B., Baudrit, J. R. V., Alkahtani, M. Q., Álvarez-Torrellas, S., Varshney, R., Serra-Pérez, E., García, J., Gkika, D. A., Kyzas, G. Z., Kadier, A., Singh, R., Periyasamy, S., Gizaw, D. G., Hossain, N., & Zahmatkesh, S. (2024). Magnetic polymeric composites: Potential for separating and degrading micro/nano plastics. *Desalination and Water Treatment, 317*, 100198. DOI: 10.1016/j.dwt.2024.100198

Loto, C. A. (2017). Microbiological corrosion: Mechanism, control and impact—A review. *International Journal of Advanced Manufacturing Technology, 92*(9–12), 4241–4252. DOI: 10.1007/s00170-017-0494-8

Lovley, D. R. (2006). Bug juice: Harvesting electricity with microorganisms. *Nature Reviews. Microbiology, 4*(7), 497–508. DOI: 10.1038/nrmicro1442 PMID: 16778836

Lovley, D. R., Holmes, D. E., & Nevin, K. P. (2004). Dissimilatory Fe(III) and Mn(IV) reduction. *Advances in Microbial Physiology, 49*, 219–286. DOI: 10.1016/S0065-2911(04)49005-5 PMID: 15518832

Luciano, K., Wang, X., Liu, Y., Eyler, G., Qin, Z., & Xia, X. (2022). Noble Metal nanoparticles for Point-of-Care Testing: Recent advancements and social impacts. *Bioengineering (Basel, Switzerland), 9*(11), 666. DOI: 10.3390/bioengineering9110666 PMID: 36354576

Luef, B., Fakra, S. C., Csencsits, R., Wrighton, K. C., Williams, K. H., Wilkins, M. J., Downing, K. H., Long, P. E., Comolli, L. R., & Banfield, J. F. (2013). Iron-reducing bacteria accumulate ferric oxyhydroxide nanoparticle aggregates that may support planktonic growth. *The ISME Journal*, *7*(2), 338–350. DOI: 10.1038/ismej.2012.103 PMID: 23038172

Lum, L.-S. H., Malghan, S. G., & Schiller, S. B. (1996). Standard reference materials for particle-size analysis of ceramic powders by gravity sedimentation. *Powder Technology*, *87*(3), 233–238. DOI: 10.1016/0032-5910(96)03086-0

Luo, J., Mamat, B., Yue, Z., Zhang, N., Xu, X., Li, Y., Su, Z., Ma, C., Zang, F., & Wang, Y. (2021). Multi-metal ions doped hydroxyapatite coatings via electrochemical methods for antibacterial and osteogenesis. *Colloid and Interface Science Communications*, *43*, 100435. DOI: 10.1016/j.colcom.2021.100435

Lv, X., Wang, C., Liu, J., Sand, W., Etim, I. N., & Zhang, Y.. (2024). The microbiologically influenced corrosion and protection of pipelines: A detailed review. *Materials (Basel)*, *17*(20), 4996. DOI: 10.3390/ma17204996 PMID: 39459699

Macdonald, D. D. (2006). Reflections on the history of electrochemical impedance spectroscopy. *Electrochimica Acta*, *51*(8-9), 1376–1388. DOI: 10.1016/j.electacta.2005.02.107

Mahadevan, R., Palsson, B. Ø., & Lovley, D. R. (2016). In situ to in silico and back: Elucidating the physiology and ecology of Geobacter spp. using genome-scale modelling. *Nature Reviews. Microbiology*, *9*(1), 39–50. DOI: 10.1038/nrmicro2456 PMID: 21132020

Mahmoud, A. S., Mostafa, M. K., & Peters, R. W. (2021). A prototype of textile wastewater treatment using coagulation and adsorption by Fe/Cu nanoparticles: Technoeconomic and scaling-up studies. *Nanomaterials and Nanotechnology*, *11*. Advance online publication. DOI: 10.1177/18479804211041181

Mahto, K. U., & Das, S. (2022). Bacterial biofilm and extracellular polymeric substances in the moving bed biofilm reactor for wastewater treatment: A review. *Bioresource Technology*, *345*, 126476. DOI: 10.1016/j.biortech.2021.126476 PMID: 34864174

Maia, M. T., Noronha, V. T., Oliveira, N. C., Alves, A. C., Faria, A. F., Martinez, D. T., Ferreira, O. P., & Paula, A. J. (2022). Silica nanoparticles and surface silanization for the fabrication of water-repellent cotton fibers. *ACS Applied Nano Materials*, *5*(4), 4634–4647. DOI: 10.1021/acsanm.1c03346

Makvandi, P., Iftekhar, S., Pizzetti, F., Zarepour, A., Zare, E. N., Ashrafizadeh, M., Agarwal, T., Padil, V. V. T., Mohammadinejad, R., Sillanpaa, M., Zarrabi, A., & Rossi, F. (2021). Functionalization of polymers and nanomaterials for water treatment, food packaging, textile and biomedical applications: A review. *Environmental Chemistry Letters, 19*(1), 583–611. DOI: 10.1007/s10311-020-01089-4

Malik, N., & Kumar, R. (2024). A Comparative Study of Different Disinfection Processes, and Management Practices to Control the Formation of Disinfection by-Products (DBPs). In *Drinking Water Disinfection By-products* (pp. 59–83). Springer Nature Switzerland., DOI: 10.1007/978-3-031-49047-7_4

Manam, N., Harun, W., Shri, D., Ghani, S., Kurniawan, T., Ismail, M., & Ibrahim, M. (2017). Study of corrosion in biocompatible metals for implants: A review. *Journal of Alloys and Compounds, 701*, 698–715. DOI: 10.1016/j.jallcom.2017.01.196

Mand, J., Park, H. S., Jack, T. R., & Voordouw, G. (2014). The role of acetogens in microbially influenced corrosion of steel. *Frontiers in Microbiology, 5*, 1–14. DOI: 10.3389/fmicb.2014.00268 PMID: 24917861

Marsili, E., Baron, D. B., Shikhare, I. D., Coursolle, D., Gralnick, J. A., & Bond, D. R. (2008). Shewanella secretes flavins that mediate extracellular electron transfer. *Proceedings of the National Academy of Sciences of the United States of America, 105*(10), 3968–3973. DOI: 10.1073/pnas.0710525105 PMID: 18316736

Masroor, S. (2023). Corrosion: Fundamental and Economic Impact. In *Carbon Allotropes* (pp. 3–14). CRC Press. https://www.taylorfrancis.com/chapters/edit/10.1201/9781003323976-2/corrosion-sheerin-masroor

Matavulj, M., & Molitoris, H. P. (2009). Marine fungi: Degraders of poly-3-hydroxyalkanoate based plastic materials. *Zbornik Matice Srpske za Prirodne Nauke*, (116), 253–265. DOI: 10.2298/ZMSPN0916253M

Matin, A., Laoui, T., Falath, W., & Farooque, M. (2021). Fouling control in reverse osmosis for water desalination & reuse: Current practices & emerging environment-friendly technologies. *The Science of the Total Environment, 765*, 142721. DOI: 10.1016/j.scitotenv.2020.142721 PMID: 33129530

Ma, Y., Rissanen, M., You, X., Moriam, K., Hummel, M., & Sixta, H. (2021). New method for determining the degree of fibrillation of regenerated cellulose fibres. *Cellulose (London, England), 28*(1), 31–44. DOI: 10.1007/s10570-020-03513-y

Ma, Y., Zhang, Y., Zhang, R., Guan, F., Hou, B., & Duan, J. (2020). Microbiologically influenced corrosion of marine steels within the interaction between steel and biofilms: A brief view. *Applied Microbiology and Biotechnology, 104*(2), 515–525. DOI: 10.1007/s00253-019-10184-8 PMID: 31807887

Mazari, S. A., Mubarak, N. M., Jatoi, A. S., Abro, R., Shah, A., Shah, A. K., & Lghari, Z. (2021). Environmental impact of using nanomaterials in textiles. In *Nanosensors and nanodevices for smart multifunctional textiles* (pp. 321–342). Elsevier. DOI: 10.1016/B978-0-12-820777-2.00018-2

Mazumder, M. J. (2020). Global Impact of Corrosion: Occurrence, cost and mitigation. *Global Journal of Engineering Sciences, 5*(4). Advance online publication. DOI: 10.33552/GJES.2020.05.000618

McBeth, J. M., & Emerson, D. (2016). In situ microbial community succession on mild steel in estuarine and marine environments: Exploring the role of iron-oxidizing bacteria. *Frontiers in Microbiology, 7*, 767. DOI: 10.3389/fmicb.2016.00767 PMID: 27252686

Meeran, S., Begum, A. S., Sheit, H. K., Mohamed, M., & Mohan, K. (2025). The deep investigation of structural and electrochemical analysis on the anti – corrosion behavior of MGCO2O4 and ZNCO2O4 spinel nanoparticles in acid medium. *Results in Surfaces and Interfaces, 100416*, 100416. Advance online publication. DOI: 10.1016/j.rsurfi.2025.100416

Mehmood, S., Ahmad, A., Ahmed, A., Khalid, N., & Javed, T. (2013). Drinking water quality in capital city of Pakistan. *Open Access Scientific Reports, 2*, 637–641. DOI: 10.4172/scientificreports.637

Mehrabi, M. R., Soltani, M., Chiani, M., Raahemifar, K., & Farhangi, A. (2023). Nanomedicine: New Frontiers in Fighting Microbial Infections. *Nanomaterials (Basel, Switzerland), 13*(3), 483. DOI: 10.3390/nano13030483 PMID: 36770443

Melchers, R. E. (2005). The effect of corrosion on the structural reliability of steel offshore structures. *Corrosion Science, 47*(10), 2391–2410. DOI: 10.1016/j.corsci.2005.04.004

Melchers, R. E., & Jeffrey, R. (2012). Corrosion of long vertical steel strips in the marine tidal zone and implications for ALWC. *Corrosion Science, 65*, 26–36. DOI: 10.1016/j.corsci.2012.07.025

Michalak, I., Chojnacka, K., & Witek-Krowiak, A. (2013). State of the art for the biosorption process—A review. *Applied Biochemistry and Biotechnology, 170*(6), 1389–1416. DOI: 10.1007/s12010-013-0269-0 PMID: 23666641

Milovanović, B., & Pečur, I. B. (2016). Review of active IR thermography for detection and characterization of defects in reinforced concrete. *Journal of Imaging*, *2*(2), 11. DOI: 10.3390/jimaging2020011

Mir, U. S. (2001). *[Patent title, if available]* (U.S. Patent No. 6,241,867). United States Patent and Trademark Office.

Mishel, L., & Shierholz, H. (2013, August). *A decade of flat wages: The key barrier to shared prosperity and a rising middle class* (Briefing Paper No. 365). Economic Policy Institute. https://www.epi.org/publication/a-decade-of-flat-wages-the-key-barrier-to-shared-prosperity-and-a-rising-middle-class/

Mishra, S., Patel, A., Bhatt, P., Chen, S., & Srivastava, P. K. (2024). Perspective Evaluation of Synthetic Biology Approaches for Effective Mitigation of Heavy Metal Pollution. *Reviews of Environmental Contamination and Toxicology*, *262*(1), 21. DOI: 10.1007/s44169-024-00072-2

Miyah, Y., El Messaoudi, N., Benjelloun, M., Georgin, J., Franco, D. S. P., Acikbas, Y., Kusuma, H. S., & Sillanpää, M. (2024). MOF-derived magnetic nanocomposites as potential formulations for the efficient removal of organic pollutants from water via adsorption and advanced oxidation processes: A review. *Materials Today Sustainability*, *28*, 100985. DOI: 10.1016/j.mtsust.2024.100985

Moattari, R. M., Mohammadi, T., Rajabzadeh, S., Dabiryan, H., & Matsuyama, H. (2021). Reinforced hollow fiber membranes: A comprehensive review. *Journal of the Taiwan Institute of Chemical Engineers*, *122*, 284–310. DOI: 10.1016/j.jtice.2021.04.052

Moghadas, B. K., Esmaeili, H., Tamjidi, S., & Geramifard, A. (2022). Advantages of nanoadsorbents, biosorbents, and nanobiosorbents for contaminant removal. In *Nano-Biosorbents for Decontamination of Water* (pp. 105–133). Air, and Soil Pollution., DOI: 10.1016/B978-0-323-90912-9.00006-X

Mohammed, M. B. (2011). *Low cost nanomaterials for water desalination and purification: Final technical report.* United Nations–UNESCO. https://www.scirp.org/(S(lz5mqp453edsnp55rrgjct55.))/reference/referencespapers.aspx?referenceid=2529208

Mondal, C., Hsiao, S., & Wickramasinghe, S. R. (2008). Nanofiltration/reverse osmosis for treatment of coproduced waters. *Environmental Progress*, *27*(2), 173–179. DOI: 10.1002/ep.10271

Moore, C. A. (1967). An Economic Evaluation of Starch Use in the Textile Industry.

Moradi, M., Song, Z., & Tao, X. (2015). Introducing a novel bacterium, *Vibrio neocaledonicus* sp., with the highest corrosion inhibition efficiency. *Electrochemistry Communications*, *51*, 64–68. DOI: 10.1016/j.elecom.2014.12.007

Moradi, M., Song, Z., Yang, L., Jiang, J., & He, J. (2014). Effect of marine *Pseudoalteromonas* sp. on the microstructure and corrosion behaviour of 2205 duplex stainless steel. *Corrosion Science*, *84*, 103–112. DOI: 10.1016/j.corsci.2014.03.018

Moreira-Filho, P., & Figueiredo, P. (2021). The influence of the marine *Bacillus cereus* over carbon steel, stainless corrosion, and copper coupons. *Archives of Microbiology*, *204*(1), 9. Advance online publication. DOI: 10.1007/s00203-021-02607-w PMID: 34873663

Moura, M. C., Pontes, F. V., & Lutterbach, M. T. (2018). Biofilm formation on corroded surfaces and its influence on corrosion resistance. *Journal of Materials Science*, *53*(14), 10145–10159. DOI: 10.1007/s10853-018-2287-3

Mourdikoudis, S., & Dominguez-Benetton, X. (2025). Physicochemical vs Electrochemical Technologies for Metal Recovery – Main Insights, Comparison, Complementarity and Challenges. *Chemistry Methods*, *5*(3), e202400046. Advance online publication. DOI: 10.1002/cmtd.202400046

Mullis, K., Faloona, F., Scharf, S., Saiki, R., Horn, G., & Erlich, H. (1986). Specific enzymatic amplification of DNA in vitro: The polymerase chain reaction. *Cold Spring Harbor Symposia on Quantitative Biology*, *51*(0), 263–273. DOI: 10.1101/SQB.1986.051.01.032 PMID: 3472723

Murcar-Evans, B. I., Cabral, A. D., Toutah, K., de Araujo, E. D., Lai, A., Macdonald, P. M., Berger-Becvar, A., Kraskouskaya, D., & Gunning, P. T. (2017). ProxyPhos sensors for the detection of negatively charged membranes. *Analyst*, *142*(23), 4511–4521. DOI: 10.1039/C7AN00568G PMID: 29098228

Muresan, L. M. (2023). Nanocomposite coatings for Anti-Corrosion properties of metallic substrates. *Materials (Basel)*, *16*(14), 5092. DOI: 10.3390/ma16145092 PMID: 37512373

Muthusamy, K., Gopalakrishnan, S., & Ravi, T. K. (2008). Biosurfactants: Properties, commercial production and application. *Current Science*, *94*(6), 736–747.

Muyzer, G., & Stams, A. J. (2008). The ecology and biotechnology of sulphate-reducing bacteria. *Nature Reviews. Microbiology*, *6*(6), 441–454. DOI: 10.1038/nrmicro1892 PMID: 18461075

Najim, A. A., Radeef, A. Y., Al-Doori, I., & Jabbar, Z. H. (2024). Immobilization: The promising technique to protect and increase the efficiency of microorganisms to remove contaminants. *Journal of Chemical Technology and Biotechnology, 99*(8), 1707–1733. DOI: 10.1002/jctb.7638

Nambela, L., Haule, L. V., & Mgani, Q. (2020). A review on source, chemistry, green synthesis and application of textile colorants. *Journal of Cleaner Production, 246*, 119036. DOI: 10.1016/j.jclepro.2019.119036

Ndukwe, A. I., Nwadirichi, B., Okolo, C., Tom-Okoro, M., Medupin, R., Uche, R., Arukalam, I., Onuoha, C., Egole, C., Okorafor, O., & Nwakuba, N. (2025). *Corrosion Control in Metals: A review on Sustainable approach using Nanotechnology.* Zastita Materijala., DOI: 10.62638/ZasMat1187

Nguyen, H. M., Le, T. T. N., Nguyen, A. T., Le, H. N. T., & Pham, T. T. (2023). Biomedical materials for wound dressing: Recent advances and applications. *RSC Advances, 13*(8), 5509–5528. DOI: 10.1039/D2RA07673J PMID: 36793301

Nielsen, A. H., Hvitved-Jacobsen, T., & Vollertsen, J. (2005). Effects of pH and iron concentrations on sulfide precipitation in wastewater collection systems. *Water Environment Research, 77*(1), 113–120. DOI: 10.2175/106143005X41645 PMID: 18536490

Nitschke, M., & Costa, S. G. (2007). Biosurfactants in food industry. *Trends in Food Science & Technology, 18*(5), 252–259. DOI: 10.1016/j.tifs.2007.01.002

Nizam, N. U. M., Hanafiah, M. M., & Woon, K. S. (2021). A content review of life cycle assessment of nanomaterials: Current practices, challenges, and future prospects. *Nanomaterials (Basel, Switzerland), 11*(12), 3324. https://www.mdpi.com/2079-4991/11/12/3324. DOI: 10.3390/nano11123324 PMID: 34947673

Noreen, S., Saleem, M. H., Ali, B., Khan, K. A., Hafeez, A., & Javed, M. A. (2024). Advancing plant resilience against microplastics and metals through nanotechnology. *BioNanoScience, 14*(2), 2065–2079. DOI: 10.1007/s12668-024-01301-7

OCWD. (2025). *ORANGE COUNTY WATER DISTRICT.* https://www.ocwd.com/

OECD. U. S. M. to the. (2023, October 31). About the OECD. *U.S. Mission to the Organization For Economic Cooperation & Development.* https://usoecd.usmission.gov/about-the-oecd/

Okabe, S., Odagiri, M., Ito, T., & Satoh, H. (2007). Succession of sulfur-oxidizing bacteria in the microbial community on corroding concrete in sewer systems. *Applied and Environmental Microbiology, 73*(3), 971–980. DOI: 10.1128/AEM.02054-06 PMID: 17142362

Okorie, I. E., & Nwokorie, R. C. (2021). A review of fungal influenced corrosion of metals. *Zaštita Materijala*, *62*(4), 333–339. DOI: 10.5937/zasmat2104333O

Oliveira, D. R., Zaiat, M., & Foresti, E. (2017). Biopolymer-based materials for the control of biofouling in water systems: A review. *Environmental Technology*, *38*(13-14), 1599–1613. DOI: 10.1080/09593330.2017.1334708 PMID: 27735215

Organisation for Economic Co-operation and Development (OECD). (2008).

Organisation for Economic Co-operation and Development (OECD). (2008). *Manufactured nanomaterials: List of manufactured nanomaterials and list of endpoints for phase one of the OECD testing programme* (ENV/JM/MONO(2008)13/REV). https://www.olis.oecd.org/olis/2008doc.nsf/LinkTo/NT000034C6/$FILE/JT03248749.PDF

Osman, E. (2020). Nanofinished medical textiles and their potential impact to health and environment. Nanoparticles and their biomedical applications, 127-145.

Padilla-Rivera, A., Hannouf, M., Assefa, G., & Gates, I. (2023). A systematic literature review on current application of life cycle sustainability assessment: A focus on economic dimension and emerging technologies. *Environmental Impact Assessment Review*, *103*, 107268. DOI: 10.1016/j.eiar.2023.107268

Palit, S., & Ranjit, P. S. (2024). Large Scale Applications of Nanomaterials for Water Treatment. In *Sustainable Production Innovations* (pp. 137–162). DOI: 10.1002/9781119792888.ch4

Pal, P., Pandey, B., Prakash, A., Sarangi, P. K., Rathore, S. S., Agarwal, V., Gupta, A. K., & Singh, A. K. (2025). *Exploration of Biosensors in Innovative Applications and Sustainable Solutions for Environmental Monitoring and Management.*, DOI: 10.1007/978-981-97-7221-6_19

Parkes, R. J., Cragg, B. A., & Wellsbury, P. (2011). Recent studies on bacterial populations and processes in subseafloor sediments: A review. *Hydrogeology Journal*, *8*(1), 11–28. DOI: 10.1007/PL00010971

Parthipan, P., Kumar, A. S., Dhandapani, P., Elumalai, P., & Thirumurugan, D. (2024). Sustainable design and production of nano-hybrid smart coatings. In *ACS symposium series* (pp. 121–138). https://doi.org/DOI: 10.1021/bk-2024-1469.ch006

Paul, S., & Ghosh, S. (2024). Nano-finishing of Natural Fibres. In *Nanotechnology in Textile Finishing: Advancements and Applications* (pp. 313–352). Springer Nature Singapore. DOI: 10.1007/978-981-97-2696-7_11

Peng, C., Zhang, W., Gao, H., Li, Y., Tong, X., Li, K., Zhu, X., Wang, Y., & Chen, Y. (2017). Behavior and potential impacts of Metal-Based Engineered nanoparticles in aquatic environments. *Nanomaterials (Basel, Switzerland)*, *7*(1), 21. DOI: 10.3390/nano7010021 PMID: 28336855

Periyasamy, A. P. (2023). Microfiber Emissions from Functionalized Textiles: Potential Threat for Human Health and Environmental Risks. *Toxics*, *11*(5), 406. Advance online publication. DOI: 10.3390/toxics11050406 PMID: 37235219

Peydayesh, M., & Mezzenga, R. (2021). Protein nanofibrils for next generation sustainable water purification. *Nature Communications*, *12*(1), 3248. DOI: 10.1038/s41467-021-23388-2 PMID: 34059677

Pillai, S. B., & Thombre, N. V. (2024). Coagulation, Flocculation, and Precipitation in Water and Used Water Purification. In *Handbook of Water and Used Water Purification* (pp. 3–27). DOI: 10.1007/978-3-319-78000-9_63

Pimenov, N. V., & Neretin, L. N. (2006). Composition and activities of microbial communities involved in carbon, sulfur, nitrogen and manganese cycling in the oxic/anoxic interface of the Black Sea. In *Past and Present Water Column Anoxia* (pp. 501–521). Springer., DOI: 10.1007/1-4020-4297-3_19

Pinck, S., Ostormujof, L. M., Teychené, S., & Erable, B. (2020). Microfluidic microbial bioelectrochemical systems: An integrated investigation platform for a more fundamental understanding of electroactive bacterial biofilms. *Microorganisms*, *8*(11), 1841. DOI: 10.3390/microorganisms8111841 PMID: 33238493

Pistone, A., Scolaro, C., & Visco, A. (2021). Mechanical properties of protective coatings against marine fouling: A review. *Polymers*, *13*(2), 173. DOI: 10.3390/polym13020173 PMID: 33418953

Poirier, K., Lotfi, M., Garg, K., Patchigolla, K., Anthony, E. J., Faisal, N. H., Mulgundmath, V., Sahith, J. K., Jadhawar, P., Koh, L., Morosuk, T., & Al Mhanna, N. (2023). A comprehensive review of pre- and post-treatment approaches to achieve sustainable desalination for different water streams. *Desalination*, *566*, 116944. DOI: 10.1016/j.desal.2023.116944

Pokrajac, L., Abbas, A., Chrzanowski, W., Dias, G. M., Eggleton, B. J., Maguire, S., Maine, E., Malloy, T., Nathwani, J., Nazar, L., Sips, A., Sone, J., van den Berg, A., Weiss, P. S., & Mitra, S. (2021). Nanotechnology for a Sustainable Future: Addressing Global Challenges with the International Network4Sustainable Nanotechnology. *ACS Nano*, *15*(12), 18608–18623. DOI: 10.1021/acsnano.1c10919 PMID: 34910476

Popescu, C., Alain, S., Courant, M., Vardelle, A., Denoirjean, A., & Cavarroc, M. (2022). Thermal spray copper-based coatings against contamination of thermoplastic surfaces: A systematic review. *Engineering Science and Technology, an International Journal, 101194*, 101194. Advance online publication. DOI: 10.1016/j.jestch.2022.101194

Popescu, M., & Ungureanu, C. (2023). Green nanomaterials for smart textiles dedicated to environmental and biomedical applications. *Materials (Basel), 16*(11), 4075. DOI: 10.3390/ma16114075 PMID: 37297209

Pousti, M., Zarabadi, M. P., Amirdehi, M. A., Paquet-Mercier, F., & Greener, J. (2018). Microfluidic bioanalytical flow cells for biofilm studies: A review. *Analyst, 144*(1), 68–87. DOI: 10.1039/C8AN01526K PMID: 30394455

Pozzer, S., Azar, E. R., Rosa, F. D., & Pravia, Z. M. C. (2020). Semantic segmentation of defects in infrared thermographic images of highly damaged concrete structures. *Journal of Performance of Constructed Facilities, 35*(1), 04020131. Advance online publication. DOI: 10.1061/(ASCE)CF.1943-5509.0001541

Prajapati, D., Jabborova, D., Saharan, B. S., Singh, N., Patani, A., Singh, S., & Joshi, C. (2024). Bionanotechnology: A Paradigm for Advancing Environmental Sustainability. *Indian Journal of Microbiology*. Advance online publication. DOI: 10.1007/s12088-024-01389-1 PMID: 40371027

Procópio, L. (2020). Microbial community profiles grown on 1020 carbon steel surfaces in seawater-isolated microcosm. *Annals of Microbiology, 70*(1), 13. Advance online publication. DOI: 10.1186/s13213-020-01547-y

Prosser, J. I. (2015). Dispersing misconceptions and identifying opportunities for the use of 'omics' in soil microbial ecology. *Nature Reviews. Microbiology, 13*(7), 439–446. DOI: 10.1038/nrmicro3468 PMID: 26052662

Purwasena, I. A., Astuti, D. I., Fauziyyah, N. A., Putri, D. A. S., & Sugai, Y. (2019). Inhibition of microbial influenced corrosion on carbon steel ST37 using biosurfactant produced by *Bacillus* sp. *Materials Research Express, 6*(11), 115405. DOI: 10.1088/2053-1591/ab4948

Qian, P.-Y., Cheng, A., Wang, R., & Zhang, R. (2022). Marine biofilms: Diversity, interactions and biofouling. *Nature Reviews. Microbiology, 20*(11), 671–684. DOI: 10.1038/s41579-022-00744-7 PMID: 35614346

Qu, X., Alvarez, P. J. J., & Li, Q. (2013). Applications of nanotechnology in water and wastewater treatment. *Water Research, 47*(12), 3931–3946. DOI: 10.1016/j.watres.2012.09.058 PMID: 23571110

Rabaey, K., Boon, N., Siciliano, S. D., Verhaege, M., & Verstraete, W. (2004). Biofuel cells select for microbial consortia that self-mediate electron transfer. *Applied and Environmental Microbiology*, *70*(9), 5373–5382. DOI: 10.1128/AEM.70.9.5373-5382.2004 PMID: 15345423

Radjenović, M., Petrović, M., Ventura, F., & Barceló, D. (2008). Rejection of pharmaceuticals in nanofiltration and reverse osmosis membrane drinking water treatment. *Water Research*, *42*(13), 3601–3610. DOI: 10.1016/j.watres.2008.05.020 PMID: 18656225

Rahaman, M. T., & Khan, M. S. H. (2024). Applications of green nano textile materials for environmental sustainability and functional performance: Past, present and future perspectives. *Nano-Structures and Nano-Objects*, *40*, 101332. Advance online publication. DOI: 10.1016/j.nanoso.2024.101332

Rahidul Hassan, H. (2023). A review on different arsenic removal techniques used for decontamination of drinking water. *Environmental Pollutants and Bioavailability*, *35*(1), 2165964. Advance online publication. DOI: 10.1080/26395940.2023.2165964

Rahman, S. a. A., Priyadharsini, P., Deeksha, R. V., & Arun, J. (2023). Grafted Biopolymer Composites and Nanocomposites as Sustainable Corrosion Inhibitors. *Grafted Biopolymers as Corrosion Inhibitors: Safety, Sustainability, and Efficiency*, 397–416. https://doi.org/DOI: 10.1002/9781119881391.ch18

Raman, A., & Sankar, A., SD, A., Anilkumar, A., & Saritha, A. (2022). Insights into the sustainable development of lignin-based textiles for functional applications. *Macromolecular Materials and Engineering*, *307*(8), 2200114. DOI: 10.1002/mame.202200114

Ramasamy, B. S. S., & Palanisamy, S. (2021). A review on occurrence, characteristics, toxicology and treatment of nanoplastic waste in the environment. *Environmental Science and Pollution Research International*, *28*(32), 43258–43273. DOI: 10.1007/s11356-021-14883-6 PMID: 34173146

Rana, D., Matsuura, T., Kassim, M. A., & Ismail, A. F. (2023). Reverse Osmosis Membrane. In *Handbook of Membrane Separations* (pp. 6–32). CRC Press., DOI: 10.1201/9781003285656-2

Randis, R., Darmadi, D. B., Gapsari, F., Sonief, A. A., Akpan, E. D., & Ebenso, E. E. (2023). The potential of nanocomposite-based coatings for corrosion protection of metals: A review. *Journal of Molecular Liquids*, *390*, 123067. DOI: 10.1016/j.molliq.2023.123067

Rathod, P. B., Singh, M. P., Taware, A. S., Deshmukh, S. U., Tagad, C. K., Kulkarni, A., & Kanagare, A. B. (2024). Comprehensive insights into water remediation: Chemical, biotechnological, and nanotechnological perspectives. *Environmental Pollutants and Bioavailability*, *36*(1), 2329660. Advance online publication. DOI: 10.1080/26395940.2024.2329660

Rathour, A., & Pathania, S. (2025). Sensing of microplastics using advanced materials: A comprehensive review. *Sensing and Imaging*, *26*(1), 80. Advance online publication. DOI: 10.1007/s11220-025-00613-6

Rawlings, D. E. (2005). Characteristics and adaptability of iron- and sulfur-oxidizing microorganisms used for the recovery of metals from minerals and their concentrates. *Microbial Cell Factories*, *4*(1), 13. DOI: 10.1186/1475-2859-4-13 PMID: 15877814

Rayasam, S. D. G., Koman, P. D., Axelrad, D. A., Woodruff, T. J., & Chartres, N. (2022). Toxic Substances Control Act (TSCA) Implementation: How the Amended Law Has Failed to Protect Vulnerable Populations from Toxic Chemicals in the United States. *Environmental Science & Technology*, *56*(17), 11969–11982. DOI: 10.1021/acs.est.2c02079 PMID: 35980084

Ray, S., & Vashishth, R. (2024). From water to plate: Reviewing the bioaccumulation of heavy metals in fish and unraveling human health risks in the food chain. *Emerging Contaminants*, *10*(4), 100358. DOI: 10.1016/j.emcon.2024.100358

Razzak, S. A., Faruque, M. O., Alsheikh, Z., Alsheikhmohamad, L., Alkuroud, D., Alfayez, A., Hossain, S. M. Z., & Hossain, M. M. (2022). A comprehensive review on conventional and biological-driven heavy metals removal from industrial wastewater. *Environmental Advances*, *7*, 100168. DOI: 10.1016/j.envadv.2022.100168

Reguera, G., McCarthy, K. D., Mehta, T., Nicoll, J. S., Tuominen, M. T., & Lovley, D. R. (2005). Extracellular electron transfer via microbial nanowires. *Nature*, *435*(7045), 1098–1101. DOI: 10.1038/nature03661 PMID: 15973408

Regulation (EC) No. 1907/2006 of the European Parliament and of the Council concerning the Registration, Evaluation, Authorisation and Restriction of Chemicals (REACH) [2006] OJ L396/1 (ECHA, 2022). Toxic Substances Control Act, 15 USC §2601 et seq (EPA, 2022). Cabinet Decision No. 10 of 2022 on the Control of Substances Depleting the Ozone Layer and their Alternatives (United Arab Emirates). Emirates Authority for Standardization and Metrology (ESMA), ESMA Standard 5025:2021 – Safety Requirements for Chemical Products (ESMA, 2021).

Ruggiero, E., Vilsmeier, K., Mueller, P., Pulbere, S., & Wohlleben, W. (2019). Environmental release from automotive coatings are similar for different (nano) forms of pigments. *Environmental Science. Nano*, *6*(10), 3039–3048. DOI: 10.1039/C9EN00227H

Safaei, M., Taran, M., & Imani, M. M. (2019). Preparation, structural characterization, thermal properties and antifungal activity of alginate-CuO bionanocomposite. *Materials Science and Engineering C*, *101*, 323–329. DOI: 10.1016/j.msec.2019.03.108 PMID: 31029325

Safe Drinking Water Act, 42 U.S.C. §§ 300f–300j (1986).

Sagar, K., Vergis, B. R., & Devendra, B. K., & SampthKumaran, P. (. (2024). Electrochemical corrosion influencing the al beryl alloy system subjected to extrusion process. *Results in Surfaces and Interfaces*, *100399*. Advance online publication. DOI: 10.1016/j.rsurfi.2024.100399

Sahu, A., & Poler, J. C. (2024). Removal and degradation of dyes from textile industry wastewater: Benchmarking recent advancements, toxicity assessment and cost analysis of treatment processes. *Journal of Environmental Chemical Engineering*, *12*(5), 113754. Advance online publication. DOI: 10.1016/j.jece.2024.113754

Sahu, A., Swain, L. P., Behera, I. D., & Basak, G. (2025). Regulatory Framework and Policies to Achieve Zero Hydrocarbon Waste Management Through Circular Economy. In Behera, I. D., & Das, A. P. (Eds.), *Environmental Hydrocarbon Pollution and Zero Waste Approach Towards a Sustainable Waste Management* (pp. 385–408). Springer Nature Switzerland., DOI: 10.1007/978-3-031-81615-4_16

Sahu, J. N., Karri, R. R., Zabed, H. M., Shams, S., & Qi, X. (2021). Current Perspectives and Future Prospects of Nano-Biotechnology in Wastewater Treatment. *Separation and Purification Reviews*, *50*(2), 139–158. DOI: 10.1080/15422119.2019.1630430

Saleem, H., & Zaidi, S. J. (2020). Sustainable use of nanomaterials in textiles and their environmental impact. *Materials (Basel)*, *13*(22), 5134. DOI: 10.3390/ma13225134 PMID: 33203051

Salehi, E., Khajavian, M., Sahebjamee, N., Mahmoudi, M., Drioli, E., & Matsuura, T. (2022). Advances in nanocomposite and nanostructured chitosan membrane adsorbents for environmental remediation: A review. *Desalination*, *527*, 115565. DOI: 10.1016/j.desal.2022.115565

Salgar-Chaparro, S. J., Darwin, A., Kaksonen, A. H., & Machuca, L. L. (2020). Carbon steel corrosion by bacteria from failed seal rings at an offshore facility. *Scientific Reports*, *10*(1), 69292. DOI: 10.1038/s41598-020-69292-5 PMID: 32703991

Samson Prince, H., Macarius, N. M. N., Chandrashekhar Walkikar, S., Das, S., Rajan, A. P., & Deepankumar, K. (2025). *Harmonizing Nature and Innovation: Sustainable Environmental Solutions Through Nano-biotech Synergy* (pp. 35–83). DOI: 10.1007/978-3-031-78483-5_2

Saravanan, A., Senthil Kumar, P., Jeevanantham, S., Karishma, S., Tajsabreen, B., Yaashikaa, P. R., & Reshma, B. (2021). Effective water/wastewater treatment methodologies for toxic pollutants removal: Processes and applications towards sustainable development. *Chemosphere*, *280*, 130595. DOI: 10.1016/j.chemosphere.2021.130595 PMID: 33940449

Savage, N., & Diallo, M. S. (2005). Nanomaterials and water purification: Opportunities and challenges. *Journal of Nanoparticle Research*, *7*(3), 331–342. DOI: 10.1007/s11051-005-7523-5

Sawyer-Beaulieu, S., Tam, E., & Hussein, A. (2022). Measuring corrosion on vehicles, in Real-Time, using digital imaging and analysis techniques. *Materials (Basel)*, *15*(9), 3053. DOI: 10.3390/ma15093053 PMID: 35591387

Saxena, N. (2023). Bio-nanotechnology in waste to energy conversion in a circular economy approach for better sustainability. In *Bionanotechnology towards Green Energy* (pp. 253–274). CRC Press., https://api.taylorfrancis.com/content/chapters/edit/download?identifierName=doi&identifierValue=10.1201/9781003316374-13&type=chapterpdf DOI: 10.1201/9781003316374-13

Schillinger, J., & Özerol, G. (2025). For better or worse: The influence of conflict-driven decentralization on the resilience of urban water supply infrastructure in the Middle East. *Sustainable and Resilient Infrastructure*, *10*(1), 1–21. DOI: 10.1080/23789689.2024.2355768

Schulz, M. J., Ruff, B., Johnson, A., Vemaganti, K., Li, W., Sundaram, M. M., Hou, G., Krishnaswamy, A., Li, G., & Fialkova, S. (2014). New applications and techniques for nanotube superfiber development. In *Nanotube superfiber materials* (pp. 33–59). Elsevier. DOI: 10.1016/B978-1-4557-7863-8.00002-5

Selim, S., Ahmad, B., & Uddin, M. M. (2024). Guardians of the Depths: Managing Groundwater Contamination in Developing Countries. In *Water Crises and Sustainable Management in the Global South* (pp. 595–623). Springer Nature Singapore., DOI: 10.1007/978-981-97-4966-9_19

Sesia, R., Spriano, S., Sangermano, M., & Ferraris, S. (2023). Natural Polyphenols and the Corrosion Protection of Steel: Recent Advances and Future Perspectives for Green and Promising Strategies. *Metals*, *13*(6), 1070. DOI: 10.3390/met13061070

Seyeux, A., Zanna, S., Allion, A., & Marcus, P. (2015). The fate of the protective oxide film on stainless steel upon early stage growth of a biofilm. *Corrosion Science*, *91*, 352–356. DOI: 10.1016/j.corsci.2014.10.051

Shafey, A. M. (2020). Green synthesis of metal and metal oxide nanoparticles from plant leaf extracts and their applications: A review. *Green Processing and Synthesis*, *9*(1), 304–339. DOI: 10.1515/gps-2020-0031

Shafique, M., & Luo, X. (2019). Nanotechnology in Transportation Vehicles: An Overview of its applications, environmental, health and safety concerns. *Materials (Basel)*, *12*(15), 2493. DOI: 10.3390/ma12152493 PMID: 31390752

Shahalaei, M., Azad, A. K., Sulaiman, W. M. W., Derakhshani, A., Mofakham, E. B., Mallandrich, M., Kumarasamy, V., & Subramaniyan, V. (2024). A review of metallic nanoparticles: Present issues and prospects focused on the preparation methods, characterization techniques, and their theranostic applications. *Frontiers in Chemistry*, *12*, 1398979. Advance online publication. DOI: 10.3389/fchem.2024.1398979 PMID: 39206442

Shahcheraghi, N., Golchin, H., Sadri, Z., Tabari, Y., Borhanifar, F., & Makani, S. (2022). Nano-biotechnology, an applicable approach for sustainable future. *3 Biotech*, *12*(3), 65. DOI: 10.1007/s13205-021-03108-9

Shaheen, K., Shah, Z., Asad, A., Arshad, T., Khan, S. B., & Suo, H. (2020). Synthesis, characterization, and multifunctional applications of CU-FE and NI-Fe nanomaterials. *ACS Omega*, *5*(26), 15992–16002. DOI: 10.1021/acsomega.0c01259 PMID: 32656420

Shah, M. A., & Ahmed, T. (2011). *Principles of nanoscience and nanotechnology*. Narosa Publishing House.

Shantaram, A., Beyenal, H., Raajan, R., Veluchamy, A., & Lewandowski, Z. (2005). Wireless sensors powered by microbial fuel cells. *Environmental Science & Technology*, *39*(13), 5037–5042. DOI: 10.1021/es0480668 PMID: 16053108

Shao, B., Shen, L., Liu, Z., Tang, L., Tan, X., Wang, D., Zeng, W., Wu, T., Pan, Y., Zhang, X., Ge, L., & He, M. (2023). Disinfection byproducts formation from emerging organic micropollutants during chlorine-based disinfection processes. *Chemical Engineering Journal*, *455*, 140476. DOI: 10.1016/j.cej.2022.140476

Sharma, S., Sharma, K., & Grover, S. (2024b). Real-Time Data Analysis with Smart Sensors. In *Application of Artificial Intelligence in Wastewater Treatment* (pp. 127–153). DOI: 10.1007/978-3-031-69433-2_5

Sharma, S., Yadav, A., & Ahmad, W. (2018). The classification, characterization, and application of ion exchange resins: A general survey. SSRN *Electronic Journal.* https://doi.org/DOI: 10.2139/ssrn.3299226

Sharma, A., Goel, H., Sharma, S., Rathore, H. S., Jamir, I., Kumar, A., Thimmappa, S. C., Kesari, K. K., & Kashyap, B. K. (2024a). Cutting edge technology for wastewater treatment using smart nanomaterials: Recent trends and futuristic advancements. *Environmental Science and Pollution Research International, 31*(48), 58263–58293. DOI: 10.1007/s11356-024-34977-1 PMID: 39298031

Sharma, A., Kumar, N., Sillanpää, M., Makgwane, P. R., Kumar, S., & Kumari, K. (2022). Carbon nano-structures and functionalized associates: Adsorptive detoxification of organic and inorganic water pollutants. *Inorganic Chemistry Communications, 141*, 109579. DOI: 10.1016/j.inoche.2022.109579

Sharma, M., Liu, H., Chen, S., Cheng, Y. F., Voordouw, G., & Gieg, L. M. (2018). Effect of selected biocides on microbiologically influenced corrosion caused by *Desulfovibrio ferrophilus* IS5. *Scientific Reports, 8*(1), 1747. DOI: 10.1038/s41598-018-34789-7 PMID: 30413730

Sharma, S., Gupta, S., Kaur, S., Kumar, D., Banerjee, P., & Nadda, A. K. (2023). *Polymeric Membranes for Water Treatment.*, DOI: 10.1007/978-981-19-9176-9_1

Sheng, G. P., Xu, J., Luo, H. W., Li, W. W., Li, W. H., Yu, H. Q., & Hu, F. C. (2013). Thermodynamic analysis on the binding of heavy metals onto extracellular polymeric substances (EPS) of activated sludge. *Water Research, 47*(2), 607–614. DOI: 10.1016/j.watres.2012.10.037 PMID: 23159005

Shentu, H., Wei, J., Lin, X., Zhou, H., Wang, S., & Liu, J. (2023). Effect of microbial community on iron release under low velocity in drinking water supply pipes. *BIO Web of Conferences, 60*, 1008. DOI: 10.1051/bioconf/20236001008

Sherar, B. W. A., Power, I. M., Keech, P. G., Mitlin, S., Southam, G., & Shoesmith, D. W. (2011). Characterizing the effect of carbon steel exposure in sulfide-containing solutions to microbially induced corrosion. *Corrosion Science, 53*(3), 955–960. DOI: 10.1016/j.corsci.2010.11.027

Shi, F., Tieu, P., Hu, H., Peng, J., Zhang, W., Li, F., Tao, P., Song, C., Shang, W., Deng, T., Gao, W., Pan, X., & Wu, J. (2024). Direct in-situ imaging of electrochemical corrosion of Pd-Pt core-shell electrocatalysts. *Nature Communications, 15*(1), 5084. Advance online publication. DOI: 10.1038/s41467-024-49434-3 PMID: 38877007

Shi, X., Yan, W., Xu, D., Yan, M., Yang, C., Shan, Y., & Yang, K. (2018). Microbial corrosion resistance of a novel Cu-bearing pipeline steel. *Journal of Materials Science and Technology*, *34*(12), 2480–2491. DOI: 10.1016/j.jmst.2018.05.020

Shi, Z., & Guo, N. (2014). *Nanomaterials and Devices.*, DOI: 10.1016/C2012-0-03204-8

Silber, J., Kramer, A., Labes, A., & Tasdemir, D. (2016). From discovery to production: Biotechnology of marine fungi for the production of new antibiotics. *Marine Drugs*, *14*(7), 137. DOI: 10.3390/md14070137 PMID: 27455283

Simões, M., Simões, L. C., & Vieira, M. J. (2010). A review of current and emergent biofilm control strategies. *Lebensmittel-Wissenschaft + Technologie*, *43*(4), 573–583. DOI: 10.1016/j.lwt.2009.12.008

Singha, K., Ray, A., Maity, S., & Pandit, P. (2021). Chitosan-based bionanocomposites in medical textile. In Bionanocomposites in Tissue Engineering and Regenerative Medicine. DOI: 10.1016/B978-0-12-821280-6.00027-1

Singh, A. V., Aditi, A. S., Gade, W. N., Vats, T., Lenardi, C., & Milani, P. (2010). Nanomaterials: New generation therapeutics in wound healing and tissue repair. *Current Nanoscience*, *6*(6), 577–586. DOI: 10.2174/157341310793348632

Singh, B., Thakur, V., Bhatia, G., Verma, D., & Singh, K. (2016). Bioremediation and biodegradation. *Journal of Bioremediation & Biodegradation*, *7*, 374. Advance online publication. DOI: 10.4172/2155-6199.1000374 PMID: 29142804

Singh, M., & Pokhrel, M. (2017). A fuzzy logic-possibilistic methodology for risk-based inspection (RBI) planning of oil and gas piping subjected to microbiologically influenced corrosion (MIC). *International Journal of Pressure Vessels and Piping*, *159*, 45–55. DOI: 10.1016/j.ijpvp.2017.11.005

Singh, N. B., Nagpal, G., Agrawal, S., & Rachna, . (2018). Water purification by using adsorbents: A review. *Environmental Technology & Innovation*, *11*, 187–240. DOI: 10.1016/j.eti.2018.05.006

Singh, S. P., Sharma, M. K., Kumar, C., & Gaur, R. C. (2022). Effect of Filter Media on Performance of Biofilm Reactor for Treatment of Domestic Wastewater. *Journal of Hazardous, Toxic and Radioactive Waste*, *26*(3), 04022014. DOI: 10.1061/(ASCE)HZ.2153-5515.0000700

Singh, S., Tiwari, H., Verma, A., Gupta, P., Chattopadhaya, A., Singh, A., Singh, S., Kumar, B., Mandal, A., Kumar, R., Yadav, A. K., Gautam, H. K., & Gautam, V. (2024). Sustainable Synthesis of Novel Green-Based Nanoparticles for Therapeutic Interventions and Environmental Remediation. *ACS Synthetic Biology*, *13*(7), 1994–2007. DOI: 10.1021/acssynbio.4c00206 PMID: 38899943

Snider, R. M., Strycharz-Glaven, S. M., Tsoi, S. D., Erickson, J. S., & Tender, L. M. (2012). Long-range electron transport in *Geobacter sulfurreducens* biofilms is redox gradient-driven. *Proceedings of the National Academy of Sciences of the United States of America*, *109*(38), 15467–15472. DOI: 10.1073/pnas.1209829109 PMID: 22955881

Soni, V., Raizada, P., Singh, P., Cuong, H. N., S, R., Saini, A., Saini, R. V., Le, Q. V., Nadda, A. K., Le, T.-T., & Nguyen, V.-H. (2021). Sustainable and green trends in using plant extracts for the synthesis of biogenic metal nanoparticles toward environmental and pharmaceutical advances: A review. *Environmental Research*, *202*, 111622. DOI: 10.1016/j.envres.2021.111622 PMID: 34245729

Sreekumari, K. R., & Ohtsu, Y. (2010). Atomic force microscopy: A powerful tool for studying microbiologically influenced corrosion. *Corrosion Science*, *52*(3), 775–781. DOI: 10.1016/j.corsci.2009.11.007

Srivastava, A., Srivastava, O. N., Talapatra, S., Vajtai, R., & Ajayan, P. M. (2004). Carbon nanotube filters. *Nature Materials*, *3*(9), 610–614. DOI: 10.1038/nmat1192 PMID: 15286755

Starosvetsky, D., Armon, R., Yahalom, J., & Starosvetsky, J. (2001). Pitting corrosion of carbon steel caused by iron bacteria. *International Biodeterioration & Biodegradation*, *47*(2), 79–87. DOI: 10.1016/S0964-8305(99)00081-5

Steidl, R. J., Lampa-Pastirk, S., & Reguera, G. (2016). Mechanistic stratification in electroactive biofilms of *Geobacter sulfurreducens* mediated by pilus nanowires. *Nature Communications*, *7*(1), 12217. DOI: 10.1038/ncomms12217 PMID: 27481214

Stetter, K. O., Huber, R., Blochl, E., Kurr, M., Eden, R. D., Fielder, M., Cash, H., & Vance, I. (1993). Hyperthermophilic archaea are thriving in deep North Sea and Alaskan oil reservoirs. *Nature*, *365*(6448), 743–745. DOI: 10.1038/365743a0

Stewart, P. S., & Costerton, J. W. (2001). Antibiotic resistance of bacteria in biofilms. *Lancet*, *358*(9276), 135–138. DOI: 10.1016/S0140-6736(01)05321-1 PMID: 11463434

Stoica, A. E., Chircov, C., & Grumezescu, A. M. (2020). Nanomaterials for wound dressings: An up-to-date overview. *Molecules (Basel, Switzerland)*, *25*(11), 2699. DOI: 10.3390/molecules25112699 PMID: 32532089

Stoler, J., Jepson, W., Wutich, A., Velasco, C. A., Thomson, P., Staddon, C., & Westerhoff, P. (2022). Modular, adaptive, and decentralised water infrastructure: Promises and perils for water justice. *Current Opinion in Environmental Sustainability*, *57*, 101202. DOI: 10.1016/j.cosust.2022.101202

Subramanian, V. (2017). *Development and testing of a framework for sustainable nanotechnology*. https://iris.unive.it/handle/10579/10350

Su, H., Mi, S., Peng, X., & Han, Y. (2019). The mutual influence between corrosion and the surrounding soil microbial communities of buried petroleum pipelines. *RSC Advances*, *9*(33), 18930–18937. DOI: 10.1039/C9RA03386F PMID: 35516885

Sun, D., Xu, D., Yang, C., Shahzad, M. B., Sun, Z., Xia, J., Zhao, J., Gu, T., Yang, K., & Wang, G. (2016). An investigation of the antibacterial ability and cytotoxicity of a novel Cu-bearing 317L stainless steel. *Scientific Reports*, *6*(1), 29244. DOI: 10.1038/srep29244 PMID: 27385507

Sunita, S., & Chakrabarty, J. (2022). Size-Dependent Catalytic Properties of Nanomaterials, Their Suitability in Terms of Efficiency, Cost-Effectiveness and Sustainability. In *Advanced Nanocatalysts for Biodiesel Production* (pp. 199–215). CRC Press., DOI: 10.1201/9781003120858-9

Sun, Y., & Sun, G. (2002). Durable and regenerable antimicrobial textile materials prepared by a continuous grafting process. *Journal of Applied Polymer Science*, *84*(8), 1592–1599. DOI: 10.1002/app.10456

Tan, C. H., Lee, K. W., & Burmølle, M. (2016). The role of quorum sensing in microbial corrosion. *Biofouling*, *32*(5), 537–547. DOI: 10.1080/08927014.2016.1143394

Tang, H.-Y., Yang, C., Ueki, T., Pittman, C. C., Xu, D., Woodard, T. L., Holmes, D. E., Gu, T., Wang, F., & Lovley, D. R. (2021). Stainless steel corrosion via direct iron-to-microbe electron transfer by Geobacter species. *The ISME Journal*, *15*(10), 3084–3093. DOI: 10.1038/s41396-021-00990-2 PMID: 33972726

Telegdi, J., Shaban, A., & Trif, L. (2017). Microbiologically influenced corrosion (MIC). In *Elsevier eBooks* (p. 191). Elsevier BV. DOI: 10.1016/B978-0-08-101105-8.00008-5

Thakur, A., Zarrouk, A., Selvaraj, M., Assiri, M. A., Khanna, V., Kumar, A., Berdimurodov, E., & Eliboev, I. (2025). Nanomaterial-based smart coatings for sustainable corrosion protection in harsh marine environments: Advances in environmental management and durability. *Inorganic Chemistry Communications, 114280*, 114280. Advance online publication. DOI: 10.1016/j.inoche.2025.114280

Thakur, S., Chandra, A., Kumar, V., & Bharti, S. (2025). *Environmental Pollutants: Endocrine Disruptors/Pesticides/Reactive Dyes and Inorganic Toxic Compounds Metals*. Radionuclides, and Metalloids and Their Impact on the Ecosystem., DOI: 10.1007/978-981-97-7221-6_3

Thangadurai, D., Prakash, L., Sangeetha, J., Al-Tawaha, A. R. M. S., David, M., Islam, S., . . . Adetunji, J. B. (2020). Sustainable synthesis of greener nanomaterials: Principles, processes, and products. Handbook of Nanomaterials and Nanocomposites for Energy and Environmental Applications, 1-23.

Thomas, P., Sahoo, B. N., Thomas, P. J., & Greve, M. (2024). Recent advances in emerging integrated anticorrosion and antifouling nanomaterial-based coating solutions. *Environmental Science and Pollution Research International, 31*(60), 67550–67576. Advance online publication. DOI: 10.1007/s11356-024-33825-6 PMID: 38831147

Tirandaz, H., Dastgheib, S. M. M., & Hamedi, J. (2024). Sub-minimum inhibitory concentration of tetrakis(hydroxymethyl)phosphonium sulfate enhances biocorrosion of carbon steel by *Pseudomonas aeruginosa. Scientific Reports, 14*(1), 28918. Advance online publication. DOI: 10.1038/s41598-024-70157-4 PMID: 39572582

Tiwari, A., Tiwari, V., & Banik, B. K. (2025). Waste Management: Nano Photocatalysis as an Efficient Future Pathway. In *Emerging Trends in Photoredox Synthetic Transformation* (pp. 305–339). DOI: 10.1007/978-981-97-8205-5_11

Tong, T., Liu, X., Li, T., Park, S., & Anger, B. (2023). A Tale of Two Foulants: The Coupling of Organic Fouling and Mineral Scaling in Membrane Desalination. *Environmental Science & Technology, 57*(18), 7129–7149. DOI: 10.1021/acs.est.3c00414 PMID: 37104038

Tran, T. T. T., Kannoorpatti, K., Padovan, A., & Thennadil, S. N. (2021). Sulphate-reducing bacteria's response to extreme pH environments and the effect of their activities on microbial corrosion. *Applied Sciences (Basel, Switzerland), 11*(5), 2201. DOI: 10.3390/app11052201

Trevors, J. T. (1989). The role of microbial metal resistance and detoxification mechanisms in environmental bioassay research. *Hydrobiologia, 188*(1), 143–147. DOI: 10.1007/BF00027779

Tripathi, A. K., Thakur, P., Saxena, P., Rauniyar, S., Gopalakrishnan, V., Singh, R. N., Gadhamshetty, V., Gnimpieba, E. Z., Jasthi, B. K., & Sani, R. K. (2021). Gene sets and mechanisms of sulfate-reducing bacteria biofilm formation and quorum sensing with impact on corrosion. *Frontiers in Microbiology*, *12*, 754140. DOI: 10.3389/fmicb.2021.754140 PMID: 34777309

Turakhia, B., Turakhia, P., & Shah, S. (2018). Green synthesis of zero valent iron nanoparticles from *Spinacia oleracea* (spinach) and its application in wastewater treatment. *Journal of Advanced Research in Applied Sciences*, *5*(1), 46–51.

Tyagi, P., Ahmed, S., & Lai, C. W. (2023). Future prospects of biobased materials. In *Advanced Applications of Biobased Materials* (pp. 749–764). Elsevier. DOI: 10.1016/B978-0-323-91677-6.00013-1

U.S. Bureau of Reclamation & Sandia National Laboratories. (2003). *Desalination and water purification technology roadmap: A report of the executive committee*. Water Purification.

U.S. Environmental Protection Agency. (2011). *Drinking water and ground water statistics* (EPA 816-R-13-003). Office of Water. Retrieved from http://water.epa.gov/scitech/datait/databases/drink/sdwisfed/upload.epa816r13003.pdf

Uchiyama, T., Ito, K., Mori, K., Tsurumaru, H., & Harayama, S. (2010). Iron-corroding methanogen isolated from a crude-oil storage tank. *Applied and Environmental Microbiology*, *76*(6), 1783–1788. DOI: 10.1128/AEM.00668-09 PMID: 20118376

United Arab Emirates Legislations | Cabinet Resolution On the Technical Regulations for the Control of the Quantity of Product in Pre-packages. (2025). https://uaelegislation.gov.ae/en/legislations/2594

United Arab Emirates—Standards for Trade. (2025, August 26). https://www.trade.gov/country-commercial-guides/united-arab-emirates-standards-trade

United States Environmental Protection Agency (USEPA). (2013). *Nanotechnology and radiation cleanup*. Retrieved from https://www.epa.gov/radiation/cleanup/nanotechnology.html

Urošević, T., & Trivunac, K. (2020). Achievements in low-pressure membrane processes microfiltration (MF) and ultrafiltration (UF) for wastewater and water treatment. In *Current Trends and Future Developments on (Bio-)* (pp. 67–107). Membranes., DOI: 10.1016/B978-0-12-817378-7.00003-3

Usher, K. M., Kaksonen, A. H., & MacLeod, I. D. (2014). Marine rust tubercles harbour iron corroding archaea and sulphate reducing bacteria. *Corrosion Science*, *83*, 189–197. DOI: 10.1016/j.corsci.2014.02.014

Van der Bruggen, B., & Vandercasteele, C. (2003). Removal of pollutants from surface water and groundwater by nanofiltration: Overview of the possible applications in the drinking water industry. *Environmental Pollution, 122*(3), 435–445. DOI: 10.1016/S0269-7491(02)00308-1 PMID: 12547533

Van Eck, N. J., & Waltman, L. (2010). *VOSviewer [Computer software].* Centre for Science and Technology Studies, Leiden University. https://www.vosviewer.com

Van Eck, N. J., & Waltman, L. (2022). *VOSviewer* manual (Version 1.6.18). Leiden University, CWTS. Retrieved from https://www.vosviewer.com/documentation/Manual_VOSviewer_1.6.18.pdf

Venkidusamy, K., & Megharaj, M. (2016). A novel electrophototrophic bacterium, *Rhodopseudomonas palustris* strain RP2: Application in microbial fuel cells. *Bioresource Technology, 218*, 602–608. DOI: 10.1016/j.biortech.2016.07.007

Venzlaff, H., Enning, D., Srinivasan, J., Mayrhofer, K. J. J., Hassel, A. W., Widdel, F., & Stratmann, M. (2013). Accelerated cathodic reaction in microbial corrosion of iron due to direct electron uptake by sulfate-reducing bacteria. *Corrosion Science, 66*, 88–96. DOI: 10.1016/j.corsci.2012.09.006

Videla, H. A., & Herrera, L. K. (2005). Microbiologically influenced corrosion: Looking to the future. *International Microbiology, 8*(3), 169–180. DOI: 10.2436/im.v8i3.9490 PMID: 16200495

Vigneron, A., Alsop, E. B., Chambers, B., Lomans, B. P., Head, I. M., & Tsesmetzis, N. (2016). Complementary microorganisms in highly corrosive biofilms from an offshore oil production facility. *Applied and Environmental Microbiology, 82*(8), 2545–2554. DOI: 10.1128/AEM.03842-15 PMID: 26896143

Vigneron, A., Head, I. M., & Tsesmetzis, N. (2016). Damage to offshore production facilities by corrosive microbial biofilms. *Applied Microbiology and Biotechnology, 100*(6), 2527–2537. DOI: 10.1007/s00253-016-7323-0 PMID: 29423635

Vlasopoulos, A. (2024). *Assessment of life cycle impacts of sustainable technologies in industrial environments* [PhD Thesis, Brunel University London]. https://bura.brunel.ac.uk/handle/2438/31458

Vrinceanu, N., Bucur, S., Rimbu, C. M., Neculai-Valeanu, S., Ferrandiz Bou, S., & Suchea, M. P. (2022). Nanoparticle/biopolymer-based coatings for functionalization of textiles: Recent developments (a minireview). *Textile Research Journal, 92*(19-20), 3889–3902. DOI: 10.1177/00405175211070613

Wagner, W. E., Gold, S. C., & McGarity, T. O. (2024). Breaking the Corporate Stranglehold over Toxics Regulation: A Possible Path Forward. *DePaul L. Rev., 74*, 723. https://heinonline.org/hol-cgi-bin/get_pdf.cgi?handle=hein.journals/deplr74&section=25

Wang, F., Gu, Y., O'Brien, J. P., Yi, S. M., Yalcin, S. E., Srikanth, V., Shen, C., Vu, D., Ing, N. L., Hochbaum, A. I., Egelman, E. H., & Malvankar, N. S. (2019). Structure of microbial nanowires reveals stacked hemes that transport electrons over micrometers. *Cell, 177*(2), 361–369.e10. DOI: 10.1016/j.cell.2019.03.029 PMID: 30951668

Wang, J., Du, M., Li, G., & Shi, P. (2022). Research progress on microbiological inhibition of corrosion: A review. *Journal of Cleaner Production, 373*, 133658. DOI: 10.1016/j.jclepro.2022.133658

Wang, J., Wang, L., Gardner, D. J., Shaler, S. M., & Cai, Z. (2021). Towards a cellulose-based society: Opportunities and challenges. *Cellulose (London, England), 28*(8), 4511–4543. DOI: 10.1007/s10570-021-03771-4

Werber, J. R., Osuji, C. O., & Elimelech, M. (2016). Materials for next-generation desalination and water purification membranes. *Nature Reviews. Materials, 1*(5), 1–15. DOI: 10.1038/natrevmats.2016.18

Widmer, M., & Meili, C. (2010). 12 Approaching the nanoregulation problem in chemicals legislation in the EU and US. *International Handbook on Regulating Nanotechnologies*, 238. http://innovationsgesellschaft.ch/wp-content/uploads/2013/11/Hodge-et-al-2010-Chapter-12-Widmer-and-Meili.pdf

Wohlleben, W., Bossa, N., Mitrano, D. M., & Scott, K. (2024). Everything falls apart: How solids degrade and release nanomaterials, composite fragments, and microplastics. *NanoImpact, 34*, 100510. Advance online publication. DOI: 10.1016/j.impact.2024.100510 PMID: 38759729

Woolf, A. D., Stierman, B. D., Barnett, E. D., Byron, L. G., Woolf, A. D., Stierman, B. D., Barnett, E. D., Byron, L. G., Bole, A., Balk, S. J., Huerta-Montañez, G. M., Landrigan, P. J., Marcus, S. M., Nerlinger, A. L., Patel, L. H., Philipsborn, R., Zajac, L., Gray, K. A., Briskin, J., & Frantz, J. M. (2023). Drinking Water From Private Wells and Risks to Children. *Pediatrics, 151*(2), e2022060645. Advance online publication. DOI: 10.1542/peds.2022-060645 PMID: 36995188

Wu, J., Zhang, D., Wang, P., Cheng, Y., Sun, S., Sun, Y., & Chen, S. (2016). The influence of Desulfovibrio sp. and Pseudoalteromonas sp. on the corrosion of Q235 carbon steel in natural seawater. *Corrosion Science, 112*, 552–562. DOI: 10.1016/j.corsci.2016.04.047

Xie, R., Wei, Y., Luo, K., Lin, N., Yu, Y., Cheng, P., & Zeng, Q. (2025). Effect of environmental factors on the corrosion behavior of CoCrNi and CoCrFeMnNi alloys in 3.5% NaCl solution. *Journal of Materials Research and Technology*, *35*, 2994–3007. Advance online publication. DOI: 10.1016/j.jmrt.2025.01.227

Xie, W., Liang, X., Wang, H., Zhao, X., Tang, Y., Wu, M., & Yang, H. (2023). Structurally Tailoring Clay Nanosheets to Design Emerging Macrofibers with Tunable Mechanical Properties and Thermal Behavior. *ACS Applied Materials & Interfaces*, *15*(2), 3141–3151. DOI: 10.1021/acsami.2c19295 PMID: 36598369

Xu, D., Li, Y., Song, F. & Gu, T. (2013). Laboratory investigation of microbiologically influenced corrosion of C1018 carbon steel by nitrate-reducing bacterium Bacillus licheniformis. *Corrosion Science*, *77*, 385–390. DOI: https://doi.org/ DOI: 10.1016/j.corsci.2013.08.007

Xu, D., & Gu, T. (2014). Carbon source starvation triggered more aggressive corrosion against carbon steel by the Desulfovibrio vulgaris biofilm. *International Biodeterioration & Biodegradation*, *91*, 74–81. DOI: 10.1016/j.ibiod.2014.03.014

Xu, D., Gu, T., & Lovley, D. R. (2023). Microbially mediated metal corrosion. *Nature Reviews. Microbiology*, *21*(11), 1–14. DOI: 10.1038/s41579-023-00920-3 PMID: 37344552

Xu, L., Campos, L. C., Li, J., Karu, K., & Ciric, L. (2021). Removal of antibiotics in sand, GAC, GAC sandwich and anthracite/sand biofiltration systems. *Chemosphere*, *275*, 130004. DOI: 10.1016/j.chemosphere.2021.130004 PMID: 33640744

Xu, L., Khan, A., Kijkla, P., Kumseranee, S., Punpruk, S., & Gu, T. (2024). Prevention of severe pitting corrosion of 13Cr pipeline steel by a sulfate reducing bacterium using a green biocide cocktail. *Frontiers in Materials*, *11*, 1407655. Advance online publication. DOI: 10.3389/fmats.2024.1407655

Yadav, J., Rani, M., Shanker, U., & Sillanpaa, M. (2024). Forging the advances of iron-based nanomaterials by functionalizing charge carriers regions for eradication of heavy metal ion contamination. *Inorganic Chemistry Communications*, *164*, 112440. DOI: 10.1016/j.inoche.2024.112440

Yadav, S., Raman, A. P., Singh, M. B., Massey, I., Singh, P., Verma, C., & AlFantazi, A. (2024). Green nanoparticles for advanced corrosion protection: Current perspectives and future prospects. *Applied Surface Science Advances*, *21*, 100605. DOI: 10.1016/j.apsadv.2024.100605

Yanardag, T. (2019). Corrosion behavior of noble metals in different oral media. *International Journal of Scientific and Technological Research.* https://doi.org/DOI: 10.7176/JSTR/5-4-14

Yaraki, M. T., Nasab, S. Z., Zare, I., Dahri, M., Sadeghi, M. M., Koohi, M., & Tan, Y. N. (2022). Biomimetic metallic nanostructures for biomedical applications, catalysis, and beyond. *Industrial & Engineering Chemistry Research, 61*(22), 7547–7593. DOI: 10.1021/acs.iecr.2c00285

Yeoman, S., Lester, J. N., & Perry, R. (1992). Phosphorus removal and its influence on particle size and heavy metal distribution during wastewater treatment. *Environmental Technology, 13*(9), 901–924. DOI: 10.1080/09593339209385227

Yingngam, B. (2024). Nanocoatings for cosmetotextiles. In Sustainable Approach to Protective Nanocoatings. https://doi.org/DOI: 10.4018/979-8-3693-3136-1.ch006

Ying, S., Guan, Z., Ofoegbu, P. C., Clubb, P., Rico, C., He, F., & Hong, J. (2022). Green synthesis of nanoparticles: Current developments and limitations. *Environmental Technology & Innovation, 26*, 102336. DOI: 10.1016/j.eti.2022.102336

You, S. (2024). Cosmopolitan pathways from the Global South: How non-middle-class students become desirable Fulbright applicants. *Global Networks, 24*(2), e12458. Advance online publication. DOI: 10.1111/glob.12458

Yuan, S., Liang, B., Zhao, Y., & Pehkonen, S. O. (2013). Surface chemistry and corrosion behaviour of 304 stainless steel in simulated seawater containing inorganic sulphide and sulphate-reducing bacteria. *Corrosion Science, 75*, 171–181. DOI: 10.1016/j.corsci.2013.04.058

Yu, L., Duan, J., Du, X., Huang, Y., & Hou, B. (2013). Accelerated anaerobic corrosion of electroactive sulfate-reducing bacteria by electrochemical impedance spectroscopy and chronoamperometry. *Electrochemistry Communications, 26*, 101–104. DOI: 10.1016/j.elecom.2012.10.022

Zaghloul, M. M. Y., Zaghloul, M. M. Y., & Fuseini, M. (2023). Recent progress in Epoxy Nanocomposites: Corrosion, structural, flame retardancy and applications — A comprehensive review. *Polymers for Advanced Technologies, 34*(11), 3438–3472. DOI: 10.1002/pat.6144

Zeng, Q., Wang, B., Lai, X., Li, H., Chen, Z., Zeng, X., & Zhang, L. (2024). A multifunctional flame-retardant TA-MXene based nanocoating for cotton fabric. *Progress in Organic Coatings, 189*, 108333. DOI: 10.1016/j.porgcoat.2024.108333

Zhang. (2008). *Physical Fundamentals of Nanomaterials.* https://doi.org/DOI: 10.1016/C2012-0-02139-4

Zhang, G., Li, B., Liu, J., Luan, M., Yue, L., Jiang, X., Yu, K., & Guan, Y. (2018). The bacterial community significantly promotes cast iron corrosion in reclaimed wastewater distribution systems. *Microbiome, 6*(1), 222. Advance online publication. DOI: 10.1186/s40168-018-0610-5 PMID: 30545419

Zhang, J., Yu, S., Wang, J., Zhao, Z.-P., & Cai, W. (2023). Advanced water treatment process by simultaneous coupling granular activated carbon (GAC) and powdered carbon with ultrafiltration: Role of GAC particle shape and powdered carbon type. *Water Research, 231*, 119606. DOI: 10.1016/j.watres.2023.119606 PMID: 36680821

Zhang, P., Xu, D., Li, Y., Yang, K., & Gu, T. (2015). Electron mediators accelerate the microbiologically influenced corrosion of 304 stainless steel by the Desulfovibrio vulgaris biofilm. *Bioelectrochemistry (Amsterdam, Netherlands), 101*, 14–21. DOI: 10.1016/j.bioelechem.2014.06.010 PMID: 25023048

Zhang, P., Zhang, Z., & Wang, J. (2014). Nanoscale investigation of microbiologically influenced corrosion by atomic force microscopy. *Electrochimica Acta, 133*, 303–310. DOI: 10.1016/j.electacta.2014.04.051

Zhang, T., Fang, H. H., & Ko, B. C. (2018). Corrosion of mild steel in the presence of marine biofilms. *Biofouling, 34*(1), 11–24. DOI: 10.1080/08927014.2017.1409345

Zhang, T., Fang, H. H., & Ko, B. C. (2019). Methanogen population in a marine biofilm corrosive to mild steel. *Journal of Applied Microbiology, 115*(4), 1071–1081.

Zhang, W. X. (2003). Nano-scale iron particles for environmental remediation: An overview. *Journal of Nanoparticle Research, 5*(3–4), 323–332. DOI: 10.1023/A:1025520116015

Zhang, W., Ding, W., Li, Y., Tam, C., Bougouffa, S., Wang, R., Pei, B., Chiang, H., Leung, P., Lu, Y., Sun, J., Fu, H., Bajic, V. B., Liu, H., Webster, N. S., & Qian, P. Y. (2019). Marine biofilms constitute a bank of hidden microbial diversity and functional potential. *Nature Communications, 10*(1), 84–463. DOI: 10.1038/s41467-019-08463-z PMID: 30705275

Zhang, Y., Ma, Y., Duan, J., Li, X., Wang, J., & Hou, B. (2019). Analysis of marine microbial communities colonizing various metallic materials and rust layers. *Biofouling, 35*(4), 429–442. DOI: 10.1080/08927014.2019.1610881 PMID: 31109195

Zhang, Y., Zhai, X., Guan, F., Dong, X., Sun, J., Zhang, R., Duan, J., Zhang, B., & Hou, B. (2022). Microbiologically influenced corrosion of steel in coastal surface seawater contaminated by crude oil. *NPJ Materials Degradation, 6*(1), 35. DOI: 10.1038/s41529-022-00242-4

Zhao, H.-X., Zhang, T.-Y., Wang, H., Hu, C.-Y., Tang, Y.-L., & Xu, B. (2022). Occurrence of fungal spores in drinking water: A review of pathogenicity, odor, chlorine resistance and control strategies. *The Science of the Total Environment*, *853*, 158626. DOI: 10.1016/j.scitotenv.2022.158626 PMID: 36087680

Zhao, S., Zhao, Y., Cui, Z., Zhang, H., & Zhang, J. (2024). Effect of pH, Temperature, and Salinity Levels on Heavy Metal Fraction in Lake Sediments. *Toxics*, *12*(7), 494. DOI: 10.3390/toxics12070494 PMID: 39058146

Zhou, E., Lekbach, Y., Gu, T., & Xu, D. (2022). Bioenergetics and extracellular electron transfer in microbial fuel cells and microbial corrosion. *Current Opinion in Electrochemistry*, *31*, 100830. DOI: 10.1016/j.coelec.2021.100830

Zhou, E.-Z., Ren, G.-Y., Sun, Y.-L., Fan, Y.-Q., Yang, Y., Sun, M.-Y., Xu, D.-K., Lu, Y.-P., Wang, J.-J., Li, H.-B., & Wang, F.-H. (2022). Antibacterial activities of a novel Cu-bearing high-entropy alloy against multi-drug-resistant Acinetobacter baumannii and Staphylococcus aureus. *Rare Metals*, *41*(2), 570–579. DOI: 10.1007/s12598-021-01837-7

Zhou, E., Zhang, M., Huang, Y., Li, H., Wang, J., Jiang, G., Jiang, C., Xu, D., Wang, Q., & Wang, F. (2022). Accelerated biocorrosion of stainless steel in marine water via extracellular electron transfer encoding gene phzH of Pseudomonas aeruginosa. *Water Research*, *220*, 118634. DOI: 10.1016/j.watres.2022.118634 PMID: 35691192

Zhou, M., Wang, H., Hassett, D. J., & Gu, T. (2013). Recent advances in microbial fuel cells (MFCs) and microbial electrolysis cells (MECs) for wastewater treatment, bioenergy and bioproducts. *Journal of Chemical Technology and Biotechnology*, *88*(4), 508–518. DOI: 10.1002/jctb.4004

Zuo, R. (2007). Biofilms: Strategies for metal corrosion inhibition employing microorganisms. *Applied Microbiology and Biotechnology*, *76*(6), 1245–1253. DOI: 10.1007/s00253-007-1130-6 PMID: 17701408

About the Contributors

Shrikaant Kulkarni, Ph.D., is currently working as an Adjunct Professor at the Victorian Institute of Technology, Melbourne, Australia. Dr. Kulkarni has been a senior academician and researcher for over four decades. He has delivered invited lectures and conducted sessions at national and international conferences and faculty development programs. He has guided many major and minor projects in subject areas like engineering chemistry, green chemistry, nanotechnology, analytical chemistry, catalysis, chemical engineering materials, industrial organization, and management. He has published over 100+ research papers in national and international journals and conferences of repute in Green, Analytical, Coordination Chemistry, Green Nanotechnology, Sustainability, Advanced AI, and Educational Technology. He has edited 25+ books, and authored 50+ book chapters in books published by CRC, Springer Nature, Apple Academic Press, Elsevier, Wiley, and IET. He authored four textbooks in engineering chemistry too. Dr Kulkarni possesses MSc, MPhil, and PhD degrees in Chemistry and master's degrees in Economics, Business Management, and Political Science. He has expertise in green chemistry and engineering, analytical chemistry, green nanoscience and nanotechnology, wastewater treatment, advanced areas in Chemical Engineering and Material Science, and Advanced AI. Two of his Patents have been published and two more are applied for grant. He was invited by UNESCO for an invited talk on " Green Chemistry Education for Sustainable Development " at the International Conference on Green Chemistry, Bangkok, and was well received by international Jury members and dignitaries from across the world.

Susanta Das is a seasoned educator with a proven record of teaching and mentoring diverse students, working at multiple institutions, and providing service to the various initiatives of organizations. He received his Ph.D. and M.A. degrees from Western Michigan University (WMU), USA, and M.Sc. degree from Banaras Hindu University (BHU), India all in Physics. He continued his research as a Marie-

Curie post-doctoral fellow at Stockholm University, Stockholm, Sweden on beam diagnostics for the DESIREE (Double ElectroStatic Ion Ring ExpEriment) in the project DITANET (Diagnostic Techniques for particle Accelerators – a Marie-Curie initial training NETwork), at the Indian Institute of Science Education and Research-Kolkata, India on high-pressure physics as a Project Scientist-B, and at the University of Electro-Communications, Tokyo, Japan on ion-surface interactions as a Post-doctoral fellow. Throughout his career, Dr. Das worked and collaborated with many researchers, Ph.D., master, and visiting students from different countries. He further visited many countries to discuss research and an international conference participant (UK, Italy, Germany, Greece, Belgium, Bulgaria, Romania, Brazil etc.). He co-authored several research articles in WoS/Scopus indexed international journals, conference proceedings, and scholarly book chapters published by renowned international publishing houses. Beyond his research endeavors, Dr. Das has a storied history of service to academic and administrative committees, exhibiting his commitment to the growth and development of educational institutions. His experience includes tenure at Central University South Bihar, Sri Sri University, and P.K. University before assuming his current role at Ajeenkya DY Patil University. He received the Marie-Curie post-doctoral fellowship in Sweden, Science Academies' Summer Research Fellowship in India, Gwen Frostic Doctoral Fellowship, Department Graduate Research and Creative Scholar Award by WMU, and Leo R. Parpart Doctoral Fellowship by Dept. of Physics, WMU, among many others, throughout his academic journey. At present, Dr. Das continues to delve into cutting-edge fields, with a keen interest in nanotechnology, UN SDGs, and data science. His multifaceted contributions, spanning teaching, research, and administrative leadership, showcase a dedicated professional who is instrumental in advancing the vision and mission of the institutions he serves.

Ibrar Ahmad is an accomplished LL.M. graduate with a research focus on Public International Law, Environmental law, and Energy regulation, particularly concerning climate change and sustainable development. He is the author of the book "Legal Framework of Green Hydrogen Production and Hydrogen Economy for Sustainable Development" and has published several SSCI/SCOPUS-indexed research articles . Academically, he holds an LL.M. from South West University of Political Science and Law (China) and an LL.B. from Abdul Wali Khan University Mardan (Pakistan), with his master's thesis focusing on restructuring the UN's role in ocean acidification governance. Professionally, he has served as an Associate Advocate in China and an Advocate in Pakistan, gaining practical experience in civil, criminal, constitutional, and environmental law. He is a recipient of the CSC Scholarship and an Editorial Board Member.

Farouq Sabre Al-Shibli joined the University of Kalba as an Associate Professor of Law in September 2024, bringing over a decade of expertise in administrative law, constitutional law, environmental law, and arbitration. He earned his Ph.D. in Administrative Law from the University of Manchester, United Kingdom, with a thesis that focused on the role of arbitration in resolving administrative contract disputes through a comparative study of the UK and Arab countries. Dr. Al-Shibli has extensive experience in teaching, research, and academic supervision at leading institutions, including the University of East Anglia and the University of Essex in the UK, Sarajevo University, Dubai University, the Judicial Institute of Jordan, the American University in the Emirates, the British University in Dubai, and Philadelphia University. He has taught courses in administrative law, constitutional law, and research methodology and has supervised numerous doctoral and master's theses. In addition to his academic career, Dr. Al-Shibli has been a practicing lawyer and arbitrator since 2008, representing clients in administrative, civil, and commercial cases. He has also served as a legal consultant for various academic institutions and private-sector organizations, specializing in administrative law, labor law, government contracts, and investment law. A prolific researcher, Dr. Al-Shibli has published extensively in Scopus-indexed journals, exploring topics such as administrative law, administrative judiciary, public-private partnerships, the legal frameworks of Islamic finance, and the impact of Shari'a/Islamic law. He is an active contributor to the academic community, serving as a peer reviewer for prestigious international law journals and as a member of various scholarly and editorial committees. At the University of Kalba, Dr. Al-Shibli is committed to advancing legal education and mentoring students in their pursuit of legal expertise.

Sudarshana Banerjee, Assistant Professor in the Department of Physics and Assistant Dean (Curriculum Development) at Srinath University, Jamshedpur, specializes in nanostructures for photovoltaic and environmental applications. A PhD from IIEST, Shibpur, she contributed to a DST-sponsored solar project. With expertise in curriculum development, she is committed to advancing education and research in Physics and Engineering sciences.

Aftab Haider is a highly accomplished PhD Scholar of International Law at Southwest University of Political Science and Law in China, specializing in the complex intersection of Environmental Law, International Law, and Human Rights, with a specific focus on the legal impact of climate change and global governance issues like data protection and cybersecurity. A dedicated scholar with a Master of Laws degree from Abdul Wali Khan University, he is also an active contributor to the global legal discourse, currently serving as an International Arbitrator at the Chongqing Arbitration Commission (CQAC) and holding prestigious editorial and reviewer roles, notably as a Reviewer for PLOS ONE (SCIE, Q1). Mr. Haider's extensive publication record includes numerous peer-reviewed articles in highly-ranked journals (SSCI, SCOPUS, SCI), addressing critical topics from the geopolitical impact of BRICS and International Trade and Plastic Waste in Oceans to the implementation challenges of environmental laws in developing countries. Formerly a practicing Lawyer with the KP Bar Council, he combines real-world legal advocacy with rigorous academic research, earning multiple distinctions including the Chongqing Mayor Scholarship Award for his PhD

Yadnyesh Khapekar is a student of Ajeenkya D. Y. Patil University, pursuing a Bachelor of Technology degree in Computer Science and Engineering. With a sharp analytical mind and a natural inclination for problem-solving, he thrives in tackling complex challenges with precision and creativity. His proficiency in C++ programming enables him to develop efficient algorithms and software solutions, making him an asset in the world of coding. Beyond programming, He has honed his skills in Computer-Aided Design (CAD) software, allowing him to conceptualize and create innovative designs with technical accuracy. His enthusiasm for 3D printing technology further complements his expertise, as he seamlessly translates digital models into tangible prototypes. His ability to design in three dimensions showcases his blend of technical prowess and artistic vision. An active participant in academic and technical competitions, he has proudly represented his university on multiple occasions, demonstrating his capability to perform under pressure and excel in competitive settings. His dedication to research and knowledge dissemination is evident in the numerous review papers he has authored and published in books and journals relating to concepts like artificial intelligence, lead nanoparticles, climate risks, and many more, contributing valuable insights to the fields. With a strong foundation in engineering principles and a drive to push boundaries, he aspires to innovate and lead in the realm of technology. His blend of coding expertise, design acumen, and research experience positions him as a future visionary in the ever-evolving field of computer science and engineering.

Diwesh Kumar, Assistant Professor at Srinath University's Civil Engineering Department, specializes in Water Engineering and Management. With expertise in teaching, project design, and water treatment, he integrates QGIS-based vegetation analysis into research. His contributions advance sustainable civil engineering solutions, blending technical proficiency with academic excellence.

Ashish Kumar Nayak is an Assistant Professor at PVP Siddhartha Institute of Technology, Vijayawada, Andhra Pradesh. He earned his Ph.D. in Civil Engineering from IIT Kharagpur, his M.Tech. in Environmental Engineering from IIT Guwahati, and his B.Tech. in Civil Engineering from NIT Rourkela. He has published numerous research articles and book chapters with reputed international publishers. His research focuses on water and wastewater treatment, physico-chemical treatment processes, sustainable environmental technologies, nanomaterial-based pollutant remediation, process optimization using soft computing techniques, solid waste management, waste recycling, waste minimization, composting technology, and anaerobic digestion. He also serves as an editor for 19 international journals published by Springer, Taylor & Francis, Wiley, Elsevier, Bentham Science, Nature, BMC, the Discover series, and MDPI, and actively reviews manuscripts for various SCI/SCIE/Scopus-indexed journals.

Amudhanila Padmanaban is an Assistant Professor and biotechnology researcher with a strong focus on marine bioresources, nanotechnology, and anticancer studies. She has published several research articles and book chapters, guided postgraduate students, and actively participates in academic conferences. She is currently pursuing her Ph.D. in Biotechnology.

Sunil Sankathala is pursuing B. Tech Computer Science and Engineering at Ajeenkya DY Patil University, with a diverse range of interests that bridge technology, sustainability, and innovation. His academic focus lies in computer science fundamentals like programming, data structures, algorithms, and software development. He actively supplements this with hands-on experience in web development and robotics. Beyond the core curriculum, He has a strong research inclination in emerging and impactful fields. His current research interests include Artificial Intelligence in Healthcare, bioplastics, nano bioplastics, e-waste management, and solid waste management—areas where he believes technology can make a meaningful difference in solving real-world problems. He has been publishing research articles and book chapters in reputed national and international journals and books. He is also part of an ambitious Hyperloop project, which involves interdisciplinary collaboration and cutting-edge engineering aimed at revolutionizing transportation. Working on this project has enhanced his skills in teamwork, innovation, and systems thinking. In the tech domain, he has developed proficiency in web development using technologies such as HTML, CSS, JavaScript, and React. His interests in robotics further reflect building smart, responsive systems that interact with the physical world. Throughout his course, he has helped strengthen problem-solving and collaborative skills. He is proficient in languages such as Python, C, C++, MySQL, Ubuntu, Machine Learning and familiar with frameworks and tools like React, Git. He actively participates in hackathons, coding competitions, and tech events to challenge himself and learn from peers. He is also a quick learner and highly adaptable, with an eagerness to stay updated with the latest trends in the tech world. He aspires to become a skilled software engineer and contribute to building innovative solutions that have a positive impact on society. The long-term goal is to work in a dynamic environment where knowledge, grow professionally, and make meaningful contributions to the field of technology. Driven by a spirit of exploration and sustainability, he aims to contribute to technological solutions that address both human and environmental needs. The long-term goal is to become a versatile engineer and researcher, working at the intersection of computer science, environmental innovation, and human well-being.

Aditya Shrivastav is an aspiring technologist currently pursuing a Bachelor of Technology (B. Tech.) in Computer Science and Engineering at Ajeenkya D. Y. Patil University. He holds solid programming expertise in C, C++, and Python, alongside strong command over data and business tools such as Microsoft Excel and Tally software. Aditya is a member of Team Vac-Vectoor Hyperloop and participated with the team in the Global Hyperloop Competition organized by IIT Madras, contributing to cutting-edge transportation solutions. He actively engages in hackathons and various technical events to expand his skills and contribute to innovative problem-solving. He has authored and co-authored book chapters in the domains of solid waste management, bioplastics, nanotechnology applications, and artificial intelligence, showcasing his multidisciplinary approach and research acumen. Aditya is also working on an innovative project on predictive e-waste management. The project is targeted toward IT departments, schools, and colleges, aiming to support sustainable practices and resource optimization. In parallel with his academic journey, Aditya is serving as a faculty intern at Disha Computer Institute, where he teaches C, C++, basic Python, basic Java, Excel formulas, and Tally software. His dedication and teaching proficiency earned him a Letter of Appreciation and the opportunity to conduct a workshop on Advanced Excel at G. H. Raisoni College of Arts, Commerce and Science. With a strong foundation in technology, a passion for sustainability, and hands-on experience in both teaching and project development, Aditya Shrivastav is committed to driving impact through research, education, and innovation.

Surya Pratap Singh, Assistant Professor and Head of Civil Engineering at Srinath University, Jamshedpur, specializes in Civil and Environmental Engineering. A PhD from Manipal University, Jaipur, he excels in project handling, biofilm reactor research, and sustainable water management. He holds an Indian patent and has presented at India Water Week 2024.

Index

Symbols

A

B

C

E

F

G

www.ingramcontent.com/pod-product-compliance
Lightning Source LLC
LaVergne TN
LVHW080902300926
849162LV00003B/518